Bridging Mobilities

Langaa &
African Studies Centre

Bridging Mobilities

ICTs appropriation by Cameroonians in South Africa and The Netherlands

Henrietta Mambo Nyamnjoh

Langaa
Langaa Research and Publishing Common Initiative Group
PO Box 902 Mankon
Bamenda
North West Region
Cameroon
Phone +237 33 07 34 69 / 33 36 14 02
LangaaGrp@gmail.com
www.africanbookscollective.com/publishers/langaa-rpcig

African Studies Centre
P.O. Box 9555
2300 RB Leiden
The Netherlands
asc@asc.leidenuniv.nl
http//:asc.leidenuniv.nl

Photos: Henrietta Nyamnjoh
Cover photo: Pinyin and Mankon members at a cry-die celebration in Cape Town

ISBN: 9956-791-51-2

Contents

Contents

List of photographs, maps and diagrams

Photographs

Map

Diagram

Acknowledgements

This study stems from my mobility and the way I have criss-crossed different spaces to gain an understanding of the lives of other mobile communities. These journeys of mine have been taken with many people, and various institutions have contributed to the realization of this PhD. I am particularly grateful to my supervisors, Prof. Mirjam de Bruijn and Prof. Michael Rowlands, for their insightful and thought-provoking comments. Mirjam, our regular meetings were of immense benefit and you constantly challenged me and my way of thinking. I appreciated your indefatigable reading of my texts and your detailed remarks. Your constant emails asking me for 'anything to read' were motivating and encouraged me to keep going. Thanks so much for your patience and enthusiasm. Our regular meetings and Skype conversations when I was not in Leiden provided us a platform where we could come up with useful analytical concepts. Michael, although you were in the UK, distance never stood in our way or prevented regular Skype contact. My brief stay in London, as well as your visits to Cameroon, Leiden and Cape Town, helped to sharpen my thinking, and your critical and astute questions and challenging comments led me to see the transcultural aspect of this study. Our discussions went beyond this thesis and culminated in our joint conference presentation in Edinburgh and a subsequent publication in *Critical African Studies*. Thank you for your confidence in me. Also, I would like to thank your families for the time I spent in your homes in Utrecht and London respectively. When I was not working on my thesis and needed some 'relaxation', I even tried to bake some bread thanks to the book Sue Rowlands gave me.

This study is about the mobile communities that opened up their lives to me and let me in. I would like to thank all my respondents. You made yourselves available and allowed me to follow all your activities, and this gave me the opportunity to understand your everyday lives. You introduced me to your families in Pinyin and Mankon who welcomed me into their homes not knowing that I would pay numerous visits and continuously ask them so many questions that 'can even make one tell a lie'. I enjoyed every moment of the time I spent with you all. Your stories form the pillar of this book. I hope I have not disappointed you and that I have accurately documented your stories and that they truly reflect your lives. Thank you and I will forever be indebted to you, my informants. In particular, I would like to thank Jibraeel and Margarette Kiyang, Kenneth and

Stella Fru, Ignatius Ticha, Ernest Pineteh, Javis and Relindis Nkam, Cletus Muluh, Elvis Kiyang, Victoire Ticha, Eric and Miranda Anunchu, Sylvia Tasi, Pascaline, Anah George Fonjah, Martin Chifor, George Mbuya, Napoleon and Marcelline Ngwasi, Solange Awa, Samuel Ayaba and Richard Khan. And to the late Ernest Ticha, Elvis Anah and Ernest Fon; our meetings were brief but you have had a lasting impact on this study. My gratitude also goes to Cyprian Nsuh, Blaise & Mirabel Anye, Cyprian & Billet Ndeh, Collins & Dylis Tangie, Cletus & Pamela Foben and George Che. Your input was immense and I really appreciated it. My profound gratitude equally goes to Augustine Konje who introduced me to most of my informants and volunteered to drive me around The Hague to meet informants there, as well as being a key informant.

I had a wonderful time in Pinyin and Mankon! I am so very grateful to Esther and Brigit Kiyang my host mothers and their families in the village, and to Paul, Miriam & Martha Anunchu, Jacob Nkam, Ticha Anyam, Susan Muluh, Alfred Fonkwa, Daniel & Stella Tumasang, Lucas Tasi Tang, Eugene Tamungang, Martin Tangu, Isaiah Tamungang, Martin Tamungang, Evodia Tamungang, Cecilia Lum, Angelina Manka, Simon & Elizabeth Monikang, Trephine Fomeh and the late Papa Kiyang. Limited space here means that I cannot mention everyone who helped me and I hope you understand this but I do acknowledge the valuable contribution you all made to this study. Francis Fokwang, your time-consuming transcription of my interviews sped up the writing process enormously and I thank you for so diligently transcribing so many interviews.

I was awarded a WOTRO grant as part of the Mobile Africa Revisited project for the duration of this PhD research and also benefitted from the WOTRO-sponsored writing workshops, seminars and conferences I attended. Special thanks go to Inge Brinkman who was always on hand to ensure the smooth running of the programme and for her comments. These occasions offered the opportunity for stimulating intellectual debate that sharpened my ideas and opened up my mind to new ones. Writing a thesis can be a lonely undertaking but the African Studies Centre (ASC) provided a serene working space and I am grateful to my colleagues there. In particular, my sincere thanks go to Ineke van Kessel, Wim van Binsbergen & Patricia, Peter Geschiere, Ann Reeves for editing the English in this volume and Dick Foeken. I am extremely grateful to Ton Dietz for his warm welcome and words of encouragement and for having faith in me. In addition, I would like to thank Maaike Westra, Gitty Petit, Trudi Blomsma, Kim van Drie, Marieke van Winden, Lenie van Rooijen and Jan Binnendijk for their assistance and cheerfulness. The library staff were always ready to give me a helping hand too: thank you Jos Damen, Ella Verkaik, Monique Kromhout and Elvire Eijkman.

The PhD room at the ASC was a warm place filled with laughter and stimulating debates, and attempts at putting the world to rights allowed us to escape from the solitary process of writing. To Evelyne Tegomoh, Fatima Diallo, Martin van Vliet, Michiel van den Bergh, Margot Leegwater, Serena Adede and all the others who passed through the PhD room: thank you for your friendship and the wonderful times we spent together. I would also like to thank Sophie Feyder and Inge Ligtvoet for their moral support.

Piet Konings and Lenie Schoenmakers offered me endless encouragement and moral support. The weekends I spent with you in Posterholt in the Netherlands gave me a break from writing and some time to make mental notes. Thank you for always being there for me.

My friendship with Amina and Joorst van Rossum started on the day we met in 2007 and has blossomed ever since. You did all you could to make my stay in Leiden comfortable and tried to ensure that I did not miss home too much. Thank you. Writing this thesis made me 'disconnect' from so many friends and to all the many friends who would phone with encouraging words to find out about my progress, a big thank you, especially to Divine Fuh, Moses & Stella Nintai, Rose Mbawa, Jude Fokwang and Charles Mbella. Many thanks to Martin Evans for providing me with his edited volume and articles on home-town associations that enriched my understanding and analysis in Chapter 7, and also to Walter Nkwi for assisting me with some of the archival research. I would also like to thank Afe and Esther Adogame for welcoming me into their home in Edinburgh and the invitation you (Afe) gave me to present my research in one of your seminars at the School of Divinity at the University of Edinburgh, as well as for the conference invitations. My stay in the Netherlands was at times lonely but various friends invited me into their homes: thanks Edwin & Nicole Akum, Henry & Victorine Nforbin and Maureen Minjo!

This project has come to fruition as a result of the moral support and love I received from my family. I would especially like to thank my mother Theresia Ngu for her love, for always being there for me and for her continuous support over the years, especially during my MPhil studies that led to this PhD. You constantly showed interest in my work with your many phone calls asking about my progress and your encouragement to forge on lifted my spirits. And my deepest heartfelt thanks go to Charles and Florence Awasom, Magdalene Asaah and Emilia Ambe-Niba as well. My gratitude equally goes to the other family members who showed interest in my research.

And to you Francis, Caro, Anye, Emmanuella and Sue, you kept the home fires burning. Immense thanks to Sue for transcribing some of my interviews, Anye you helped with the layout and Emmanuella you did some editing. And you importantly managed to revive my back by massaging it when it was giving way due to the long hours I was spending working. You all endured my extended periods of absence with love and encouragement and I know that even when I was at home, I tended to sometimes be an 'absent presence'. Thank you for your continuous support and for believing in me. This PhD would not have been possible without such immense love and I dedicate this book to you all.

1 Introduction

Setting the scene

As a child growing up in the Grassfields of Bamenda, the provincial capital of the North West Region, I recall vividly how I would be woken up early in the morning by one of my parents to go and see a certain person to deliver a short note, a verbal message or simply to inform them that one of my family would pass by at a particular time. The 1970s was a time when even fixed phones were not a common phenomenon, and there was no inkling of the mobile phone that was to come so many years later. There was just a reliance on a plethora of traditional means of communication to pass on messages. Physical mobility and face-to-face communication were the predominant means. And then there was letter writing, which is fast eroding today. The world has been hijacked by Information and Communication Technologies (ICTs) and, by the same token, the world has hijacked ICTs, thus creating an avalanche of information that has led to a network society. This network society has emerged as a result of phenomenal strides in ICTs and the wider expectations of societal and livelihood transformation that were forecast by earlier studies on ICTs (Castells 2004; de Bruijn *et al.* 2009, 2013; Katz 2008; Ling & Donner 2009). Information and knowledge are the prerequisites that drive mobility in the network society. In the process, ICTs have seemingly brought together families and other far-flung migrant communities, thereby creating virtual and social spaces and making transnational migration and mobility a way of life for many. This new lifestyle of transnationalism (Glick Schiller *et al.*1992; Levitt 1995; Basch *et al.* 1994; Vertovec 2007, 2009) defies the age-old tradition found in receiving countries that idealized assimilation as opposed to a dual pattern that would make the host country and the home country a single arena. However, even if the latter were to be the case (double engagement), earlier forms of communications, such as slow pace of posting letters or the dependence on emissary would not have fully supported double engagement.

The perception of assimilation, as such, was due to the tardiness of the process of communication, in particular the length of time it took mail to go back and forth. If accelerated communication and the living of dual lives were the catalyst to strides in communication technologies as well as to road and air transport (de Bruijn *et al.* 2009; Nieswand 2011), the question is what these societies were like before the arrival of ICTs.

This thesis is an ethnographic study about the working of the social fabric of mobile communities. It aims to establish whether there has been a change or transformation in the mobility dynamics of the Pinyin and Makon societies since the ICT revolution and whether its appropriation by the Pinyin and Mankon mobile communities in Cape Town, South Africa and the Netherlands have brought about transformation in the ways in which they relate back home and the changes in the livelihoods of those in the home country. Based on a triangular study that spans South Africa, the Netherlands and the respective Pinyin and Mankon home villages in the Grassfields of Bamenda in Cameroon, this research considers the consequences of ICT appropriation in relation to people's mobility. The study articulates how mobile communities criss-cross the globe virtually or in person in order to create linkages not only with the home country but with wider migrant communities mediated by the Internet and wireless communication. It focuses on the wide array of ICTs used by these mobile communities to maintain or sever links with relations and other mobile communities and how kin back home relate to them via these means.

Communication is all about building bridges and this study sheds light on (i) how communication facilitates the building of bridges; and (ii) whether these bridges are strong enough to maintain relationships. The bridge thus becomes both a metaphor for connection and a practical instrument with regard to mobility and ICTs. This means that, in today's mobile world, relations are simultaneously being built, maintained and broken, but by adopting the dual concept of a bridge it is easier to understand the 'importance and the transformative power embedded in linkages'. Equally, one can ascertain the continuities and changes, given mobility, in communication, and connectivity is 'symbolized by the bridge' (de Bruijn & van Dijk 2012: 2). Bridges and connections are mutually reinforcing as they connect communities, places and distances as well as providing virtual connections via the Internet and mobile phone.

Although the arrival of the mobile phone and other technologies is relatively recent, this study goes beyond this 'newness' and is embedded in the construction of historical narrative more than in the recent advances in communication technologies themselves. Historical narratives help us to understand how the mobile trajectories of the past link it to the present and elicit the mutual bene-

Map 1.1 Research sites

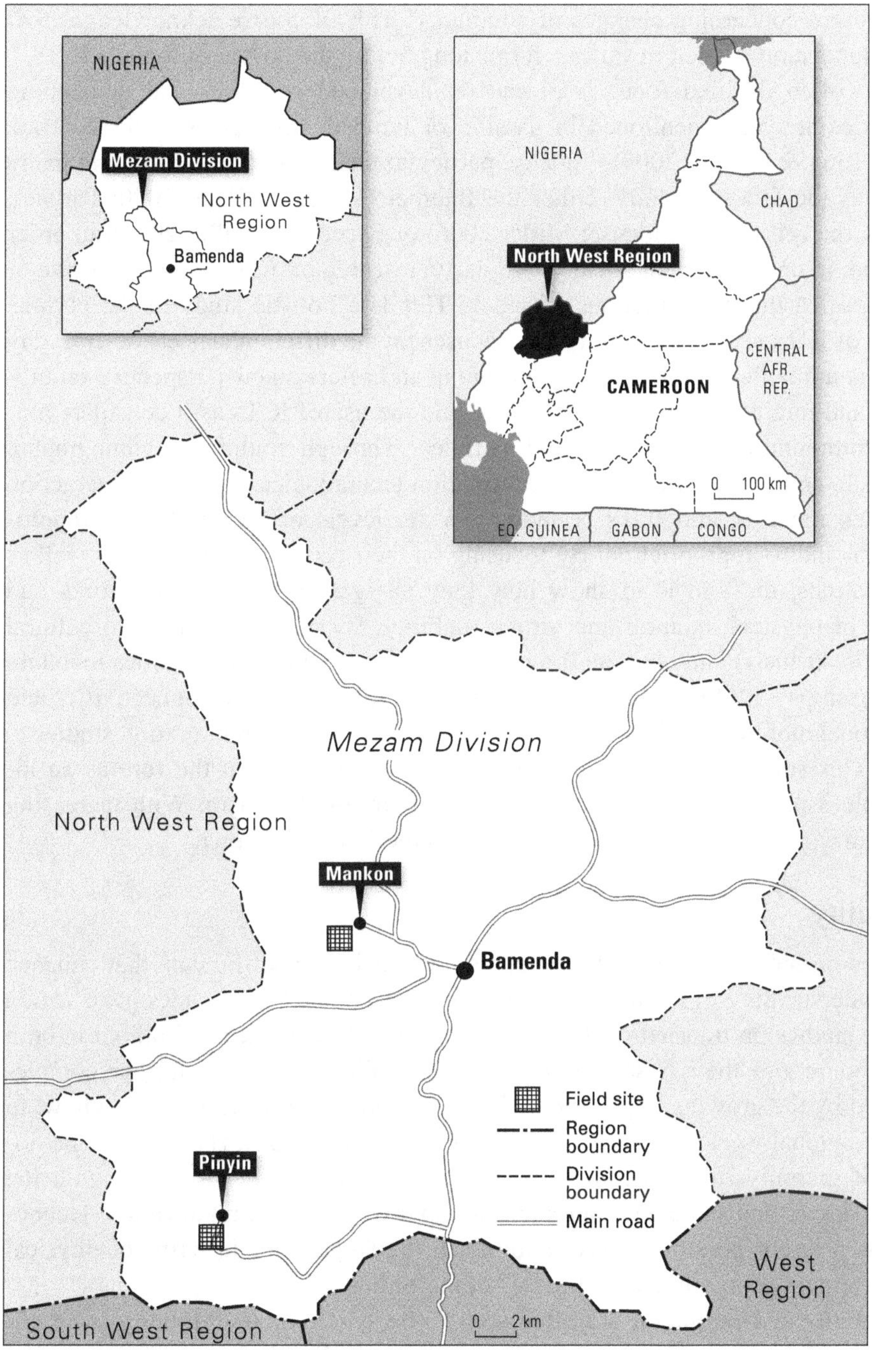

fit that has occurred with regard to improved ICTs, or how the past has informed the present concerning changes or continuity. This of course acknowledges that communication existed in various forms long before the arrival of today's ICTs.

Too often detailed focus is placed on developed countries and developing countries are only mentioned in passing or lumped together as 'Africa' (Katz 2008; Ling & Donner 2009). Equally, particular technologies are chosen in many cases as the focus of study, either the Internet (Miller & Slater 2000; Parham, 2005), the cell phone (Horst & Miller 2006) or Facebook (Miller 2011), in order to zoom in on a particular society. Similarly, research on ICTs is not extensive or broad, often due to limitations of space. This is a holistic study of the various forms of ICTs *vis-à-vis* mobility that focuses on the different aspects of everyday life that a mobile community may engage in and offers a new perspective on mobility and migration – the in-betweens – and the use of ICTs as it considers mobile communities, individuals and families. Through traditional ethnographic research, coupled with the use of information technologies, the social interaction between mobility and ICTs is shown, as are levels of social shaping, appropriation and re-appropriation. By focusing on their economic activities as well as associations, the aim is to show how they navigate the various networks and forms of physical, cultural and virtual mobility. By paying attention to cultural activities (rituals) and how the Pinyin and Mankon mobile communities maintain relationships with kin in the home country, attention is paid to connectivities and emotional mobility, notions that have been largely ignored in migration studies to date. This study aims to fill this gap by placing emotions at the fore as an intangible aspect of migration that accounts for transculturalism. With increasing mobilities, there is a tendency for increased emotion to be shown.

Mobility

Studies on mobility and migration contain extensive scientific data that enhance our understanding of connectivities and flows. The last two decades have seen a rise in studies on transnationalism that shows migrants having a foothold in both their home and their host countries. These studies have been further complemented by the growing appeal of studies on mobility and ICTs that show how, in today's global world, mobility is being enhanced and informed by ICTs. The notion of mobility often encompasses physical displacement. However, with the world increasingly being interconnected by wireless communication technologies, it is not possible to see mobility in the singular or in terms of physical displacement anymore. Hence the talk of mobilities.

Mobility and migration are often seen to be synonymous. Mobility is an all-encompassing form of movement that is not mutually exclusive to persons but seen as the perceptible and simultaneous shifts in both spatial (place) and social locus. It presupposes that one kind of movement is not the focus, given that there is a mobility of ideas. And as humans, they are mobile with their habitus – practices, ideas, beliefs and social and material culture. In conformity with Adey's (2010) notion of being "mobile with" migrants seek to reproduce or re-invent customs that they are familiar with to suit the host context. These include cultural practices as birth (born house) and funeral (cry die). The notion of mobility has gone past the idea of physical displacement. It is in this light that de Bruijn *et al.* (2001), with reference to nomads, talk of 'travelling cultures'. If social mobility, as espoused by Elliot & Urry (2010), 'becomes the de facto stratification of migration between migrants and non-migrants', can we also talk of stratification among the mobile communities in host countries?

The term mobility, as such, cannot simply be used to denote the movement of persons but rather by looking at it from the notion of constellations of mobilities – persons, goods, ideas, social, cultural, virtual, and emotional mobilities. These forms of mobilities have been enhanced by the new communication technologies. They 'connect the analysis of different forms of travel, transport, communications with the multiple ways in which economic and social life is performed and organized throughout time and across various spaces' (Urry 2007: 6). I also use the concept of mobility to refer to the itinerary of migrants, as well as the various things they carry to their host country – the movement of food, photos, objects and ideas to reinforce the notion of cultural mobility. By so doing, this permits me to adopt, on the one hand, an historical geographical approach that will enable an in-depth analysis of the 'formation of movements, narratives about mobility and mobile practices', by looking at mobility holistically (Cresswell 2010: 17). On the other hand, this also allows me to ascertain to what extent communication technologies have enhanced mobility and connectivity.

Conversely, migration refers mainly to a person's geographical movement from one place to another, permanently or temporarily. While mobility is the preferred *modus operandi,* the terms will be used interchangeably, denoting their various forms as conceptualized here. This study goes beyond a simple description of the use and impact of ICTs and human mobility, or the myopic focus on the communication practices that migrant communities establish and maintain in either the host or home country with no regard to the other. It critically foregrounds an in-depth analysis of the multifaceted ways in which mobile communities interact in the host and home countries and with the wider migrant network.

In this study, I maintain that ICTs have been incorporated to enhance migrants' co-presence in interesting ways in the familiar on-going patterns of everyday social life, as well as to insulate them from the excessive demands of their families. I begin with the hypothesis that ICTs, especially mobile phones, shape and are shaped by particular relationships in particular ways and that these are mutually reinforcing. Many authors, such as Horst (2006); Horst & Miller (2006); de Bruijn *et al.* (2009, 2013); de Bruijn & van Dijk (2012); Nyamnjoh (2005, 2011) and Vertovec (2004), have demonstrated the profound importance of ICTs, especially the mobile phone, showing how its introduction in rural communities has catapulted the process of 'linking up' and has transformed everyday lives.

The trajectory of mobile communities here are those of a community that wants to reassert itself in the host society by travelling with whatever tangible or intangible objects they can take from home. This travelling culture or cultural mobility denotes the historical mobility of the people and their 'codes, structures, and definitions that enabled the massive transfers of prestigious cultural norms' (Greenblatt 2010: 11). Such mobility is embedded in current trends of flexible mobility that are 'facilitated through a wholesale reinterpretation of history and a change in its valence' (*Ibid*: 13; de Bruijn *et al.* 2001, 2013).

Transnationalism or ... ?

Recalling Greenblatt's notion of flexible mobility (2010) leads us to also interrogate the concept of transnationalism in the present context of accelerated flows, mobilities and connectivities. Should we not now be talking of 'transculturalism' and 'transcultural habitus' as a way of understanding the new complexities of mobilities and connectivities? The era of migrants being uprooted from their societies and cultures and moving to a host country is fast becoming a thing of the past, as migrants today seek to maintain a life of double engagement and a foothold in both the host society and the home country (Mazzucatu 2008). This is equally reflected in current social sciences studies where the study of migration seeks to understand the places of origin and destination (Brettel 2000: 98). Stemming from the seminal studies by Glick Schiller *et al.* (1992), various studies[1] have provided insights into the concept of transnationalism. Acknowledging these contributions, but also being critical of the prioritization of the nation-state over the home village in today's globalizing age where the global and the local have been merged into a single arena thanks to strides in ICTs, I propose enhancing this debate by entering into dialogue with both transnational and transcultural concepts as I see them as different sides of the same coin. I go further by bringing in a second concept, namely 'transcultural habitus'. What do these concepts represent? Borrowing from Bourdieu's notion of habitus (1990) and Guarnizo's (1997) concept of transnational habitus, I take transcultural habitus as the conscious and unconscious habitual and internalized cultural practices that migrants

[1] Basch *et al.* (1994); Glick Schiller *et al.* (1995); Levitt & Glick Schiller (2004); and Vertovec (1999, 2004, 2007).

engage in and cultivate to express their identity and belonging to the home village while at the same time maintaining links with their home and host countries. What this suggests is that there is a paradigm shift from rootedness (assimilation) and deterritorialized subjects (Basch *et al.* 1994) to one that is increasingly accompanied by the re-inscription of identity into the territory of the homeland. Vertovec (2004) offers more clarity by claiming that transnationalism does not cause modes of transformation but rather that transformations are caused by each practice that contributes significantly to the on-going processes of transformation embedded in the historical trends of a long-distance social network. Such trends I note are attributed to the concepts of transculturalism and transcultural habitus. Transculturalism, especially in the host country, has received less attention and, when mentioned, this is usually only briefly. This study, while it takes as point of departure the concepts of transnationalism discussed above, shows the increasing trend of trans-culturalism amongst Mankon and Pinyin communities. The study fills this lacuna as regards transculturalism. Culture here is conceived as an:

> … evaluative conversation constructed by actors out of the raw materials afforded by tradition and on-going experience. It is continually modified by them in processes of social interaction and their behaviour is guided by anticipation of such cultural behaviour. (Hammel, 1990 cited in Brand 2000: 10)

I would also add that their behaviour is socially and culturally constructed, and is produced and reproduced by actors. If this is the case, should we not consider mobilities as the way migrants form their own habitus based on their experiences of mobility and a mixture of local, national and cosmopolitan cultural references, and how rooting occurs? Following this path signals a gradual shift from transnational theories and emphasizes instead the 'dialogic dimension of the migrants' transcultural and transnational ways of being and belonging' (Landolt 2001), which enhances the understanding of the cosmopolitan concept and sociocultural transformation. While this approach is quite useful, we should remember the fact that a cosmopolitan lifestyle is not a *sine qua non* for all migrants' connectivity given the inaccessibility of the Internet in rural African communities (including Pinyin and Mankon). Transnationalism and modes of connectivity are thus often useful in explaining the global trends of connectivity. Bringing in transcultural habitus, the aim is to show that migrants' activities in the home and host countries are guided by a sense of what they already know and a sense of learnt practice because it defines who they are and therefore re-enforces their belonging to the home village.

The path mapped by Bryceson & Vuorela (2002: 7-8) as regards transnational families underscores how 'families shape and are shaped by movement, separation, and reunion, and the boundaries and vistas they establish for themselves'. Like other transnational communities, transnational families are guided by relational ties, with welfare and mutual support being fundamental to their existence. However, their overt focus on the transnational trajectory of transnational family inhibits them from linking these trajectories to the emotional transculturalism that is intertwined with it. This is a linkage that I capture in this study. In the same vein, I position my study from an historical perspective and unequivocally state that transnational migration, though the concept may be relatively new, is not new but goes back to earlier mobile migrants, especially those who were engaged in bush trade, rural-rural migration or went to work on the plantations during and after colonialism. This idea has also been proven by Thomas & Znaniecki (1958: 303-15)[2] in their study of Polish peasants in Europe and America decades before the coinage of the term 'transnationailsm' when showing how letter writing was of prime importance between migrants and relations in the home country in the absence of the telephone (and of course the Internet). Mobile communities have today moved beyond letter writing to voice communication and other forms of faster speech communication (the Internet).

These concepts allow us to understand the new trends of dual engagement, as well as permitting me to delve into various ethnographic spaces in an attempt to understand the social interaction and everyday life enmeshed in mobilities and the appropriation of ICTs. By introducing these concepts, my study contributes to the development of further frameworks that offer an understanding of the complexity of mobilities in the age of ICTs. ICTs have become the interface between the wider mobile communities and families who stay behind (Horst & Miller 2006; Vertovec 2004).

So far, I have recognized that links exist between mobilities, ICTs and network/network societies (communities), as has been underscored by several authors.[3] This presupposes that in today's age of mobilities, technology and greater interconnectedness, ICTs, mobilities or network societies could be studied in isolation. It is in this respect that Panagakos & Horst (2006: 115) call for studies that subscribe to literal and horizontal accounts that invite us to look at the 'process of production', consumption, appropriation and re-appropriation, highlighting the ways in which ICT use are 'bundled'. I contend that it is only by studying mobilities, including transnationalism, transculturalism and transcultural habitus, ICTs and networks in a holistic manner that we can highlight the everyday practices of mobile communities. In doing so, we start to realize how ICTs have come to propel mobility, which is embedded in the daily lives of the people we are studying. Equally, by focusing on the historical pattern of mobility, we understand trends in contemporary mobility and society.

---

2 Cf. Poitier (2000).

3 Urry (2003, 2004, 2007); Castells (1996, 2004); Castells *et al.* (2007); de Bruijn *et al.* (2009, 2013); Brettel (2000); and Mazzuccato (2008).

Pinyin and Mankon as frontier people

Pinyin is a mountainous area and the village is traversed by small streams that can turn into rivers during the rainy season. Pinyin people, according to oral history, migrated from Widikum sometime in the 17th century following a chieftaincy struggle and then settled in their present site (Tasi, unpublished manuscripts).

As a result of wars and in a quest for good arable land, the Mankon people, who originated from *Nsahnyoum* in northeast Africa, migrated to the Mbam Valley, the land of the Tikars, in the 13th century (Warnier 2012: 42; Ncheawah & Tanibang 2004: 3-4). After a series of migratory movements, they settled where they are today.

Pinyin is in the North West Region of the Grassfields, about 45 km southwest of the regional capital of Bamenda. Pinyin is the largest clan in Santa Subdivision in terms of population and land, covering an area of about 750 km^2. It falls under the jurisdiction of Santa Council.

Mankon is in the Mezam Division of North West Region. The entire region is referred to as the Bamenda Grassfields and Mankon lies on the Mezam River that rises on the *mushiig* (Bamenda Escarpment). Unlike the geographically remote Pinyin, Mankon is only about 8 km from the centre of Bamenda and does not experience the telephone problems that Pinyin has. In contrast to Pinyin, mobility in Mankon was rural-urban, with some migrants going to the Coast to work on the plantations.

While most villagers, Mankon included, practised rural-urban mobility, the people from Pinyin were among the exceptions that focused on rural-rural migration.[4] This took them to Nkambe, Mesaje, Sabong-gari, Fonfuka, Fundong and Njinikom as bush traders and cattle owners/traders. Its geographic location in the highlands of the Grassfields overlooking Bali, Batibo, Santa and Bamubu may explain the Pinyin people's mobility in an attempt to discover what could benefit them financially.[5] Their mobility was characterized by weekly visits to neighbouring markets in Bali, Ashong, Santa, Guzang, Bawaju and Widikum and this later evolved into cross-border trade between Cameroon and Nigeria and West and East Cameroon during the colonial period.

Trade with East Cameroon saw the introduction of coffee planting in Pinyin as traders came back with coffee seeds. This perhaps explains why a customs post was set up at Santa during the colonial era and is still operational today. After independence, Pinyin men used to trade palm oil in Nigeria and come back with cigarettes, gunpowder, liquor and textiles; some of which were sold in neigh-

4 Conversations with papa Gabriel (bush trader), Pinyin: 27/05/2010 and 14/01/2011 and papa Anyam, Bamenda: 21/09/2011.

5 Personal conversation with papa Nkeng, Bamenda: 07/10/2011.

bouring villages. Most of the textiles were sold in the West Region to the Bamilekes.

In Mankon, I can still recall my paternal grandfather and some of his peers who were long distance traders between Calabar and Cameroon taking palm oil from Metta and bringing back kerosene, dane guns and gunpowder as well as cloth. As mentioned by Warnier (2012: 51) and corroborated by Pa Muluh, a bush trader who visited markets in Boyo and Donga & Mantung divisions, 'regional and long-distance trade was carried out through elaborate institutions including professional traders and porters, synchronized markets, currencies (cowries and brass rods), savings associations and a credit system'. Chapter 6 discusses these practices that have simply been transposed to Cape Town by both the Mankon and the Pinyin communities.

Their mobility was rudimentary as it was marked by long-distance trekking, head porters and indigenous forms of communication with kin networks and those forged in the process and providing accommodation and respite. Such long-distance mobility and trade brought matrimonial alliances and diplomatic relations with other chiefdoms (Warnier 2012).

Accounts of the histories of Pinyin and Mankon have always included mobility, which was facilitated then by indigenous forms of communication such as messengers and village criers and influenced contemporary migrants to embrace mobility. This mobility has continued unabated until today but in various forms: rural/rural, rural/urban and international. And nowadays it is embedded in the notion of 'bush-faller', where the loci of migration are synonymous with bush/farm and the current generation of migrants are expected to work hard and send home remittances and lavish goods and to help develop the home village (Nyamnjoh 2011; Pelican 2008; Alpes 2011).

As one of the villages that leapfrogged into wireless communication, Pinyin was connected to mobile telephony in the mid-2000s and faced challenges regarding network reception. On account of the village's rocky terrain, communication by road and mobile phone (due to network reception problems) presented difficulties, especially in the wet season. People still had to go to the tops of hills and/or climb trees to make calls. To date, Pinyin is only connected to one of the country's three mobile telephone service providers, namely Mobile Telephone Network (MTN). In addition, the problem used to be further compounded by a lack of electricity for users needing to charge their phones. This had to be done, for a fee, in bars and shops at night or in homes that boasted a generator. The scenario gradually changed when the village was connected to the national grid and phones could be regularly charged. This had a huge impact on mobility and connectivity between those who stay and migrants. It led to an increase in traffic in communication and connectivity.

The rapid encroachment of the city towards Mankon village has resulted in the distinction between village and city life becoming blurred. Most Mankon residents commute into town now for their daily activities. Mankon had the lone landline that connected the Mankon Palace until the early 2000s but the village experienced wireless landlines when Cameroon Telecommunication (CAMTEL) introduced this new service. However, mobile telephone services reached the village when the rest of the province was connected in 2000. The arrival of telephony has connected the village to the rest of the world and increased the speed of connectivity and mobility. Compared to 1999 when there were about 6 Pinyin migrants in Cape Town, the number today is over 100. In the same vein, being connected has seen mobility amongst those who stay at home. ICT has given them the opportunity to 'go places' through the calls they make. This was previously unimaginable. It has not changed the notion of the bush market trade but it has offered them the opportunity to make more informed decisions.

Contrary to the rudimentary forms of communication that marked earlier mobility and migration patterns, mobility today is largely defined and informed by ICTs as well as by old forms of communication (village/town crier). Juxtaposing these trade trends to contemporary socioeconomic activities among Pinyin migrants in Cape Town, trade patterns are identical but have been facilitated by communication (information) and knowledge about the goods in demand as people now often make business transactions without having to migrate or may initiate a first visit and subsequent transactions are done electronically and/or by phone.

It is critical to understand present-day mobility patterns and reconstruct the past to comprehend and assess how patterns then may have informed the present trend of mobility. From this brief sketch and historical account of Pinyin and Mankon migrations, we can infer that the aim of migration has been mainly in search of fertile farm land and to be strategically located geographically, with natural buffers warding off imminent threats from neighbouring villages.

Similarly, the search for greener economic pastures in new destinations is central in the current trend of migration. Prior to migrating to post-apartheid South Africa, favoured destinations were mainly Europe and North America. However migrants are now selecting destinations in the new South Africa, particularly Cape Town and Johannesburg, which are the preferred destinations for most Pinyin and Mankon migrants. Other emerging African countries with economic prowess are also popular with these migrants too.

The Pinyin and Mankon are represented in Cape Town by their respective associations: the Pinyin Family Movement (PIFAM) and the Mankon Cultural and Development Association (MACUDA). Like their predecessors, the essence of mobility is not about severing ties, it is about maintaining relationships across borders, the search for belonging and creating a new life. When people move, they do not do so in isolation as they carry tangible and intangible objects that remind them of the life they have left behind in their home village. They also follow the paths taken by others before them and maintain connections with those they leave behind. This presupposes that communication is possible between migrants and family in the home country, between migrants and would-be migrants, and amongst migrants in the host country. There has been an increase in mobility patterns. Beyond this dialectic of home and host country, migrants have carved out a space for themselves and are seen as a mobile community that is deeply embedded in its cultural practices through the daily social and cultural activities that they participate in. Such formations in a mobile community are in contrast with findings in earlier studies that posit that informational society creates a fourth world of marginalized people who have been bypassed by communication technologies (Castells 1996, in Donner 2008: 29-31). In fact, to conceive communication in this light is to obfuscate the interpersonal communication and connectivity that have made it meaningful to regions like Pinyin that have leapfrogged to wireless mobile communication. As such, connectivity has become the norm not the exception and has led to a mobile community and society.

Although Pinyn and Mankon migrants may be considered as residing in the margins, or as marginal, I see them as 'frontier' people (Kopytoff 1987) because they display the characteristics of those perpetually on the move and who are in search of a niche for themselves. They are constantly exploring frontiers. Given that they are mobile as a community, they have become each other's keepers. How then does one extrapolate from their frontier nature the 'workings' of earlier societies to make sense of the contemporary one? To do so, I focus on social and kinship networks that have become fundamental to the Pinyin and Mankon migration and mobility process and how the new technologies are enhancing these networks. I look at the different practices that underpin the societal fabric and how it defines ideas of belonging and home.

In the remainder of this section, I focus on the concepts that are seen as the main issues speaking to mobilities in relation to the formation of mobile communities: social and kinship networks, society and technology, habitual practices and the notion of belonging and home.

'Tuyau' and 'lines': Social and kinship networks

Mobility occurs due to the interaction, multiple networks and community solidarity that members are able to forge with one another and amongst other migrant communities. In this respect, social and kinship networks provide an avenue to unpacking the constellations of mobilities, given that they are propelled by networks. My interest in personal/kinship networks, linkages and network society was heightened by the respondents in the villages and how they see themselves as part of the information superhighway and how they are connected with their children and other relations far away by the click of the call button on their mobile phones. Equally I was intrigued by the ways in which migrants are connected at various levels in the host country and with other migrant and ethnic migrant groups and Internet-mediated forums, and how these levels of interconnection make them belong to a shifting network society that would have otherwise been impossible. Communication technologies in this regard are creating a network society that helps to expand mobility and the migration industry. However, whatever the different levels of connection and networks that are forged, they still continue to maintain strong social and cultural ties with the home country.

A network society and community is an important prerequisite for migration as well as enhancing the survival of migration and mobilities respectively. If migration is a dynamic system that takes into account sending and receiving countries, then this also implies that it centres on linkages. The social network and a network society are important within the framework of this study as they provide an understanding of the different social field as it is seamed by networks.

Migrants often have a defined terminology to refer to particular networks. In this regard, 'line' is used to refer to migration syndicates that assist in the migration process. It is thus common to hear a prospective migrant say that they have found a 'line' for a particular destination. And for these 'lines' to function, one needs to pay for them. 'Line' in this sense is synonymous with *tuyau*, a French word (English: pipe) that is used by migrants in Cape Town.

Following Fawcett (1989: 673), networks/linkages do not only refer to the persons involved that make up this network. Instead, I take into account the elements that are dependent on the people, such as communication between migrants and would-be migrants and families trying to use their networks to secure migratory passage for their children. Networks are, according to Castells (2004: 3), 'process flows that channel information between nodes'. The node is 'significant in its ability to contribute to the network's goal' (*Ibid.*). This implies that linkages/networks are not simply examined at one location but in the entire migration system both at home and in the host country. And at the centre of these interactions are social networks that give meaning to every action in the process. Networks thus occur at multiple levels and invariably tend to cooperate or compete with each other, thus enhancing our understanding of mobilities, the mobile community and its intricacies. By extending the framework of networks to include home and host countries, the communication nodes that facilitate these linkages and networks permit me to move away from anachronistic single-site research to multi-sited research, and to simultaneously capture community solidarity and that of kinship as well. In other words, it sheds light on the multifaceted networks, be they personal, kinship, community/ethnic or agency/migration industry, and when and how they are played out at the different ends. However kinship and personal networks, also known as strong ties (Granovetter 1973), include various forms of assistance. What marks these networks are the lasting impacts the networks have on migration given obligation towards kin relation are of an abiding nature (see also Nieswand 2011). The speed, power and connectivity of networks have increased due to improvements and transformations in communication technologies, especially in mobile and wireless technology, that have given users the chance of interaction. Could this be the imminent change brought about by ICTs on network? Of late, the possibilities of accessing the Internet on the move with smart mobile phones have increased interaction and, by extension, virtual mobility. Arguing along similar lines Mitchell (2003: 144, in Castells 2004: 11) notes that 'wireless connections and portable access devices create continuous fields of presence that may extend throughout buildings, outdoors and into public space as well as private'. This has profound implications for the relocations and spatial distributions of all human activities that depend ... upon access to information'. As regards access to migration information, weak ties (Granovetter 1973) play an enhancing role. These are ties that are not strong enough to be considered within the frame of regular networks (kin and kith) yet are not excluded nor included and are quite useful. In Chapters 6 and 7, I refer to some of them as legal networks – lawyers, staff at the department of Home Affairs (charge with issuing/extending refugee permits) and migration syndicates that assist migrants with legalizing their stay.

Networks do not necessarily entail strong ties with lasting impact on relationships. While we talk of strong ties, the reverse can also be true. There are weak ties that in some instances perform tasks that strong ties cannot, such as the migration syndicate. Often this tie is orchestrated by family and personal relationships that set the process in motion financially. The same can be said for refugee permits, where an unholy alliance has been formed between migrants, Home Affairs officials and lawyers (legal ties) to perpetuate the process. Granovetter (1973: 1360) notes that it is through these 'networks that small-scale interaction becomes translated into large-scale patterns, and that, these in turn feed back into small groups'. Not limiting the network approach to strong ties/networks because these are standard ties, the weak ties have the ability to bridge both strong and weak ties. However, contrary to Granovetter's notion that 'no strong tie is a bridge', I would maintain that strong ties are a bridge in themselves, although not exclusively because they depend on weak ties to negotiate the migration process of relations. They are the financial facilitators. Weak ties in this case would refer to the migration industry/syndicates and agencies that facilitate the migration process from Cameroon to South Africa. How does the notion of strong and weak ties inform the population ratio of Pinyin and Mankon in Cape Town?

Society and technology

The interaction and conviviality between ICTs and society has enhanced migrants' navigation between networks. This section considers society and technology, with particular reference to the Actor Network Theory (ANT) widely associated with John Law and Bruno Latour. Latour (1996: 369) maintains that the ANT 'aims at accounting for the very essence of societies and natures without limiting itself to human actors only but extends to the word actor – or actant – to non-human, non-individual entities'. Succinctly put by Kien (2009: 43), 'ANT seeks to expose the intimate relations between humans and technologies, actors and actants and network', and how they manifest themselves in practice. It is this intimate relationship that I seek to examine here and how the relationships and appropriation propel networks (de Bruijn & van Dijk 2012). I argue that actors and actants cannot be placed on a par because, on their own, phones (actants) would never ring or take photos without someone (the actor) giving the command. But as actors, humans are endowed with the know-how to give mobile phones the specific task of performing. However, concurring with Latour (1988: 159), their primary focus is building and bridging networks and to maintain this position, an actant continuously builds relationships with other actants. The relationship between ICTs and mobility therefore is highly intertwined as ICTs today play a determining role in mobility (dis)connection and networking. Connections thus become central to ICT insofar as relationships are (dis)connected, configured and reconfigured.

An actant can be anything provided it is 'granted to be the source of an action' (Latour 1996: 373). Although Latour holds that the main criticism for this theory is the 'complete indifference for providing a model of human competence' (cf. Mutch 2002: 487), or by not recognizing the higher agency and power of humans that, in my opinion, point to the limitation of this theory as well. Although power is not tangible, it is an effect of the capacity of actors on the networks – a capacity that communication technologies possess. I imagine that the arrival of mobile and communication technologies could invite us to re-examine this critique and the whole notion of ANT and give adequate importance to human agency with regard to how it relates to communication technologies (Rose & Jones 2005: 27). Throughout this chapter, I have underscored the relationship between mobility, networks and ICTs and how their speed and connectivity have increased as a result of the inextricable relationship and appropriation of technology by society.

This infiltration of ICTs into our daily lives and the use of technologies I term the appropriation of ICTs by mobile communities. In doing so, a convivial relationship has been created whereby ICT is shaping the lives of the users and by the same token the users are equally shaping and adapting ICTs in their lives not just to serve new adaptations but principally to serve their old motives by developing innovative ways of doing it (Brynin & Kraut 2006; 8). As a result of their social shaping, appropriation and re-appropriation, they have formed symbiotic relationships that have become actors, and actants on each other, something Rose & Jones (2005) refer to as the 'double dance of agency'. Technology is meaningful insofar as it serves the purpose for which it is meant.

It follows that technologies and humans build relationships and networks over time – actor-network – that may be unstable but are held together by the continuous engagement of the actors (Kien 2009; William-Jones & Graham 2003). However, actors at times may resist being involved with a network because of their independence (agency) that can lead to (dis)connection. There is a 'glue' that attracts them to each other, something Latour (1996) refers to as 'translation'. In other words and as noted by William-Jones & Graham (2003: 275), 'each actor has its own diverse sets of interest, thus a network's stability will result from the continual translation of interests'. Often, this is to have a continued sense of connecting with family though far away. This suggests that by the very essence of their translation (social shaping and re-shaping), networks mutate in multifaceted ways as they are appropriated by different users in unintended ways. They go beyond the normative values and function built into them, as they are not mere passive entities.

This is what I consider as the social shaping and re-shaping of technology and society respectively, and their interaction, when appropriation and re-appropriation occur, which is the focus of how we see the interaction between society and communication technologies. As Lehoux *et al.* argue (1999: 440, cited in William-Jones & Graham 2003: 277), the 'meaning of a technology is not entirely a given a priori, but rather, progressively constructed (*and reconstructed, I may add*) through social practices'.

ICTs come in a conglomerate of what Slater & Tacchi (2004: 3) refer to as 'communicative ecology' that embodies the totality of ICTs: radio, print media, computer, mobile phones and more recently social media and co-presence interaction. We should not lose sight of the choice, role and power of distinctive technologies, with some more accessible and easier to use than others. This availability provides users with informed choices about which technology to adopt and adapt to, and which to use for specific purposes. Arguing along similar lines, Ling (2004) shows that there is preference by others for a technology that enables a co-presence experience, thus making Skype calls by far the most widely used form of communication besides the mobile phone among transnational migrants with Internet-literate family and friends.

With regard to this study, the ANT provides mobile communities with the savvy of bridging networks given that these technologies are shaped by consumers in how they translate them for their own usage. In the process of translation, technologies may meet the standards of consumers and be incorporated into their everyday lives and affect collective wellbeing through its sustainability and multiplicity of networks. However, it could also lead to inequalities in society with the haves and have-nots (between those with 3G smart phones and those still using 1G and 2G phones). This effectively means that they are excluded from accessing the apps that are available to 3G users. This also shows the coexistence of old and new technologies, giving actors the agency to choose which network to use. It also helps to see both the actions of machines and those of humans simultaneously and to 'understand how the agency of humans and machines are mutually dependent and intertwined' (Rose & Jones 2005: 33). This does not take away their agency as they have found ways of translating the technology to suit their needs. While maintaining networks is thanks to the conviviality between humans and technology, it is also informed by learnt habits that have been facilitated by ICTs.

Habitual practice

According to Bourdieu (2009), such kinship relational networks and other social networks are informed by the internalized culturally accepted norms and habits that they have learnt over the years. This leads us to the notion of habitus and loci and the logic of practice of everyday life. He claims that habitus highlights structured responses and structured behaviour as informed by rules of behaviour that have been internalized. But at the same time one has measures and possibilities (agency) to transform these rules. The environment or context that informs behaviour allows the possibility that what is internalized is transformable within the logic of everyday practice. Thus habitus makes sense only within a given locus or context. How does the notion of habitus inform mobilities, social and kinship networks? Although Bourdieu draws on Durkheim's notion of collective consciousness in the sense that it regulates behaviour, the former's idea of habitus is not as frozen as Durkheim's. As such, habitus is different from one field to the other, and thus structured. Again Bourdieu argues that even if one is committed to a habitus, that everyday context depends on the extent to which one can reproduce it effortlessly and this translates to a sort of everyday practice of that habitus. It is in the process of reproduction that habitus is transformed. The notion of habitus is thus born out of Bourdieu's social reproduction – the process by which a society maintains itself over time. It is 'the capacity each player of a game has to improvise the next move, the next play, and the next shot' (Calhoun 2003: 3). This manoeuvring does not occur in space but rather in a social field where each actor tries to exert coercion on others by playing the game that is specific to that field and what is at stake. With specific reference to knowledge of the rules of the game, it requires that 'every field of social participation demands of those who enter it a kind of preconscious adherence to its ways of working. This requires seeing things certain ways and not others, and this will work to the benefit of some participants more than others' (Calhoun 2003: 18). In this respect, when jostling for the extension of a refugee permit for instance, migrants are astutely versed in the game and its rules (shrouded in secrecy and mediated by middlemen). Hence, practical knowledge of the rules in any game in a given field, as well as being shrewd in understanding the values and rules of the game is of particular importance. In relation to the fields of pitching (economic activities, Chapter 6) and that of the various associations (PIFAM and MACUDA, see Chapter 7), these are two different fields where play is executed by the same actors but the objective structures are different. To reconcile both the power of the structure and the subjective choice that gives social agents agency, Bourdieu saw social agents not simply as following rules but as dynamic agents who could improvise and adapt to changing circumstances that called for going beyond rules and for acting strategically and interestedly (*Ibid.*: 19-20).

The structure of this field is often controlled by those with economic capital (wealth) who use it to acquire cultural and social capital. On the whole, the defining resource is social capital as it has the ability to bridge and bond (de Bruijn & van Dijk 2012) and strategies are of significant importance for participants in living up to the games in the different social fields. Strategies are inert and derive from a *sens pratique* (practical sense or logic). In migrants' associations, the structure is ruled by the elders and those in positions of power, such as the president and the vice president, although subversive power is held by younger people and newly arrived migrants, especially if they want to thwart attempts that impinge on their financial stability without any direct reward. The ethnographic chapters in this volume discuss the various forms of networking, ethnic enclaves and multiple alliances that underpin actors' actions and play out in the various fields amid constant improvisation and strategic creativity. This also means that games are subjected to continual change. As Calhoun (2003: 4) puts it, 'when we improvise our actions, we respond both to the social and cultural structures in which we find ourselves and to our own previous experiences'. To what extent are Pinyin and Mankon mobile communities able to strategize and improvise in their daily activities?

From this perspective, my reading of the interaction among community and families, amongst other migrant communities and with ICTs is based on the practice of everyday habitus and social fields. It allows me to analyze whether interaction and connectivity have increased with the arrival of ICTs and whether the migrant communities have changed their everyday practices. This approach emerges throughout the chapters and complements the analytical framework.

This approach helps me to map the various social fields and provide an analytical framework with which to categorize the different levels of relationships and interaction that occur within each level of the game Chapters 6 and 7), the power dynamics and hierarchization embedded in such relational interaction. I am also able to assess the ease with which migrants improvise and straddle various social fields by reminiscing and emphasizing their belonging to home.

Belonging and home

Increased global mobility in the wake of revolutions in ICTs has rekindled interest in the notion of belonging. Ethnographic data show how migrants' quest for home and belonging is heightened, reconfigured and reinterpreted by their appropriation of information and communication technologies. In analyzing the notion of home and belonging, Heidegger (1993: 361) provides the framework of 'dwelling' to unpack this concept where dwelling means to 'reside or to stay, to dwell at peace, to be content or at home within a place'. Mobile communities' continuous 'mobile with' (Adey 2010) suggests that they do not consider Cape Town as home and, as such, do not belong because the notion of home carries with it a powerful connotation of belonging.

Using the bridge as a metaphor for connecting people and places and bringing streams and banks and land to each other's neighbourhood (de Bruijn & van Dijk 2012), I go even further and consider the new forms of ICTs as a bridge that is changing the notion of home and belonging, and links/delinks and reinforces/despises the notion of home and belonging in the case of migrants, while in the host country, they are disconnected and are instead connected to the home country, notwithstanding the geographical distance that separates them. Increasingly, ICTs are cutting them off from the host communities and linking them with home. In this regard, while the notion of home may speak to the physical archetype, it extends here to the ideational notions of home that migrants travel with and reconstruct over time. In addition, objects have become an integral part of mobility and belonging as they help soothe emotions of longing and separation. Objects transported from home include photos, architectural objects and 'sacred icons or consumer goods – fetishes attest to the family's virtue, their status and their stability' (Wright 1991: 214). They have become purveyors of nostalgia and the emotionalized ideas of home. These objects become a connecting device between the home and host countries, between relations and between the living and the dead, as some are given by default for having been crowned successor: they bridge and bind communities, an aspect that has been expounded on by van Dijk (2012) with regard to marriage.

The idea of home and belonging is inextricably linked to feelings of nostalgia and longing, although these may not be mutually shared. In this respect, home to the individual migrants serves as a mirror that reflects the family's portrait and binds them to it (Wright 1991). The home thus becomes the tangible and intangible container that allows both home and host communities to draw loyal support from to maintain a continuous sense of family and gain solidarity and strength from it. The home could well be physical land in the plantain grove behind the house where one's *nitong* (umbilical cord) is buried, where one retires or where one is repatriated for burial because 'home' to migrants is where one was born and thus where they belong. Belonging, above all, is marked by where one is buried (Chapter 7). In this regard, by self-consciously creating a home away from home through associations and sub-associations and taking part in different rituals, migrants are simply affirming their belonging. The commitment by both Pinyin and Mankon communities is rooted in the fact that they consider themselves as sojourners whose aim is to invest back home for future return, and they have no desire to actively integrate in the host society. Could such determinism be more fuelled by the policies of promoting legal integration than social integration that has created this sense of apathy among migrants?

Migration among both communities thus continues to echo the notion of home and is seen in what van de Veer (1995) calls the 'dialectics of belonging and longing' (cited in Oonk 2007: 17). Belonging juxtaposes rootedness with uprootedness, and establishment with marginality, while longing is related to the desire for change and movement. These dialectics of belonging and longing reignite a sense of identity that Oonk (2007) contends is dual or paradoxical in nature. Belonging is reminiscent of discrimination and exclusion, and also identifies positively with the highly esteemed Pinyin and Mankon heritage. Longing creates an awareness of multilocality that mutually reinforces belonging to the 'here and there' as well as sharing the same 'roots' and 'routes'. While belonging is considered in the dichotomy of autochthones and allochthons (Geschiere & Nyamnjoh 2001), I demonstrate a community trying to regroup in a new territory to recreate a sense of home.

The fact that members of these communities have consciously chosen to recreate their home communities in the host community points to how they want to maintain their personhood. If getting married is seen as a sign of integration, then it has to be reconsidered with respect to the communities in question. Marriage is clearly not a prerequisite to integration, instead those who are married to South African women or are in long-term relations with migrants have been co-opted into the community but they accrue the benefits derived from relationships to excel in business.

To sum up, I argue that while ICTs continue to enhance mobility for those who move and for those who stay put, they have become inextricably linked in forging networks and reconfiguring existing ones. Contrary to earlier studies that predicted change and transformation in societies that leapfrogged into ICTs, these new ICTs have been appropriated to enhance the workings of Cameroonian society. We should not see this as having changed society fundamentally or as having divided it, but as improving people's mobility and keeping them in touch with their homes and their cultural practices.

Research questions and outline of the book

The concepts above provide me with a roadmap with which to understand current trends in mobility and migration, the interactive nature between humans and ICTs, the level of connectivity among mobile communities and that between migrants and families back in the home country.

In the light of this analysis, the underlying question in this study is:

- How do people in mobile worlds navigate their social fabric and make sense of their individual being?

And as follow-up to the main question, I also pose the following questions:

- ICTs have been hailed by earlier studies for bringing about change and transformation in societies. To what extent is this true for Pinyin and Mankon communities?
- To what extent has the introduction of ICTs reduced the perception of mobility and migration as rupture between host country and home village?
- How do increased mobility and improved ICTs change the dynamics of transnational social spaces?
- To what extent has increased connectivity in ICTs and mobility informed communication between migrants and the home country as well as feelings of home and belonging?

In an attempt to provide answers to the questions above, each chapter in this volume focuses on the workings of the Pinyin and Mankon mobile communities and each is a piece in the puzzle. The volume consists of ten chapters including this introduction. The historical, theoretical and conceptual frameworks in this first chapter set the scene and show that migration and mobility are as old as the history of any given society. Chapter 2 introduces the fieldwork site and explains the process of data collection (methodology) and analysis. This chapter provides an account of how the field and I mutually engaged with each other and allows me room to reflect on my role as a researcher as well as charting my own personal navigation of mobility, connectivity and networking, both as a migrant and as an researcher providing a practical and empirical understanding to the stories told.

The analytical framework is guided by the concepts of mobility, network, society and technology, habitus and belonging.

Chapters 3, 4 and 5 are the core chapters and offer a greater understanding of the theoretical concepts of mobility, Information and Communication Technology (ICT) and network before the rest of the chapters (Chapters 6, 7, 8 and 9) present the data analysis. The final chapter, Chapter 10, summarizes the study and draws some conclusions. In an age of accelerated mobility and ICTs, it is increasingly difficult to extricate the themes described in Chapters 3, 4 and 5 and to place them under separate headings as I have opted to do here because of their overlap and the intricate web of connectivity. However, this separation is a deliberate choice related to a decision to further the contextual understanding of contemporary trends in mobility, ICTs and networking. Chapter 3 seeks to draw on the past patterns of migration to understand the current trends of Pinyin and Mankon migration. To do so, I have adopted an historical approach to explore this and go on to look at mobility and migration and the way they are, with communication technologies mutually reinforcing each other and creating a new mobile culture of virtual presence and co-presence. Chapter 4 questions how and to what extent the change in communication technologies has impinged on social relations. By focusing on Information and Communication Technologies (ICTs), with a historical note on how Cameroon has leapfrogged from its embryonic stage of poor accessibility of ICTs among the population at large to today's wide availability, we can comprehend the relationship between mobility and ICTs. In addition, the provision of basic amenities, such as electricity, has helped the process of connectivity as cell phones no longer go for days without being charged, thereby increasing reachability and connectivity between home and host country and among mobile communities. Chapter 5 examines the multi-stranded networks that migrants accumulate and how they navigate these forms of network. Together, these three chapters provide answers to all the questions while also laying the foundations for the subsequent chapters. The chapter also deals with network and the emergence of a network society with respect to the PIFAM and MACUDA and all the other associations that have emerged as Cameroonian in a show of solidarity. The accumulation of network underpins mobile culture and networks become important factor in social structure relationships as well as in the formation of snowball networks. Besides existing ethnic enclaves and social relational networks, there is a tendency to forge new ones and consolidate others. One such new network, which is of prime importance to the mobile communities in Cape Town, is the legal network of lawyers whose role it is to bail out migrants from police cells, arrange contractual co-habitation (life partnerships) for documentation and other forms of legal papers that permit migrants to stay in the country.

The following four chapters (Chapters 6, 7, 8 and 9) synchronize the 'being there' and 'being here' (Geertz 1988), which simultaneously informs the writing process and engages my 'ethnographic skills into textual analysis in order to persuade you (the readers) of the authenticity of the accounts' (Segall 2001: 581).

Chapters 6 and 7 discuss the ways migrants form different levels of bonding within their intra-groupings. Chapter 6 dwells on migrants' economic activities, elucidating on how migrants are constantly breaching laws and operating on the fringes of the long arm of the law when engaging in business activities. Compared to the economic activities of their parents in the past, ICTs have facilitated communication and knowledge and created faster business transactions with lower overhead costs. Chapter 7 offers in-depth accounts of both associations and their workings, with a particular focus on how they manage ROSCAs. These have come to be the backbone of the associations but, unlike in the past when these associations were engaged in home village activities, less out of fear of witchcraft than for the prestige it brings to those involved. To these new groups of migrant communities, participation in village development is considered as the 'noble' thing to do and participation is rewarded with cultural capital and respect from peers and the village community.

Chapter 8 explores the transcultural activities that migrant communities engage in and the various rituals they perform in the host country. These are essentially life-crisis rituals, with the most prominent being death. How ICTs have helped to transform the way these rituals are performed is also discussed. As a result of mobility, some rituals have become fluid and dynamic but performing them helps mobile communities develop as a fortress and reifies their identity and belonging, disconnecting them from the host country.

Chapter 9 presents additional aspects of transculturalism with a greater focus on the 'I' as opposed to the works of the associations, shifting from the sense of 'we' to that of 'I' (he/she), and it zooms in on the individual migrants and their relationships with their respective families and friends. Emphasizing individualisation illustrates the way individual migrants relate to and (dis)connect with family. It shows the in-betweens and challenges they deal with, the emotions that are generated and how they respond to them. While studies of transnationalism have focused more on transnational links with the home country than on the transcultural, this chapter acknowledges the influence of the transcultural, taking both perspectives into consideration.

With the exception of the introduction and the conclusion, all the chapters illustrate and provide answers to the research questions posed above. They show how mobilities and ICTs are linked and permeate the social lives of communities and the practice of everyday life. Through this permeability, we are able to appreciate the social processes of the appropriation of ICTs by societies and the social shaping and reshaping that influence both society and ICTs.

2

Methodological considerations and data collection

Introduction

This study is part of a larger research project entitled 'Mobile Africa Revisited' that has been examining the appropriation of ICTs and how mobile migrant communities are shaping and being shaped by them. The study is an ensemble of conversations, my observations and interaction in the various field sites as well as extensive interviews and the employment of other ethnographic tools that together have resulted in this book. This chapter is divided into two main sections. The first section focuses on my methodological reflections, paying particular attention to my role as a researcher, ethics and how I selected my informants, while the second section offers detailed insights into the methods used for collecting data and my analysis.

The study adopted a multi-sited approach and concentrated on migrants from the Pinyin and Mankon areas of the Bamenda Grassfields in Cameroon, following them to their new localities in Cape Town, South Africa and The Hague and Amsterdam in the Netherlands. The study adheres to the call by MacClancy & Fuentes (2011), Marcus (1995) and Gupta & Ferguson(1997) for a spatial and temporal fieldwork method that takes into account both the point of departure and of arrival and, in this case, a multiplicity of points of departure and arrival. Its particular ethnographic contribution is supported by a holistic approach and the intensive fieldwork and data collection are grounded in techniques of qualitative research. In the process, trust and confidence were gradually built up.

By drawing on Bourdieu's experiences of ethnographic fieldwork in Kabylia, Algeria, it was considered important to 'combine intimate knowledge (in this case of being Mankon and from the Grassfields) of practical activity with more abstract knowledge of objective patterns, and using the dialectical relation between the two to break with the familiar ways in which people understand their own everyday actions' (Calhoun 2003: 8; Bourdieu 2012). It is in this light that theoretical concepts, in conjunction with empirical findings, have led to a study of everyday life. These daily practices, as suggested by de Certeau (1984), are tactical in character and thus require observations from a bird's eye view that include perceptions not only the sense of sight but one that also involves all the senses of specific action be they physical or vocal. What gives this action validity is the meaning attached to the action (Holy & Stuchlik 1983: 17) and, in this vein, the observer needs to be conversant with the meaning. It is in this respect that Förster (2001) proposes that observation is and should be intentional, encompassing the whole world and any ensuing actions that emerge. Attending various tribal functions and meetings required careful observation and involved unpacking or paying attention to the various forms of speech, such as debates and comments made on the side. Conversations and interviews were constantly interrupted by phone calls and text messages and these meant that the observer needed to be alert to the different layers of interaction and connectivity to make sense of them. In the course of my interview with Clovis, for instance, he informed me that he had been issued with a 'you must go' message (30 days to leave South Africa as his asylum permit had not been renewed) and so he was actively looking for different options to legalize his stay and had contacted a number of persons who had promised to get back to him. The beeps, texts and calls were all geared towards providing him with solutions and so whenever there was a beep, we would stop for him to call someone back. By the end of the interview, he had settled on going for a life partner that was proposed by a friend's friend in the township (a weak tie). This scene speaks of the level of connectivity to the host society and the speed with which weak ties can function in bridging social relations, as well as any opportunistic tendencies.

As will be seen in Chapter 6, the Pinyin and Mankon migrants live in clusters and have thriving communities in which I participated, attending both public and private events. Being part of various communities in Cape Town and Cameroon as well as in the Netherlands obviously called for a multi-sited research approach. And I spared no energy in highlighting this multi-sited dimension of the research, as will be shown below.

An important way of trying to understand the migrants and communities I was researching was by adopting a reflexive/inter-subjective approach. In this respect, a reflexive approach that charts my own personal navigation of mobility, connectivity and networking, both as a migrant and as an interviewer, provides a practical and empirical understanding to add to the stories told here.

Methodological reflections

Reflexitity/inter-subjectivity

Being a migrant myself, having lived apart from my spouse and then finally leaving my country of origin thirteen years ago, my encounters with these mobile communities is a reminder of how our lives are inter-related. They are anchored in mobility and seamed by ICTs and networking on the one hand. And on the other, doing research amongst such communities was not matter-of-fact, even though, like them, I too come from the Grassfields of Bamenda, from Mankon, from a similar culture and, as a woman, understand the meaning of transnational parenting. Our narratives are interlinked in multifaceted ways, also as a mother who was constantly away from home during the duration of this study. My choice to do research among migrants and immigrants (Nyamnjoh 2010) was embedded not only in the self as a migrant, but culturally too. If a reflexive approach, according to Mason (1996, in Guillemin & Gillam 2004: 274), entails that the 'researcher should constantly take stock of their actions and their role in the research process and subject these to the same critical scrutiny as the rest of their "data"', then the desired effect has been achieved as there were moments in the field when I had to refrain from allowing my experiences from obfuscating those of the researched. Instead, I tried to maintain a reflexive distance. Arguing along similar lines but bringing in emotions as well, Holmes (2010: 140) perceives reflexivity as an 'emotional, embodied and cognitive process, in which the social actors have feelings about and try to understand and alter their lives in relation to their social and natural environment and to others'. In this respect, the choice of a reflexive approach concurrently helps to understand my own personal migrational trajectories and to better understand those of my informants, and is equally reminiscent of my emotional interaction and contours during my periods of separation both as an interviewer as well as a migrant and mother. It informs too the ethical decisions and the stance I took in and from the field. In this respect, this study is less focused on 'othering' than on incorporating a reflexive focus on 'own' and breaking down a specialized metaphor of the field (Gupta & Ferguson 1997: 2-5). Enhanced by memory – 'autonoetic memory' – that 'involves the ability to recall personal experiences and its different forms' (Tulving 1972, in Collins & Gallinat 2010: 11), our life paths interlink in multifaceted ways. It is in this light that Schank (1990) argues that 'memory organises experiences into retrievable units on which the self draws during social interaction', the baseline on which people (the researched or researchers) form social units. Taking the cue from Tulving (1972), Collins & Gallinats (2010: 11), opine that 'the self in consort with memory is implicated in ethnography; not only in the field but also in the study ... the self thus becomes the incubator of ethnography'. However, the question that arises is about the sort of memory that is triggered.

What does the position of researcher create in terms of data? How does it influence the production of data?

While this approach has been foregrounded by Collins & Gallinat (2010), Holland (1999) and Mauthner & Doucet (2003), it is by no means without its critics. As noted by Dyck (2010), reflexivity has been 'confined to the introduction and/or preface; the text proceeds as before and little is gained, and the information provided becomes too eclectic, limited and little commented upon'. With this critique in mind, I hope by the end of this volume to have challenged these critiques and not to succumb to such pitfalls by concentrating on the stories of those whose life histories form the basis of this thesis, taking into consideration the 'effects of my presence on the investigation', as is called for by Holland (1999: 476) before challenging Dyck's assertion by postulating that 'reflexivity enhances the most essential quality of our human existence: it is therefore the best criterion of a truly human science or practice derived therefrom'.

Ethics

Social anthropologists such as MacClancy & Fuentes (2011) and Powles (2004) have recommended that ethical observers take the utmost care during fieldwork research, but at the same time there is a call for researchers not to turn into philanthropists or aid workers. While admitting this may, in a way, taint the data that one collects, by the same token it points to anthropologists who, despite avoiding being seen as parachutists or even coming close to being seen as one, when the focus is less on the pursued information than on forging a relationship with the researched. It is, however, difficult to dethatch oneself from informants outside of the fieldwork and to pretend to focus only on the interviews held and be blinded to the physical and emotional stress that one's informants are going through. One has to be prepared for the unexpected. How then do we as researchers deal with 'ethically important moments' (Guillemin & Gillam 2004: 265) in the field or the unexpected that would not jeopardize or compromise our research and data? As a newcomer to the field, I have come to rely on my own savvy and ethics that are morally accepted within the community in negotiating such issues. What (Guillemin & Gillam 2004: 260) refer to as the 'home-grown' notion of reflexivity that actually encapsulates and extends the concerns of procedural ethics. Reflexivity, as such, not only produces knowledge or enhances but, in terms of ethics, it does sensitize us to 'ethically important moments' (*Ibid.*: 276). In the same vein and as several authors, for example Macdonald (2011), have indicated in an effort to strike a balance, de Bruijn (1998: 81) suggests that 'more room must be created for moments in fieldwork that do not convey information: moments of silence, of emotions' but also moments when the researcher goes out of his/her way to assist the community. She contends that these gestures are the right thing to do and not making them would be tantamount to unethical behaviour and there would be nothing left to bind the researcher and the researched. However, to have been indifferent to informants when they broke down during interview sessions would not have made me any different from a 'parachutist' whose major concern is to harvest information. Gifts and emotional empathy, while they brought us closer, concretized our relationship, allowed me to delve into the individual's psychic and experience their emotional problems but hidden challenges, which are rare moments in fieldwork.

Such moments in fieldwork I would not consider as strategies but as serendipity. However, on the spur of the moment, I would chose to switch off my recorder and, by turning it off, I was being sensitive to their challenges. Often I offered advice and/or the option of returning later, but the latter was turned down by informants as they said it was a way for them to accept a situation or make closure. It was a healing process for them. Visiting informants when they had new babies in the family offered me a chance to share their joy and such visits were less about interviewing than socializing. I was always alert in the process but we must guard against overindulging as this may lead to a total immersion that could create a situation where it would be difficult to distinguish between researcher and philanthropist.

Selection of participants

The Pinyin and the Mankon people from the North West Region of the Bamenda Grassfields are the target groups in this research for multiple reasons. First, Pinyin has a larger immigrant group than Mankon and it was intriguing as to why they would have such a sizeable group compared with others from the Grassfields. As such, bringing in Mankon would provide a more comparative analysis and allow for us to understand why the Pinyin continue to make up a bulk of the migrant community from Bamenda. The aim was to find out, being both from the Bamenda Grassfields and sharing a similar culture, what their modus operandi is in South Africa and whether they see themselves as separate entities. While distinctive identities are maintained, conviviality is highlighted as seen from the extensive community, cultural and economically based network they have created that is self-sustaining and equally maintains strong economic and social links back home. Secondly, compared to other immigrants groups, such as Zimbabweans, Mozambicans and Nigerians, these immigrant groups are less well known and the form of business they practice in Cape Town needs to be given some visibility. Thirdly, their dynamism in Cape Town is varied and while the Pinyins are more adventurous in terms of their business ventures, the Mankon are very conservative. And finally, Pinyin was chosen because people from here started migrating to Cape Town much earlier than the other groups.

From the list of names I was given (cf. Chapter 1) and the contacts I made during my first meetings with Pinyin Family Movement and Mankon Cultural and Development Association, a series of conversations and interviews ensued. In the process, a survey was carried out to examine the composition of the Pinyin and Mankon societies in Cape Town. It soon became clear that their stories differed remarkably depending on their time of arrival in the country. From the survey and without random sampling, respondents were selected based on their date of arrival and the accessibility of their families back home, given that some of their parents had not returned from rural-rural migration localities that they had gone to long ago. This selection strategy was intentional. It was necessary to be able to select respondents that captured the evolutionary and variegated stages of mobility, the availability of ICTs and their experiences with ICTs, as well as fitting one of the following years of arrival in South Africa:1996-1999, 2000-2002, 2003-present. The migrants' different years of arrival were seen as the proxies for independent variables that, I believed, would give varying outcomes in the mobility, availability, accessibility and appropriation of ICTS by the migrants. There are major differences in the migrants' experiences before the availability of ICTs for earlier groups, and the experiences of later groups are confronted with a multiplicity of choice and rapid advances in these technologies. What would such advances mean regarding mobility and the appropriation of ICTs for later migrants? And how does it depict the mobile community? One important criterion for selection was that respondents' families could be traced back to the home village or its vicinity to complete the circle.

During the selection process, I realized that women only featured in the later period (2003 onwards). What accounted for this late arrival? The first women came to join their husbands and it was only after this that they began to bring other female siblings. This also informs their transcultural family engagements, which were higher compared to those of males due to the conventional gender roles of caring that are expected of them. They are more family oriented and less individualistic.

Snowball sampling was an important aspect during and after the selection process. The few persons I spoke to drafted in their friends, especially those who travelled together to South Africa from Cameroon or those they believed had an interesting story they would want me to hear. I was also directed to meet those who were considered to be well versed in certain areas to corroborate earlier information or to get more information. This was not necessarily because they were part of the intended group of informants. I refer to these persons as 'casual informants' and to those who were regularly followed as 'regular informants'. Their role in the data collection process is 'non-negligible' as noted by Nyamnjoh (2009: 474) who maintains that 'some of the most valuable information in research comes from the most unlikely sources, times or place'. In this regard, chance or casual meetings with members from both communities were often informative and brought out instances of everyday practices. For instance, whenever I visit informants for born-house (see Chapter 8) or to offer my condolences, I always stayed alert and it was during one such visit that I learnt about changes in PIFAM's constitution and the ban on Pidgin being spoken as the language of operationalization during meeting sessions.

Without any prior knowledge of the composition of the communities, I had no planned number of persons to interview. This only emerged after I had decided who to follow on a regular basis. Even after selection, some were dropped and new names were added because of informants' lack of time or them not being as readily available as I would have liked them to be. I spoke with 26 Pinyin migrants in Cape Town, 2 in the Netherlands and 18 families back in Pinyin village. In most cases, I had at least 3 sessions with them, and I spoke to others on numerous occasions. However, in the Netherlands, I managed to have just one lengthy meeting with each of them due to their work schedules. As for the Mankon community, I spoke with 14 people in Cape Town, 5 in the Netherlands and 7 families in Mankon. There is a disparity between the numbers spoken to in Cape Town and in the respective villages. In Cape Town I spoke to siblings and in some cases to both spouses. However, when I arrived in the village some mothers (widows) had gone to stay with their children in other cities or were ill. Nevertheless, sitting at various Pinyin junctions (*shware* in Kwidegli, Metayen, Ngali and Ndapang) provided useful information not only from parents but also from the people with whom I interacted.

Some informants are highlighted more than others. This was for the purpose of in-depth studies and analysis, and was guided by three factors. First, the information given by them was corroborated by most other informants, secondly they turned out to be very articulate and capture what others expressed with difficulty and, lastly, some of them were the core informants who readily made themselves available for conversations and interviews without appointments or when I wanted clarification on previous interviews.

In the course of data collection, some themes recurred and it is this thematic reflection that informed the various methods. In the sections that follow, I reflect further on them and, in the process, a combination of informants' voices as well as the researcher's engagement with them form the core of the section (Collins & Gallinat 2010: 7). In the remainder of the sections, I guide you through my research sites and show how I gingerly made my way into the field, introducing the various methods I used in data collection.

Methods

Negotiating multi-sited fields

According to Marcus (1995:105),

> Multi-sited research is designed around chains, paths, threads, conjunctions, Juxtapositions of locations in which the ethnographer establishes some form of literal, physical presence, with an explicitly posited logic of association or connections among sites that in fact defines the argument of the ethnography.

In more practical terms, this approach involves following processes in motion rather than units *in situ*, away from an investigation of subalterns but towards one of the networks and the mobile community (MacGaffey & Bazenguissa-Ganga 2000). In the process, we arrive at an understanding of the functionality of the system and the in-betweens that would have been missed if we had focused solely on the subalterns from the perspective of a single-sited study. Multi-sited fieldwork is as much about field sites as it is about the informants that are followed. For this reason, I will offer snippets of accounts that transpired.

I arrived in Cape Town for the first time in January 2010 and it was at this time that I was conceptualizing a research proposal and thinking about applying for a WOTRO-funded grant. Having to rely on the contacts my spouse had, given that he had arrived here six months earlier, I was introduced to a few Cameroonians, one of whom was from Pinyin. Meeting him (Joe) was helpful as he told me about the Pinyin and Mankon communities and their respective associations, and gave me a handful of contacts from both associations and some phone numbers so that I could speak to them as well as to parents in the village whose children were in Cape Town. Importantly, he promised to talk to the President of PIFAM on my behalf and offered to take me along to the next meeting. I made contact with the Mankon people and had a follow-up meeting with some persons who briefed me about the Mankon association. This application stage gave me a brief insight into the community through my meetings with the people I contacted and by attending the meetings of both associations. Work only began in earnest when I was offered this research position in April 2010.

Between May and June 2010, I made two short visits to Pinyin and Mankon. Both were very informative and brought me into contact with the parents of children in South Africa. The first visit to Pinyin made me aware of the vast terrain I was working in and I had to devise a strategy in order to meet informants in the various neighbourhoods (with plans to attend two nearby neighbourhoods a day and, time permitting, three).

Joe offered me accommodation in his compound in the village. This receptive host family helped to introduce me to other parents and negotiated a guide with a motor bike (to whom I was eternally grateful) who took me around. We communicated extensively by phone to organize our pick-up and drop-off times and sometimes he would stay with me for the entire session and introduce me if necessary. His presence paved the way for me and led to my being accepted in the community. Back in Cape Town between July and August, I increased my outreach to include members of the community and attended meetings at which the associations' political views on co- development projects, social cohesion, hierarchy and power, and *njangis* became clearer.

I was in the Netherlands between September and December 2010 and combined proposal writing and research with meeting both the communities there. In contrast to the over 100 Pinyin migrants I met in Cape Town, I met only three and seven Pinyin and Mankon migrants respectively in the Netherlands. Despite the limited numbers, interaction led to useful contrasts that are important to the study. Research was limited to Amsterdam and The Hague. One of the Mankon informants who had been in Holland the longest (John) led me to all his Mankon contacts, while I received the phone numbers of Pinyins from friends and relatives in Cape Town. While recorded interviews were at the centre of data collection here, some informants were not as forthcoming with information as they would have been if it had not been recorded. In this respect, I decided on both interviews and conversations that allowed me to glean a lot in terms of relations with family back home and with the rest of the migrant community. Watching the funeral on DVD of parents or relations was an important aspect of this study as it always led to intimate conversations about migrants' relationships with their families.

My following visits to Pinyin and Mankon were in January 2011 and between September 2011 and January 2012, while in Cape Town between February 2011 and August 2011. I started to be accepted by informants during the second visit and by the third, I was well established. My relationships with people blossomed, especially as I had met their children and had photos of them on my computer that I showed their families or even hard copies that had been sent. I would also come back with gifts and money from the children for their families. I soon became a confidante and parents showed me land they had acquired for their children who were in Cape Town and the land demarcations in the hope that I would convey this information to their children if they should die without seeing them first. As a result of these cordial relations, the apprehension towards me that I had noticed in the beginning disappeared and a lot of information was divulged even though, at times, I asked just one question. For instance, when I asked Mike's mother about her daughter's health problems, her response was 'na witch for compound' and went on to tell me about the rivalries amongst the co-wives in the family and how her daughter had been bewitched. Others talked about simmering tensions that existed in the family and at times we cried together in the course of the interviews or we took walks or read poetry to diffuse the tension.

For others, seeing me and receiving news of their children was emotional and it was almost like seeing their children, especially for those whose children had never been home since they left. For most parents I became like their adoptive child and, as such, it would have been an insult if I had visited and not shared in family meals. This sketch of how research was conducted offers glimpses into the various field sites and of my main methodological approach, namely a multi-sited approach. By the same token, it provided insight into the relationship between migrants and families at home. Such 'deep play' (Geertz 2005) has been considered by some (Evans-Prichard 1951 in Hannerz 2003: 209) as surrendering to the field. It naturally depends on how one looks at it but I think that this is not a tangible enough reason. It is often only by being committed and being seen as truly integrated, as opposed to being a 'parachutist', that one gets to know a community and in turn is accepted by them. Then one can claim some degree of authenticity regarding the data. This is not, however, to negate the fact that profound ethnography cannot emerge from a casual encounter.

Kontri Sundays (rest days reserved for non-farming activities in both villages) and market days were good moments for long conversations with parents as we sat at *shware* (the main junction of each quarter in Pinyin) over a drink and talked for long hours. At the start of the research in the villages, my tendency was to record the interviews but I soon realized that the interviews were not as profound as the conversations so I abandoned my recorder and depended exclusively on conversations and note-taking, and when I retired home in the evenings, I tried to reconstruct the conversation and attach meaning to people's body language and tone of voice. This meant that even though I took notes, I had to devise a kind of shorthand to be able to record as much information as possible. I also had to depend on my own memory in the evenings to reconstruct the interviews and record them in diary form over my 15 months of fieldwork. I was careful not to limit myself to the conversations and interviews but also to remember the things I had seen. Memory and remembering was thus an important and integral aspect of data collection because, as Kohn (2010: 186) posits, 'memory's ability to construct emerges from the individual's contact with other people and places and events'. My diary became an important companion for recording any new and emerging ideas as well as a tool in data analysis.

My regular attendance at both associational meetings in Cape Town added another dimension to my interaction as I was not limited simply to individual visits. These meetings at times provided a platform for zooming in and out, thus taking on the betwixt and between that seemed to present 'grey matter'. Going virtual, offered the opportunity of having a look at various associational interactions and at the spatial organization that was mediated by the Internet. What I attempted to do throughout my travels was to show the network of relationships both locally and translocally and how they were being seamed or torn apart by their appropriation of ICT. In the process, I embarked on showing patterns of collaboration between migrant communities and their home villages and amongst migrants and the challenges that occur and how these are dealt with spatio-temporally.

Being a native of Mankon, my situation was slightly different there compared to in Pinyin. Knowing most of the parents as neighbours who had watched me grow up during my teenage years when my father retired to the village made me now more of an intruder in family matters and some people were apprehensive about talking to me and confiding in me. In this respect, 'going native' at the start seemed to be my nemesis. But in due course, they began to relax and, eventually, as in Pinyin, I became a close confidante. This notwithstanding, some informants, regardless of their familiarity, were very forthcoming and I am heavily indebted to them. Although by going native I was seen as the 'other' and this questions the notion of 'going away to the field', it is only by 'establishing a physical distance from home and field that sufficient objective be obtained' (Collins & Gallinat 2010: 8). While the challenge of carrying out fieldwork at home initially seemed to concern research ethics and shades of 'at homeness' (*Ibid.*), it turned out to be contentious because of my reception by my informants. Hence I was never truly 'at home' because the 'role of ethnography is enough to create a sense of difference' (*Ibid.*: 10). How did this estrangement play out in Mankon with regard to data collection? I chose to go along with this dual identity (of being from Mankon and being a researcher).Through this, in some instances, I used previous knowledge to pose intimate questions and at other times I played the researcher and feigned ignorance of certain family issues they expected that, as a neighbour, I would already know. But importantly, I had to learn a new way of belonging to a community that I had taken for granted that I belonged to. Learning to belong, in this case, meant not all questions were explicitly asked and I adopted the habit of 'knowing without asking'. In the process, I gave meaning to it in relation to the meanings attached to the action by the participants. For example, in the course of an interview with Jessica and her husband in Mankon, an argument broke out with Jessica accusing him of never reaching out to any of their five children abroad unless they called him. She maintained she was the only one who called, especially if she had not heard from them for a long time. Connection, here, seemed to be disjointed and perpetuated the assumption that it is only those abroad who should connect with home. Nevertheless, there were instances when being an insider paid off as I could juxtapose facts and rumours (after repeated interviews), and read meaning into them. The case of a neighbour's son sheds light on this. He lived in Cape Town and was not well regarded by most in the village as he operated a 'line' that took boys from the village to South Africa. The last group of boys were repatriated, although some later went back via other means and some asked to be reimbursed. The rumour in the village was that he was dishonest and fraudulent in his business dealings. The mother claimed the opposite and said that it was the fault of those who were repatriated and who were not smart enough to have run away from the border patrol. In South Africa, his friends thought he was not honest and was a bit shady in his business dealings. After my first interview session with him, he told me he was busy because of the nature of his business. Although his mother communicates with him regularly and he goes home as well, there seemed to be some disconnection. For this reason, I would concur with O'Reilly's (2009: 111) suggestion that it often poses a problem to 'achieve naïve distinction between insider and outsider'.

In Mankon, as in Pinyin, I focused more on conversations and held only two recorded interviews. A lot more was revealed through this form of data collection than if informants had been conscious that the conversation was being recorded and would have thought more carefully about what they were saying. For instance, on my first visit to Pinyin, I had a recorded interview with Mike's father but I could sense that this was not flowing as it did in normal conversations. I chose to keep the interview short and came back another day for a visit with only my pen and notebook. This time, we discussed matters for over three hours and it was then that he disclosed the difficulties he was having in getting financial assistance from his son in Johannesburg and from another in Cape Town (not Mike). He commended Mike for his assistance but expected more from him. After we finished, he left for a meeting and I went to meet Mike's mother in her kitchen. I wanted to believe that she was encouraged by the length of time I had spent with her spouse that she immediately asked me 'what papa told me'. In a nonchalant manner, I said 'something about the children'. Without any further questioning, she started recounting her own story about how her husband was too demanding of her son and this was causing family feuds. These conversations revealed the tensions that come with migration and the notion of entitlement parents feel they have regarding their children's finances. By the same token, this stops the children from calling regularly because of fears of demands that will be put on them. Back in Cape Town, Mike told me he usually calls when he has something to send home and mostly communicates with his mother because she is less demanding.

In Pinyin in particular, sitting around the fire in the evenings while my host prepared the evening meal was quite revealing. These were the moments when I asked the unanswered questions that had cropped up during the day, found deeper meanings in some of the figurative language used or simply told them about my activities that day. This usually led to extensive information about the

Photo 2.1 Researcher's host mother in Pinyin preparing the evening meal

people I had met. In this way, I could come to see the community from a broader perspective. I found this quite useful in my research in Pinyin where I lived with the family of a Cape Town migrant.

Despite the fulfilment I felt from interacting with the various communities and listening to their stories, fulfilment and challenge are not mutually exclusive.

The importance of the approach I adopted in this research project is embedded in the researcher's presence at multiple sites and his/her ability to analyse and interpret events that are chronologically or simultaneously linked. However, this approach does not come without its flaws. In certain instances it is difficult to be present at all sites at the same time (as in my case), especially when all the sites become sites of important activities. The researcher is thus compelled to make a choice and to rely on interviews to reconstruct the others. In other words, engagement in one area forecloses one's ability to participate in important happenings in other areas, and hence having to depend on interviews and second-hand information, whereas the physical presence of the researcher would be able to give different perspectives to the situation. For instance, two weeks after I left Cape Town for Cameroon to continue fieldwork in Pinyin and Mankon, Ron had an accident and later passed away. I did not hear about this although I was in Pinyin. Having worked for a few weeks in Pinyin, I then moved to Mankon and his body was brought back to Pinyin for burial in the meanwhile. I came back to Pinyin and learnt about it but the funeral had already taken place and I had missed out on all the events and had to rely on interviews to construct my analysis. Similarly, with families and relations that span home and the host country, it is not very easy to capture their story in a fluid form and in its entirety as the story changes depending on who is telling it. By the same token, participating in most activities turns out to be monotonous (Hannerz 2003).

At a personal level, it would be foolhardy to claim that the globetrotting involved in the course of this study was easy. It was not. Long periods of absence from my family were difficult to cope with and towards the end of my stay in any of the localities other than Cape Town, my motivation decreased as homesickness and nostalgia crept in and affected my attention span. It was difficult to write even a paragraph at such times and, faced with such emotional challenges, I could understand those of my informants only too well. Going home I felt as if I was a prisoner who had been released on parole because the writing still had to continue once I was home. Having to weave into my studies the role of wife and mother was quite a challenge and I often felt guilty when my children wanted to chat but I was busy writing.

This notwithstanding, when I returned to Pinyin and Mankon, I touched base with all my contacts and was given news about all that had happened in my absence. This was not necessarily to harvest information but was a way of being committed to the welfare of the individuals that I had come to know. I paid condolence visits to some and visited new-born babies and their parents too. The scenario was different in Cape Town where I was in constant communication with some by Skype, on Facebook with others and had email contact with yet others.

Comparative analysis

The comparative approach adopted in this project was an integral part of my research method and juxtaposing both societies helped to understand their commonalities and differences and how these traits could help to map out their mobility trajectories and unpack their appropriation of communication technologies. The aim of this approach was to gather different insights into the mobility patterns of Pinyin and Mankon migrants as well as to discover the social construct of their societies in different cases and capture their complexity. Based on historical and contextual comparisons, I was able to draw nuanced conclusions given the progressive nature of the study. This approach established the absence of a 'common cause or effect, even if all circumstances are identical' (Rihoux & Ragin 2009: 2).The approach does not deviate from the qualitative aspect of this study but instead complements it. And it allowed me to examine constellations, conjunctures and configurations (Ragin 1987). The configurations were the determinants that produced a given outcome (Rihoux & Ragin 2009) and through these, I was able to address the 'why' questions about 'outcomes resulting from multiple conjunctural causes' (Ragin 1987) that characterize the different communities. In this regard, it provided a basis for making statements about my empirical regularities and for evaluating and interpreting my findings, as well as room to go virtual and look at the Pinyin and Mankon Internet-mediated news forums. Following an historical approach, I looked at specific experiences beginning with the various mobility patterns of the Pinyin and Mankon and how differently these evolved. How have they informed the present and what identification can be discerned from them? I went on to look at the arrival of migrants in the host society and at how their arrivals coincided with the various phases of communication evolution. What comparisons was I able to make from the absence and presence of ICTs in the lives of the migrants according to the time of their migration?

Confidentiality

The issue of confidentiality was of concern to most participants, especially those who were better educated, and were constantly asked how I planned to use the data I had collected. I thought long and hard about this and about not infringing their privacy. And for this reason I decided on pseudonyms. In the same vein, most requested anonymity and so, prior to any recorded interview, I assured interviewees of their anonymity and promised them I would use pseudonyms. With some, I went back to crosscheck the information I had been given during interview(s)to be sure they were comfortable with the use of certain parts of it. Once someone explicitly specified what should not be used, which I naturally respected. This kind of attitude is crucial if one is to uphold the trust vested in the researcher and to be able to build up further confidence in the relationship. Abiding by strict ethical concerns insofar as confidentiality is concerned was less for academic reasons than for the benefit of our mutual relationship throughout the period (see Nyamnjoh 2009: 373). However, one major concern was how to submit to 'narrative hospitality' in certain instances or to be honest with information that would reveal the identities (such as recounting information about unsuccessful migrants) of some informants by virtue of their reputation within the community in spite of using pseudonyms. Choosing not to tell this side of a story would be falling prey to presenting a single story that was not balanced. Nevertheless I chose to go for a complete story and found reassurance in Rapport's (2010: 90) advice (see also Guillemin & Gillam 2004) where he propounds that:

> …. ethical and intellectual compromise are intrinsic characteristics of social research and whichever choice is made is unlikely to bring complete satisfaction ... the fieldworker is he or she who learns to live with an uneasy conscience but continues to be worried about it.

On my first trip to Pinyin, I was introduced to Mbu Joseph, who agreed to talk if I dropped by a day later. I started recording but soon realized he was not comfortable so I took notes instead as we conversed. Formal interviews and recordings in both villages seemed to make my respondents uncomfortable, hence I opted for conversations. This also meant I could spend long hours with families, although I tried to steer the discussion and sometimes digressions were apt. The preference for conversations in the village revealed more information as people tended to be relaxed. This in no way suggests that the formal interviews did not achieve their intended purpose. Returning to Mbu Joseph, he talked about his son who went to Cape Town hoping to assist him with his other children, but the father was woefully disappointed. He became sick and was unable to get proper medical attention. Suddenly, he changed the conversation to himself, telling me how he would have been in a senior position in the military today had his father not dissuaded him from going and urged him to marry. He said 'I have never really told this to anyone, I regret not having joined the military, I would not be here today suffering; I emptied my account to send this boy to South Africa and all the promises he made about helping with the rest of his siblings have come to nothing.' That said, he turned his face from me so I did not see his tears. I offered him a hug and proposed that we take a walk to *shware* (the junction in the quarter) where we had a drink. There was a similar situation with Pa Anyam in Bamenda who cried during the course of our interview because his son has gone to the US and had not come home in 25 years: he has seemingly (dis)connected from the family. His son's lifestyle in the US broke his heart and he is unable to tell his mother about it as it would kill her. I stopped recording and read some of his poems and we discussed the themes. This calmed him down and we continued.

In Cape Town, some informants talk of the simmering tensions between them and their in-laws who are constantly making demands on them or feel that their son (spouse) spends all his income on his spouse and her family. Of the over ten meetings I had with Joyce (none of which was recorded) because she had a lot to tell but was not organized in her telling, we engaged in conversations both at her business stand and at her home. I did record one conversation with her husband and three with CJ about how he came to have a South African passport. People like him, having just returned from fieldwork himself, confided in me because they understand the importance of research.

The case of Lola (in Pinyin) whose husband is in Cape Town is worth mentioning. She got married at the age of 16 while still in secondary school with promises that she would continue her education but these turned out to be empty promises as she started having children soon afterwards. Her husband travelled to Cape Town when she was six months pregnant. According to her, she has been mistreated by her in-laws ever since his departure. All the land she used to farm with him has been usurped by them and she showed me crops that she planted in front of the house and that were destroyed by one of her brothers-in-law. Money was being sent to her through another brother-in-law and he decides how much to give her or gives her money when he thinks the last consignment must have been exhausted. Her husband in Cape Town says he looks after her and the children very well and sends them a monthly allowance of FCFA 10,000 (clearly not enough) and sometimes more at Christmas, in September and at other festive times of the year. She would love to go back to school or to Bamenda to do some business but no one, including her husband, wants to give her a chance. Her parents often have to assist her with food and palm oil, her two daughters looked malnourished and untidy, and I sensed some kind of tension each time I visited them. On my last visit, one of her brothers-in-law (the person who destroyed her crops) interrupted our conversation for over an hour when he sat with us as (I correctly guessed) he wanted to know what we had been discussing. I stopped the conversation and we discussed life in the village. He tried unsuccessfully to send Lola on an errand but she retorted that it would be rude for her to leave me when I had come to visit her. During our conversation, I could feel a certain tension in her as there were long moments of silence and at one point her voice cracked. And when she talked about her destroyed crops, her eyes were full of tears. It was often during these moments of silence that I listened to the unsaid and felt the outpouring of emotions, especially of regret, and feelings of abandonment.

Feelings

In these periods of fieldwork there were difficult moments to draw a fine line between being a researcher and an emotional person who empathized with the stories being told. I was particularly touched by Lola's story because she never looked happy when I met her. On all three meetings with Lola as well as when I met Jessica and her husband, I could feel the emotions, tension and (dis)connections that were being brought to bear on families by migration and mobility. Although I had my eyes wide open, I needed the assistance of other senses to make a holistic analysis of what was unfolding in front of me. Concurring with Simmel's view (1997 in Urry 2007: 24), I too felt that the eye is a 'unique sociological achievement ... it is the most direct and purest interaction that exists ... and it produces the most complete reciprocity of person to person, face to face'. I could go further and add that observations go beyond using the sense of sight as this is just one of our senses and observation calls on us to employ the rest of our senses too to be truly enmeshed in the field. While conversations enabled me to learn more about their daily lives and facts omitted in some interviews, they also helped me to fine-tune my questions and this gave leads in other directions.

Language

One important thing that worked in my favour was my ability to speak Mankon and understand both languages, confirming Marcus' (1995: 101) point that the ability of the researcher to speak/understand the language impacts on the quality of the fieldwork. Being culturally tuned in and speaking and understanding the Mankon and Pinyin dialects respectively positively impacted on the research, as code switching was quite common, especially during intimate conversations when the interlocutor did not want to use Pidgin English. They would then adequately express their thoughts in the local dialect or in expressions that precisely captured the point. This was the case with Jessica's husband who was disgruntled when his son asked for money from South Africa, and he summarized his frustrations in his own dialect. Translated it would read as follows: 'Children go overseas but mine has gone to *ndzimabuh*', (a quarter in the village). This statement adequately captures his frustration at his son not being able to send home remittances and instead asking his father to send him money. Going to the next village means there is not much there, hence his father still has to support him financially. I was comfortable about having interviews in the local dialect but I did not go beyond basic greetings and words of welcome. This could be explained by the fact that the ethnographer they saw in me enhanced their perception of me being 'not native enough', as speaking the dialect seems not to make me as native (which I thought I was) in their eyes (Collins & Gallinat 2010).

Photos

One feature of my approach was collecting photos and listening to the stories behind the photos. My digital camera allowed me to take photos and make short recordings simultaneously. However, I soon realized that making videos shifted my attention away from social interaction (see Cooper *et al.* 2013) and I focused on taking photos, although this also meant that I had to be careful not to miss out on any key moments. The camera provided a significant service by recording a scene that made me contemplate more deeply or re-examine what had been said.

Some of the photos were scanned and the originals returned or those that were sent through me to kin in Cape Town or Cameroon were scanned with the permission of the sender. These photos were full of memories and meaning for both the sender and the recipient. For me too, it was crucial to know the way in which the data had been gathered and some served to document research sites and as mnemonic devices to reignite my memory later. It is in this regard that when I wrote each chapter, I often began by going through the photos on my computer. They were the perfect reminder of specific scenes and triggered memories, added detail to a story as well as being testament to having 'been there' (Collins & Gallinat 2010).

Some photographs also showed contrast between old and new or changes (see Chapter 4). As noted by Bourdieu (2012: 4), photographs were 'materializing and memorizing observations'. Some were not necessarily new photos, but ones which has been taken before the relation migrated and they were sent alongside recent photos. I suspect the intention was to see how those at home had grown. When the photos were delivered, I studied the attention and behaviour (mannerisms) of the receiver when looking at all the details on the photos, listening to his commentaries and the emotions expressed by the receiver. These were often of nostalgia, mixed feelings at times of joy and sadness about being away from the family but at the same time being able to see them. The photos Lola sent to her husband was the first opportunity he had had to see the child that his wife had been pregnant with when he left home. She was 14 months old on the photo. Lola's husband looked at the photos of Lola and the two children pensively, and I sensed feelings of regret and separation when, in reference to the younger one, he exclaimed that '*na ma pekin don big so wey I never take eye seeam*!' (My child has grown so big and I have not been able to see her up till now!).

Similarly, the most recent photos Joyce had of her sons in the village with her parents-in-law were over two years old and had been brought by her husband when he visited. Before I left for Cameroon, she pleaded with me to take photos of the boys while I was in Pinyin and bring them back for her. She immediately went with me to the photo shop and printed out all the photos of her family (her father, parents-in-law and children). Her joy at seeing the children was obvious and she proudly showed them to her neighbours in the market, all of who commented positively on their suits and shoes. When I later visited at her home, I found she had enlarged the photos and hung them on her wall. Through such acts much is revealed about the psychological state of mind of the receiver, the nature of their relationship and whether it is intimate or casual.

By the same token, I took photos from Cape Town to Paul in Bamenda. He lives in Cape Town but had travelled to Cameroon on a business trip for five months and his wife sent some recent photos that had been taken at the beach of her and their son. When I handed the photos to Paul, he took a long look at them and exclaimed; '*wey! my woman don dry finish because I no dey for ee corner*' ('my wife has gone pale because of my long absence'). One could sense remorse in him for staying away too long. But he was happy to see that she could distract herself, as well as keeping their son happy and that she was able to enjoy an outing to the beach. In the end, he told me that he has to go back to Cape Town by November in order to be with them. And the photos that Emile in Holland sent to his mother (Chapter 9) express her joy at having lived to see more grandchildren, even though only virtually. Paying such close attention resonates with Geertz's

(1973: 17) suggestion that 'it is through the flow of behaviour or social action that cultural forms find articulation, as well as in various sorts of artefacts'.

Similarly, I collected photographs, scanned and returned them (Chapter 9), re-snapped enlarged photos on the walls, and took lots of photos of happy occasions and the participants. These photos in themselves told stories and brought joy to families at home and in Cape Town when I showed them. Most of those in Cape Town opted to make prints. These photos form part of what Hammersley & Atkinson (2007: 212) refer to as documentary construction of reality wherein 'documentary sources construct "facts", "records", ... that are crucially involved in social activities'. Also, they are expressive forms of virtual presence and connectivity. Conversely, in such circumstances, the photos highlight the geographical distance between migrants and family as families express nostalgic feelings about when they will see the each other, 'maybe s/he will come when I am gone' was often said by parents. But the one thing that keeps them alive is the hope of seeing each other and the photos act as a substitute for missing kin.

The use of ICTs in fieldwork

The mainstay of multi-sited research is its ability to follow the researched in their daily activities. However, respecting this rule meant I was out of action at some field sites for a while due to my presence at another. I had to develop a Janus-like approach to be present at all sites. In this respect, ICTs helped to fill the void, especially when I was in Holland as I often communicated with informants in Cape Town then too. While communications within the villages were not too regular for parents who were often on their farm where there is no network reception, I nevertheless sought to be virtually present in Cape Town through phone calls, Skype and visiting informants' Facebook pages to see what they had been up to and to get news of events in Cape Town. I depended on my phone for my daily programme and to make calls and schedule or confirm appointments. I made calls to informants to clarify interviews when I was writing them up, which saved quite a lot of time. I received SMS information about upcoming events and also invitations and this was by far my best source of connection with all my informants. This was also my GPS navigator (Tom Tom) for getting to the homes of informants. And it also meant that if I was unable to reach informants, it made contacting them difficult and I took the chance of going to their homes at weekends to meet them. If I was lucky, I would find them. In summary, the phone was my 'Swiss Army Knife' (Ling & Donner 2009).

The Internet was key in data collection. Exchanging emails with informants meant that there was regular communication even when I was physically absent. Skype conversations with Joe would update me about the Cape Town community and developments in the village (Chapter 9), while phone calls with Joyce and

Paul would keep me abreast about happenings in the PIFAM and MACUDA respectively. My knowledge of Internet-mediated associations was gathered from the group emails that were posted on the Pinyin News Forum (PNF) and Mankon Forum (MF). The reason for checking these mails was to access the extent of mobility, virtual connectivity and connection with the home village.

Although I was not very regularly on Facebook, I did meet some informants there and we had brief chats. I took a keen interest in going through some of the photos and the comments posted, especially those brought from home. Facebook also enlightened me about migrants' virtual connectivity and this was by far the most frequently used means of commination by those who were well versed with the Internet and had smart phones. Most of them had parallel groups of friends. Chris, for instance, who belongs to four different Facebook groups, was interested in football and Cameroonian politics, and updated his page almost on a minute-by-minute basis.

Life histories

A detailed ethnographic study requires delving into life stories to piece together the past and the present so as to have a story in the continuum and to understand whether the present is informed by the past or if there is a disjuncture. As such, quasilife histories (that do not have all the attributes of a life history) draw extensively on memories and depend on how far back the informant can dig into his/her memory to X-ray the past. While part of the aim was to string together generational history, it goes further in 'illuminating the social and constructive elements of an individual that make him or her potent in the social context' (Marcus & Fischer 1986: 183). This allowed me to juxtapose mobility, trade and networks of the past with those of the present migrants and to come up with an analytical framework of transformation and continuity. Given that mobility and trade are dynamic processes, this approach was used to establish social practices of the past and link them to the present to ascertain whether there was any continuity or disjuncture in these patterns. The importance of the stories is a means of appreciating how various events and their relationship fit together. Collecting life history interviews with migrants in Cape Town, the Netherlands and families back in the respective villages of Pinyin and Mankon was done to understand and explain the mobility patterns of the migrants. Recording or recounting life histories was often integrated in interviews and conversations. In the host countries, I would usually begin a session by asking about the individual's life before coming to South Africa and the family background for those who had one living parent. They then recounted what they remembered about their fathers as well as stories they had been told. The information on Joe's father (Chapter 3) was, for instance, narrated to me by Joe himself. Some of it he had experienced first-hand and the rest he had been told by his father. In the home villages, I asked about people's parents' lifestyles from when they could remember. Often, they talked about how they tilled the fields and later joined the Fulani as herdsmen and from whom they 'earned' a calf each year. In recounting these stories, the parents back in the village were meticulous in reconstructing the journeys of the bush traders, the various stops that were made and their duration. Bianca, like Joe, narrated the story of her father's trade connections. He had been a cattle trader and would travel from Pinyin to the cattle markets in Mesaje, Sabongari and Nkambe to buy cows to resell in Bamenda and as far away as the South West Region, and he had also kept some in his herd. This meant that the family would go for weeks without hearing from him and he would occasionally send a message through a fellow Pinyin trader.

Juxtaposing the life-history accounts of those in the village and in Cape Town shows how most Pinyins in Cape Town have continued in the lines of trade practised by their parents. The same does not hold in the Mankon community where people have followed patterns quite diverse from those of their parents. When we review the stories recounted, the emergent theme that runs through them is the different pattern of mobility. Their stories (see Chapter 3) describe a plethora of encounters in different places and with different people. Despite a tendency to be mobile, the feeling towards home and to those there is still strong. While the availability of new technologies has removed the distinction between the village and the city, such distinctions are being challenged when they are overwhelmed by emotions.

When conducting life histories as interviews, it was difficult to be dethatched from my informants' emotions. Many families were happy to welcome me into their homes because my presence made them feel a little more connected to their migrant kin. Equally, the interviews were a painful reminder of the distance between them and their kin (especially wives left behind) or the disconnection between them and their sons. Such periods often brought out raw or subdued emotions that had been piling up and that were expressed by the shedding of tears, but they were determined to carry on with their stories. Telling their stories had a healing effect. At the end of the story sessions, some parents volunteered to show me the piece of land they had reserved for their sons and the demarcations, with instructions about what belonged to who. And I became a 'trustee' of their oral testament and had to transmit it to the children in case they did not see them again. I liked to believe that telling me their stories was less for the purpose of the research than because they viewed me as one who was close to their children and that I could adequately express how they felt. Proof of this was when I was able to reconnect a parent and their son, and the latter later completed the house the father was building and provided assistance with opening a small shop. Upon my return to the village, he continued his story from where we had left off. As a result of the relationship that we had built up, it did not take any prompting from me for him to start talking and he acknowledged my input in reconnecting him and his son, although this lasted for a limited time only.

Archival research

Archival research is important in linking the past with the present and to deciphering what changes, transformations or continuity is taking place. By so doing, this study called for ethnography that is historically oriented and that 'dealt with complex as well as primitive societies and that recognised culture contact, movement, and change' (Vincent 1991 cited in Gupta & Ferguson 1997: 20). Parts of this study were based on archival research as an important step to establishing previous mobility patterns that could shed light on the present but equally on how these patterns had been largely shaped and defined by ICTs and the revolution in transport. At the regional archives in Buea, I went through administrative reports on Bamenda and Victoria Provinces to gain insight into mobility and communication trends at the time. Prison and taxation records show that Pinyin traders were heavily engaged in long-distance trade and that one of them was arrested for evading taxes from bush trade and another was sent to jail, while yet others had to pay fines.

The juxtaposition and simultaneous use of these varying methods were all geared towards telling the stories and capturing the voices of those being researched, with the aim of highlighting their life trajectories in ways that have, until now, only been told piecemeal.

Data analysis

As already mentioned, the data collected are qualitative and include the words, silences, events, actions and experiences in the everyday lives of the actors. Interviews were done in English and Pidgin English. For the most part, the Pidgin transcriptions have been used as such and, where necessary, the English meaning of the quotes is given. Following Mauthner & Doucet (2003: 422), part of my approach in data analysis is a 'voice-centred' focus, which is equally hinged on reflexivity but is not reduced to a simplistic representation of the respondents' voices without any attempt to offer interpretation and/or analysis (*Ibid.*). This approach allowed me to measure my level of emotional relatedness towards the informant while at the same time trying to understand and interpret the informant's story without tainting it with my own perceptions and assumptions. Nevertheless, there were times when charting my own migratory story and challenges helped to reduce the shyness of the informant. This approach was particularly useful given the fact that I am not only culturally in tune with both cultures but also hail from Mankon, which could result in me overlooking certain narratives or questions. Over and above this, the method, as indicated by Mauthner & Doucet (2003: 422), holds at its core the 'idea of a relational ontology in which conceptions of the separate, self-sufficient, independent, rational "self" or "individual" are rejected in favour of notions of "selves-in-relation" or "relational beings".

Given the duration of the data-collection period, it would have been easy to fall prey to the standard three-phase formula of data analysis: data reduction; data display; and conclusion drawing or the verification process. However this study benefits from Hammersley & Atkinson's (2007) suggestions that reducing, condensing data, identifying or isolating groups of common categories and meaning making should begin as the study begins and continue through the duration of data collection to the writing of the thesis. They advise that 'in ethnography the analysis of data is not a distinct stage of the research ... it begins in the pre-fieldwork phase, in the formulation and clarification of research problems, and continues through the process of writing reports, article and books'. It is precisely for these reasons that my research dairies have become my 'Bible' and I have been reading and re-reading the notes I took to develop and analyse what I heard and saw. One additional point that was helpful in the analytical framework was memory – visual, aural and emotional feelings – that formed part of my everyday writing.

Conclusion

Methodological scripts often follow the traditional patterns of writing these sections such as observations, interviews etc. And, as I demonstrated above, I chose to adopt the themes that were outstanding during data collection and to expound on them following my interaction and experiences with the communities.

Working with mobile communities in three loci where activities are constantly on the move required more than the Malinowskian intensive model, i.e. flexible methods beyond traditional ethnographic ones. It called for a multi-sited approach to capture these mobile communities on the move. This means that much as the research was focused on ICTs, communication technologies became an integral research tool for negotiating multi-sitedness, understanding the mobile community that by definition is difficult to grasp, tracking down informants and simply 'linking up' to improve relations between the researcher and the researched. By juxtaposing Internet-mediated technology, social media and telephony, I was able to begin to grasp migrants' mobility and appreciate the intricate web of connections that they had built and were continuously forging. As a result, I established a virtual and Janus-like presence at the field sites when I was physically absent through constant communication.

I became a perfect bridge that linked mobile communities at home and in host countries as a result of my movements thanks to my research funds. Families were linked through the courier and the photos that I carried back and forth and these also became units of analysis. The same goes for the photos that I took. These flexible methods illuminated a rounded understanding of mobility in the face of advances in ICT as well as the power relations that ensued and the concept of social capital (cf. de Bruijn & van Dijk 2012).

Comparative analysis in research is not an entirely innovative technique (MacClancy & Fuentes 2011). However, by dint of the fact that this study spans Cameroon, South Africa and the Netherlands and features two ethnic communities, it called for a new and creative way of looking at comparative research. In this respect, this study has been innovative in its methodological considerations. The chapters are interspersed with ethnographic quotes from informants and I use them extensively because they are ethnographic sources that reflect my insights into the sociocultural and economic meanings of both communities.

3

Mobility and migration at the crossroads: Mobile communities

Introduction

Bianca's story

> When I finished high school I refused to study law at the University of Yaoundé. I just went into business. I was trading, buying beans from Ndu, Babessi and Ndop and I sold them in Douala. At one point I had a connection and I sent beans to Gabon and Equatorial Guinea. So that's how I worked and raised money. Then my first cousin, who is in Johannesburg, came to Cameroon and told me that it would be better to go out there than to pursue the business that I was doing ... She offered to help out in getting the documents. I said I had the money and if she could get someone who would do the documents, I'd be happy to go. It was when she returned that she got other documents processed and then I had my visa, and travelled in July 2007. When I arrived, I was in Johannesburg for two months, then I had a job there that paid me R1500 a month but I took it for four months just to see how life was going, but after some time, I converted it. Then, it was R 1450 to FCFA100,000 in Cameroon. I converted the money and saw that it was nothing because I was not even paying rent (sic). So that's the reason I decided to leave that city. Before then, another sister came from Cape Town ... she told me how life was in Cape Town … I asked her what I'd do if I moved to Cape Town before I settled. She told me that she owned a flat in Cape Town and she was studying. If I came, we could 'patch' for a time before I looked for my own place to settle. Then I talked to the former president of the PIFAM who met me in Johannesburg. I spoke with him and he encouraged me and that's also the reason why I moved. I did not like it there … I did not like the lifestyle in Johannesburg, it's too rowdy. It's just like Douala in Cameroon. A lot of crime … There's a high crime rate in Johannesburg.[1]

Unable to explain why she turned down going to university in favour of doing business, we can only attribute it to an attitude she grew up with at home – that of a father who was business-minded and would travel to different cattle markets to buy and sell cows. She says:

---

[1] Interview with Bianca, Cape Town; 17/08/2010

My father was a trader, he used to go to Wum, Nkambe and Mesaje, and he was a cattle trader.[2]

Bianca's story forms the backdrop through which we can understand the mobilities of the communities being studied in this research. Mobilities have increasingly become a chosen lifestyle for the majority of the mobile community. And this lifestyle of mobilities is propelled and sustained partly by kinship, by social networks and by ethnic enclaves that 'mobiles' have forged over the years. But there is also evidence to suggest that the mobility pattern of today's migrants is embedded and informed by those of their parents: it is born out of a learned and internalized habit that they have adopted.

The twenty-first century has seen a significant change in the context of mobility due to the age of globalization that has ushered in new dimensions of cross-border movements, flows and transnational networks. It therefore follows that the current migration flows are inextricably linked to globalization and have become the reagent in socioeconomic and political transformation in receiving countries as well as in sending countries (Castles & Miller 2003). However, we should avoid overly romanticizing globalization because mobility has always existed in terms of the movement of people and things. As Mbembe (2000: 263) explains; contrary to the bounded idea of place that is informed by classical ethnography, the world has always been composed of layers of multiple spaces 'constantly joined, displaced, and recombined through wars, conquests (as we saw in the history of mobility among the Mankon people) and the mobility of good roads and persons'. In the midst of this movement, there is also an illegal side, with trafficking that is equally embedded in historical perspectives as Schrover *et al.* (2008) have so well demonstrated. In this respect, it is arguable that, while humans have always been mobile, the ripple effects have been the creation of fluid identities, as shown by Gupta & Ferguson (1992: 9), a trend that is increasingly evident in the context of globalization. On the one hand, traditional sending countries, like Spain, Italy and Ireland, have seen a reversal in emigration and have become recipients of immigration and, on the other hand, the increasing wealth of some African countries like South Africa, Botswana, Equatorial Guinea and Angola, as well as some Gulf States such as Qatar, Saudi Arabia, Oman and Bahrain, have witnessed high levels of migration to these countries. Castles & Miller (2003: 4) claim that 'migration ranks as one of the most important factors in global change'. This, as most migration literature concurs, is thanks to advancements in the new Information and communication technologies and the complete overhaul of the transport sector, especially air travel, which has given people the advantage of incessant and multi-directional flows of mobility (Massey *et al.* 1998; Castles & Miller 2003; Urry 2000; Mazzucatu 2008). What can

[2] See previous footnote.

we therefore deduce from this accelerated mobility in terms of its impact on people? Elliott & Urry (2010: 7) shed light on this and hold that accelerated mobilities in people's lives will lead to the 'recasting of identity in terms of flexibility, adaptability and instant transformation', while holding onto the values that they consider as defining their identity. However much the advances in mobility/migration are hailed, they would, however, appear to have come at a cost to nation-states as they are gradually recasting the sovereignty and autonomy of nation-states and minimizing their ability to maintain effective border control. This is by no means meant to suggest that borders have become redundant but rather that flexible mobility seeks to make a nonsense of borders that have become contested spatiality between migrants and states. The age of globalization, which is defined by accelerated flows of people, goods and ideas and unbounded mobility, would appear not to be flexible to all. This limitation could explain why the sovereignty of states is being challenged. And with advances in new ICTs borders have become increasingly porous and the corollary is an explosion in mobility (Ferguson 2006; Nyamnjoh & Page 2002). Migration and mobility are therefore more of a seesaw, where closures are vehemently challenged and flows questioned and reconfigured thanks to migrants' networks. As such, mobility can best be understood through these networks and fluids. Furthermore, such developments appear to have led migrants to redirect their lives in a multiple society – to living a dual life – where geographic location and time are no longer issues of inter-connection and conversely change social structures and lead to cultural diversity and/or fusion. Pinyin and Mankon migrants are no exception. With such rapid evolution in mobility, the tendency is to obliterate the mobile culture of the past in favour of the new era.

Delving into the past informs the present, and mobile patterns of the communities under study can be seen in Bianca's story given that empirical data show that mobility, especially among the Pinyin, resembles the mobile culture of parents during the colonial and post-colonial era. This historical approach is also supported by Cresswell (2010: 29) who maintains that 'we cannot understand new mobilities, then, without understanding old mobilities. Thinking of mobilities in terms of constellations of movements, representations and practices helps us avoid historical amnesia when thinking about and with mobility.' The varied forms of mobilities as practised by the Pinyin and Mankon communities have thus come to represent an historical progression from traditional rural society to that of an urban and mobile society (Ferguson 1999; Andersson 2001).

This chapter sets out to understand the mobility patterns of the Pinyin and Mankon migrants and how these have evolved from when they were marked by bush trade and mobility to the plantations. In the same breath, it acknowledges that migration involves a point of departure and entry, as well as a dynamic system in which migration represents and evolves from linkages between sending and receiving countries. The chapter seeks to understand mobility from the perspectives of the sending and receiving countries in order to explain the 'flows within the context of other flows and emphasises that flows of people are part of, and often influenced by, flows of goods, services and information.' And it sees a social network (see Chapter 5) as the connecting device given that networks bind family ties, kith, community and weak ties and are the catalyst for migration. As a result of mobility, networks connect migrants and non-migrants across space and time, and migration and mobility revolve around a seamless weave. These forms of mobility and network become a complete reversal of the assimilation form of migration in the past where migrants said farewell to their country of origin and embraced the new country. Instead, they opt for a more cosmopolitan and elastic way of life that prioritizes double presence, taking into consideration their cultural attachment (cf. Chapter 1). Against this backdrop, I question how this has evolved over the course of time. How can we draw on past patterns of migration to understand the current trends of Pinyin and Mankon migration? What does mobility mean to migrants? How are receiving countries reacting to the new wave of in-flows? And what is the role of ICTs in enhancing or stifling these new trends in mobilities?

Mobility trends in Cameroon from colonial times to the current wave of migration to South Africa

Prior to the current global trends of accelerated mobilities, mobility existed in most African states and was an integral part of people's daily realities. Since the colonial period when the Germans established plantations to the post-colonial period in Cameroon, mobility was marked by movements to the plantations from Bamenda Division; with huge contingents of people moving from Wum and Kom and later from Momo (Meta and Ngi) from the Grassfields of the North West Region to the South West Regions, and to other coastal regions (Konings 2001, 2011; Delancey 1974: 181, 1988: 307; Warnier 1985; Geschiere & Gugler 1998: 314; Nkwi 2011). Such mobility patterns did not exclude rural-urban migration and rural-rural migration, and were not limited to movements to the plantations. They also revolved around economic activities, with traders crossing the country buying and selling goods. Although statistics on migration from Mankon to the coastal region are not available as they are only for other villages such as Meta, Ngi, Kom, Wum, Bali and Nkambe (Delancey 1974), migration to the plantations or the coast was commonplace. Oral accounts hold that migration to the plantations amongst Pinyin was uncommon because they are very industrious and business-oriented and thus tend to be self-employed.[3]Instead, mobile trends amongst them tilted more towards rural-rural migration and to weekly visits to neighbouring village markets as well as to Ikom, a town on the Nigeria-Cameroon border to sell pigs and foodstuffs such as pink potatoes, in exchange for salt[4] and other goods not available in their own village. Those who went as far as Nkambe and Wum became conversant with another lucrative trade, namely bush trade between Cameroon and Nigeria, and East and West Cameroons. Bush trade was exacerbated by the lack of good roads at the time and most goods depended on 'head-carry'. At the top of the list of goods that were brought back from Nigeria[5] were clothes and gunpowder, which would then be taken to Bameleke in East Cameroon or present-day West Region where traders would sell and buy gunpowder and coffee seeds and goods such as cigarettes, spirits (drinks). Pinyin's proximity to the West Region gave them an added advantage as it is widely believed that Pinyin people inherited their entrepreneurial spirit from the Bamileke.

These mobile patterns of the Pinyin are largely informed by historical patterns and the notion of 'rites of passage' that existed at the time, even though this was for economic reasons. It was a way of proving one's coming of age and readiness to take a bride amongst one's peers. By the same token, it catapulted those concerned onto a list of eligible marriage partners.[6] The men converted their economic capital into cultural capital, wielding positions of hierarchy. This trend has continued to the present. Against this backdrop and following Geschiere *et al.* (2008: 2), I propose looking at mobility from a genealogical approach that takes into account migration and mobility since the colonial period through to the advanced era of mass migration enhanced by the revolution in communication technologies and air and road transport. Migrating to Cape Town followed similar patterns of step-wise migration as those manifested by the migrants' parents who made the rounds of the neighbouring weekly markets. Migrants would leave the village for the city and later move to South Africa thanks to migration enterprises. Back then in Pinyin, coming of age and being independent was marked by the ability to belong to a peer group of 'mobiles', known as *nkang*, who visited weekly markets in the neighbouring villages of Ashong, Bali, Babajou, Bamessing, Bamumbu, Guzang, Wum, Nkambe and Mesaje and were also involved in bush trade between Nigerian and Cameroon, and the then East and West Came-

3 Interview with Pa Nkem, Bamenda; 21/09/2011.

4 File No.cb1919/1, Annual Report 1919 of Bamenda Division: Trade and Economics. p. 4.

5 File No. cb 1955, Annual Report 1955 of Bamenda Division. It asserts that bush trade increased between 1954 and 1955 as smugglers organized themselves in bands of upwards of 50 men when going on a trip, and used scouts, some of whom went ahead to check the road for customs officials. Archival data corroborates oral sources.

6 Interview with Pa Muluh, aged 71, Pinyin; 28/05/2010 and Pa Nkem, 68 years, Bamenda; 21/09/2011.

roons[7] simultaneously. The weekly markets provided opportunities for those who had migrated to these villages to receive news of relations and also for relations to send messages to them. In other words, information to and from the village had to be sent with villagers travelling from one weekly market to another. In line with Urry's 'systems', this was the only system in place at the time for moving information and ideas. Travellers took news from relations to Pinyin settled in other villages and *vice versa*. In the same vein, they came back with new farming techniques, new seeds and information about different breeds of pigs and piglets because Pinyin was famous for its pigs. With weekly mobility, those residents in other villages looked forward to market days as it gave them the opportunity to get news from home, especially from family. Traditional forms of communication were very much the practice at the time. Muluh, who had relocated to Njinikom, was urgently requested to return home because his father wanted to groom him as his successor (unknown to him then). He was sent a particular type of grass by his father through a village kin visiting the market. The significance of this grass was twofold: his presence was urgently needed in the village and the particularity of the grass signified the urgency of the message, i.e. to come with the grass still fresh. Corroborating this information, Nkem says that the recipient of the message was expected to return to the village before the grass died.

Rural-rural migration gave people the opportunity to move into rural areas and take advantage of the lack of emerging markets. This led to movements of Pinyin to villages like Nkambe and Mesaje (Pelican 2006) in Donga and Mantung Division, Fundung in Boyo Division and Bawuru in Menchum Division, all in the North West Region of Bamenda. In the process, an extensive use of kinship ties reaching across frontiers was used, especially by those involved in bush trade. A case in point is that of Mark and Nelson, whose parents migrated from Pinyin and settled in Fundong and Nkambe respectively.[8] The migration trail of Mark's parents took them first to Wum, then to Nkambe and finally to Fundong. They established themselves simultaneously as businessmen, farmers and cattle owners with substantial herds of cattle. To these families, although Fundong and Nkambe have become a second home to them, they have not cut their links with Pinyin as seen from the development projects they are carrying out in the village as well as their presence at all the funerals and family occasions.

The linking up of rural areas by mobile communication networks has made it possible for the flow of information and virtual mobility. The culture of pitching (Chapter 4) in Cape Town is what those there have internalized over the years

7 File no. CB 1955/1. Annual Report 1955 of Divisional Officer for Bamenda Division; pp. 4, paragraphs 10-11, see also File no. BA 1969/1: Economic Quarterly Report on Mezam Division for the Quarter Ended 31 December 1969.

8 Interviews with Mark 04/08/2010 and 06/04/2011, and Nelson on 06/07/2011. Parow; Cape Town.

and are now reproducing. At another level, the children have become migrants in their own right as Mark's journey to South Africa was more for transit purposes before moving to Britain, while Nelson, having established himself in Johannesburg as a businessman, relocated to Cape Town for his studies and then moved to the South West Region of Cameroon. The mobilities systems propounded by Urry (2007) would appear to be acting in their favour.

Such mobile trends saw not only the arrival of new ideas and business opportunities, especially agricultural innovations in Pinyin and Mankon, but also heralded the inflow of Fulani cattle owners into Pinyin where they found vegetation. Pinyin traders going to East Cameroon and the Mankon plantation labourers introduced coffee growing from the seeds they brought back, and some Pinyin diversified into cattle rearing, almost rivalling the Fulani. Only recently have Mankon people branched out into cattle rearing, an area that was exclusively reserved for the Fulani. The culture at the time was to work as cattle herdsman and this was rewarded with a calf each year. With this new-found wealth came social mobility as well as extended criteria for marriage eligibility amongst the men. This trend did not, however, go unnoticed as it featured in one of the Quarterly Reports of the former West Cameroon.[9] Could we, therefore, infer from these past mobility patterns something about the present fluidity and flexibility in mobility? Joe's story will elucidate on these new trends of mobility among the younger generation of migrants in South Africa.

Mobility and migration are no longer what they used to be but are guided by information and knowledge about migration from earlier migrants to prospective migrants; with the former becoming the communication out-post. And they have moved away from the old migration systems, that privileged assimilation to the new that integrates the 'here' and 'there'. This challenged and circumnavigated state borders and showed pride in cultural mobility.

A paradigm shift

Given the paradigm shift in trends of mobility, we are witnessing a new reorientation and diversification of mobility globally. Drawing on Urry's (2000) myriads of mobility, it is evident that we can no longer talk of individual migrants or individuation given their lives are inextricably linked to the home and host countries and to the global. Mobility is more about 'mobiles' who are aware of their surroundings and the global changes, and who are armed with informed choices and being assertive. Mobile communities are more the norm, with agentic powers that increasingly redefine the host and home countries. In this regard, the governmentality policies put in place to monitor borders are being challenged and

[9] File no. BA 1972/1; Economic and Political Quarterly Reports of the Division of West Cameroon, 1971-1972: 93.

contested. In this respect, it also influences the way communities seek to reorganize themselves in the face of such mounting closures. South Africa which used to be the cradle of labour migration to the mines among its Southern African neighbours (SADC), is witnessing a downturn in labour-intensive migration to the mines in favour of other forms of migration, such as student migrants and a new wave of economic and professional migrants who are self-employed (Shindondola 2007; Segatti 2011). The defining features of these migrants, as opposed to the traditional mine labourers, is that they are post-apartheid migrants. This has led to a huge influx of Congolese (DRC), Nigerians, Somalis and Cameroonians, and the reasons for migration especially from Cameroon have been well expounded by Shindondola (2007: 8-12). Be they post-colonial forms of mobility or new cultures of mobility, they are guided by what Urry calls 'systems' (see also Tyner 2009) given their ability to make possible movements and provide 'spaces of anticipation' (Urry 2007: 12-13). By the same token, they are making a tremendous impact on their family's stability and organization. His idea is backed by Licoppe (2004) who believes that communication technologies are the driving force behind the systems. But perhaps, Urry's 'system' is limited only to those that make movement possible and does not consider the 'systems' that endanger mobility. Although he alludes to the theory of 'complexity' that 'investigates emergent dynamic, self-organising and interdependent systems' (*Ibid.*: 2007: 30), it fails to be explicit when, even in the face of such complexities, it should be noted that one emergent dynamic is the breaching of social relations. This is something that Giddens (1990) terms as the perspective of mobility at 'disembedding' the social functions: it is a liberation of the individual. Hopkins *et al.* (2006) look at mobility from another vantage point to assess how social identity considerations shape spatial behaviour. Concurring with the latter, social identity has been one of the driving forces behind the high rate of Pinyin emigrating to South Africa. For instance, most of the informants (all but two) in this study recall how coming to South Africa was facilitated by already-settled relatives who later assisted them in their own businesses. In this regard, the mobility of the Pinyin and Mankon may be affected by the degree to which particular places (in this case South Africa) contribute to their sense of personal continuity, of community/distinctiveness, self-esteem and self-determination.

An attempt to understand the migration and mobile patterns of these groups will be anchored in a migration-systems approach that calls for the study of both ends (the home and host countries) in the migration flows that form the cardinal point in this study, the linkages between the places and the networks that have propelled the sustenance of migration (Massey *et al.* 1998; Fawcett & Arnold 1987; Fawcett 1989; Boyd 1989; Hagan 1998). While social networks (family, community and informal networks – weak ties) inform migration and are coping mechanisms that help migrants to settle in, they equally perpetuate it through chain migration and the migration industry or enterprises. The ripple effect is the creation of a network society, one that is underpinned by ICTs and migrants' strives to create a home-away-from-home. Comprehension of all these linkages will enhance our understanding of migration and mobility among Pinyin and Mankon migrants. This study also considers the nuances of the systems and complexities of mobility and migration flows. These same systems that make it possible could equally mar mobility when the 'in-betweens' are not adequately coordinated and managed. Mobility could best be understood by engaging it from the perspective of everyday life and assessing how it translates to the formation of a mobile community. Given that everyday life is where all events and daily activities take place, capturing the tendencies of everyday life should offer an understanding of the concept of mobility that involves the in-betweens. In summary, the analyses in this chapter are premised on complexities, migration systems and the notion of everyday life through which the in-betweens of mobility and migration are captured and understood.

Expounding on the concept of globalization and mobility, Appaduria (2000) argues the case for 'globalisation from below' as globalization can be experienced irrespective of whether one is mobile or not. Given that most of the world's population is sedentary, it is thanks to the mobile individuals that it is possible for most 'immobiles' to connect with the rest of the world (Sichone 2008). This presupposes that at either level, be they mobile migrants or the sedentary 'immobiles', the gap still exists even if we tend to look at globalization from below.

As part of the paradigm shift, there is an increasing discourse on cultures of mobility (Hahn & Klute 2007; Adey 2010; Greenblatt 2010) that is embedded, on the one hand, with mobile communities seeking to negotiate with other communities and the host communities, and on the other, in cultural mobility as migrants are mobile with their culture. Making a case for cultural mobility, Clifford (1992, 1997) and Greenblatt *et al.* (2010) maintain that mobility is not limited to people alone but that people and cultural travel and this process link people and places as well as sedentary cultures of host and home countries because the inhabitants of any such places are already embedded in the global flows of people and culture. It follows that cultural mobilities herald the formation of strong ethnic enclaves whose goals, besides the pursuance of economic interests, are replicating home-away-from-home with its hierarchical structures that seek to perpetuate the status quo. This is, however, not without its challenges from members who want to belong to the community even though they still want to have room to manoeuvre where belonging is on their own terms. De Bruijn *et al.* (2001: 64-67) discuss how population mobility in Africa is frequently a cultural phenomenon and is culturally mediated, but how mobility is also a way of dealing with climate variability.

In a way, mobility also articulates the migrant's conviction of belonging and, while striving to maintain their identity, they take on that of the host country given that one of the ripple effects of mobilities is the formation and re-formation of identities. These changes in mobilities lead to a variety of patterns of identity formation that can be flexible, adaptable or transformed (Clifford 1997). With regard to the Pinyin and Mankon communities, while they opt for flexible identities, they are still very conservative in terms of holding onto their 'Pinyin-ness' and 'Mankon-ness' as can be seen through their associations (Chapter 7) and their cultural practices (Chapter 8) that are an integral part of their existence in Cape Town. For instance, the traditional outfit of the Pinyin and Mankon is known as *ntogho*. The very act of wearing this outfit denotes the person's identity and the fact that they belong to the Bamenda Grassfields society. However, given the flexibility of their identity, there is no gain in saying that the outfit is limited to the Pinyin and Mankon exclusively. But what this denotes is that its wearing evokes a particular culture and identity, namely Grassfields. Amongst the mobile communities and in addition to the *ntogho* that is often transported from the home country to the host country, one other item frequently carried by the migrants is food. The *ntogho* carries with it a lot of prestige, thus making it a must-have for migrants. For example, at the start of this research only a hand full of Mankon owned *ntogho*. However, when they were visited by their sister association from Johannesburg; all of whom were dressed in *ntogho* on the cultural evening that was organized in their honour, it was a serious challenge to the Cape Town members because they did not have theirs on. Soon afterwards, most of the members sent money home for one to be (bought and) sent by DHL or with someone travelling back to Cape Town. Conversely, a sizeable number of PIFAM members now own one. Equally, the formation of tribal ethnic associations, where home is replicated in these associations and belonging is emphasized, is with those who do not want to be looked down on as an outcast.

Cultural mobility is not mutually exclusive to food and traditional regalia and can also encompass cultural objects or memorabilia (Baldassar 2008) and the significance attached to them. As opined by Lury (1997), the movement of objects from place to place reaffirms the notion of belonging and memory, and conversely creates an emotional wellness (Baldassar 2008). This is well exemplified in the movements of large framed photos that are a reminder of deceased relatives (especially parents) in the migrant community, new and old photos of the family that they take on their maiden trip and some artefacts reminding them of the home they have left behind.

Globalization has given rise to an unprecedented level of cultural mobility and it is mostly the cultural values, norms and goals of migrants that are altered in the process. This speaks to the ways in which migrants (especially the Pinyin in Johannesburg) have tended to brandish 'bought'[10] or inherited (as successors) cultural titles from the village to enhance their cultural capital. Such categorization is not far from what Hall (2000: 215) refers to as the 'unintended effects', that is 'subaltern formations and emergent tendencies' of cultural mobility, to which Greenblatt (2010: 14) contends that such a form of cultural mobility 'can take the form of attempted cultural murder'. Although Greenblatt (*Ibid.*) alludes to the era of Roman conquest, I would rather we see it as the reinvention or transformation of culture, given that culture is dynamic and not static. The Pinyin and Mankon in South Africa want to enjoy the benefits of royalty, thereby giving another interpretation to their culture thanks to mobility.

The discourse on mobility is not limited to the above-mentioned points *per se*. Given the dynamism of mobility, other systems and forms have given rise to propel it. These include the virtual/real and imaginative as suggested by Urry (2000) and transnational mobility that forms part of this culture of mobility. Internal and transnational mobilities have become part of a single arena of spatial mobility and social transformation that connects these communities to those back home. It should be recalled that their spatial mobility starts long before they leave the country with information fed to them by earlier migrant communities. For most Pinyin and Mankon migrants who are unable to go home due to a lack of proper documentation, virtual and imaginative migrations are the options they are left with. By asking relations to put a phone on speaker so that they can follow the events during a funeral, they can imagine what is happening and participate virtually. Similarly, through their mobile phone communication with families back home, the sending back and forth of photos, videos of events at home or in the host country, migrants and families are transposed respectively to the home village and the host country and are temporarily connected to their families and bear witness to each other's daily lives and happenings. Similarly, they learn the new cultural dance styles that are in the village and the songs that go with them, especially the *mbaghalum* dance from videos. Permit me at this juncture to share what I consider a classic example of virtual migration. The scene is at the home of John in The Hague where we were both watching the DVD of the funeral of his late mother. When we got to the section when the casket is being lowered into the grave, John said:

---

10 Interview with Bianca, Cape Town; 17/08/2010. During fundraising occasions in Johannesburg, members offer to be the highest contributor in order to be given titles such as chiefs of the quarter they hail from in the village. They take this role seriously and act accordingly.

> The day they were lowering my mum on that Saturday, I was alone in this house at home! And I called my brother and asked that he put the phone on speaker so I could follow the funeral. I was listening to the choir and also dancing at home here.
>
> Q: Why did you ask that the phone be put on speaker?
> A: Because I wanted to be present physically but it was not possible. And that was only way I could be present. I imagined how the site of the grave was full of mourners, except me. I really felt bad I was not there like the rest.[11]

After we finished watching the video, John played the ring tone on his phone and asked if I recognized the song. Apparently it was the song that was sung during his mother's interment. He had chosen to keep this song as his ring tone to remind him of his mother until he goes home to see her grave and organize his own 'cry die'. While this confirms Lury's (1997) assertion of memory and its importance to migrants, the ring tone reminds John of his mother and he wants to remember her as the 'tall, soft spoken and hardworking mum he left back home'. Similarly, it highlights the relationship between mobility, ICTs and the social shaping and appropriation that occurs at such a time.

Similar accounts from Pinyin migrants in Cape Town and family members in the village concur with John's. Often such virtual mobility is heightened in times of grief and emotional stress and these directly affect the migrant. Although they acknowledge the cushioning effects of the phone in mitigating their sorrow, they equally admit it does not come close to appeasing the emotional trauma they experience. However, the phone gives them a partial feeling of presence and they can imagine how the events were managed.

In this respect, memory and imagination are useful tools that sustain their mobility. To them, remembering the places, persons left behind and things give them a sense of virtual and imagined mobility, as well as a sense of connectivity. Although it keeps them attached to home, their knowledge of home is invariably one that is stunted or informed by the information they receive from home. For those who have regular documentation that permits them to oscillate between home and village, albeit occasionally, they are able to effect real migration and face-to-face meetings (Urry 2003).

Face-to-face meetings are often necessitated by the obligation to go home after a long period of absence, the need to re-internalize memory and also by fear. This may be fear of what the family will say and think about the migrant's long absence and/or fear of bad-mouthing by relations that the migrant has severed links with the family, something Jake does not want to happen to himself. When Jake goes home from Cape Town, which he does only infrequently, he makes a point of visiting all his relations and acquaintances. A large part of his stay at home is spent travelling from one region to another with his family. He undertakes these journeys with the aim of immortalizing the visit for years to come,

---

[11] Interview with John, Holland; 19/09/2010.

travelling with a camcorder and a digital camera. Over and above this is the need to see all his relations to dissociate himself from any form of badmouthing and to receive blessings from the elderly: '… what they will say about me when they heard I was at home and did not visit them, and to receive blessings from aging relatives'.[12] Although not all migrants pay attention to badmouthing, for those who care about it, it is fulfilling and takes on a cultural dimension. According to Jake, when relations talk too much ill about one, it gives room for illluck to befall one, an idea deeply rooted in the cultural repertoire of the Pinyin and Mankon and expressed in the following: 'If you do not open your door to the devil, the devil will not come in'. Kinship relations, however, are one of the primordial reasons for going home and taking part in funerals and family gatherings. How then do these mobile communities function and what does their trajectory look like? For the most part, the mobility of the present mobiles is informed by the habits they learned from their parents in their childhood years. Joe's story below has followed the pattern of his father's.

According to Joe, his father was a bush trader who traded between Cameroon and Nigeria, specializing in buying guns and gunpowder to sell in Pinyin and the neighbouring villages. He was eventually arrested and imprisoned for tax evasion. Could this be the lone case that is mentioned in the archives? Prior to this, he was a herder and would earn a calf every year from the Fulani owner. Once out of prison, he became a builder, a trade he acquired in prison and that he complemented with farming. Soon he gave up building and went back into bush trade, buying cloth from Nigeria and selling them in bush markets. The Nigerian route was later abandoned and he would then go to the West Region to buy goods and this brought him into contact with coffee. He became a coffee farmer and went to neighbouring villages to buy coffee and supply it to the Pinyin cooperative. When the coffee boom was over, he reinvented himself as a butcher and reared cattle, travelling to Nkambe and Mesaje to buy cows and to Douala and Kumba in the Littoral and South West Regions respectively to sell them on. As we read Joe's story today, the rate of mobility has not changed much: it is a circle of mobility.

Mobile society

Joe's story

When I finished university in 1986, I went to practise as a trainee lawyer after I wrote the entrance exam for the military academy and failed. Not satisfied with this, I moved to Limbe where I set myself up as a businessman selling *fuenge* (petrol from Nigeria). Business was going well and I later branched out into medical supplies – laboratory reagents – that I supplied to hospitals in Foumban because I had a girlfriend working there. I would go to Nigeria

12 Interview Jake, Cape Town; 04/08/2011.

by speedboat, buy things and within a week I was back to supply my clients. This went on for a while and things were promising, but my business crumbled when the boat I was travelling in almost capsized due to rough waters and I had to throw all my stuff into the ocean in order to save the boat from sinking. Back in Limbe, I changed from being a Christian to a Muslim and went on *hajj* to Mecca. My intention was to take advantage of the *hajj* and move to Saudi Arabia in search of greener pastures. I made it, and after the *hajj* I got a job but the lack of work permit made me a fugitive and I left Saudi Arabia after eight months.

Back in Cameroon, I decided to go to South Africa with a friend – who I literally pulled along – in 1995. Our whole idea of leaving Cameroon in the first place was to get into South Africa before independence and pass for exiled African National Congress members so we could benefit from South African nationality at the time as incoming exiles. It took us 16 months from Cameroon to South Africa, as almost all the journey was done by road. When we got to DRC, we were robbed and spent 12 months working, begging and hawking to get some money to continue the journey. But one important thing that helped us was Islam: it gave help to stranded traveller. For a while we slept in the mosque and when we had saved enough money, we set off for Angola, got to the small oil island of Cabinda and spent some more time there and got a visa for Luanda, from where we got a Namibian visa. While in Namibia, I got a fake South African visa but my friend did not. When we got into South Africa we were repatriated because I had a one-way ticket. We then smuggled ourselves into South Africa again and went straight to Cape Town to the refugee centre there. Here, it took me three weeks to save R40 to buy my first box of chips, which it took me three days to sell as a hawker.[13]

Joe's story expounds on the fluidity and flexibility of mobility and identity, and the porous nature of borders. Being a Christian and taking up the identity of a Muslim and name changing was a means to an end, although he has come to appreciate Islam. Joe's mobile life spans a quarter of a century and mobility is still part and parcel of his lifestyle. Even after settling in Cape Town, he still frequently travels to Johannesburg and other towns in South Africa for business, and occasionally to China to meet suppliers.

Elias from Mankon similarly changed to Islam while studying in the UK because of the benefits it offers but, unlike Joe, he has remained a Muslim even after migrating to Cape Town. Joe is the embodiment of the late-twentieth/early-twenty-first-century migrant who takes advantage of every avenue that facilitates mobility and takes on multiple identities as he goes along. But in his case, his major handicap has been a lack of communication technology as, back then, the mobile telephone was just starting its journey to Africa. His ability for smooth sailing was hampered.

By contrast, Nixon's case will shed more light on how advances in road infrastructure coupled with mobile telephony acted as a catalyst for him to navigate geographical spaces with ease. As Nixon failed to complete secondary school in 1992, he moved from Bamenda town back to Pinyin where he took up gardening and running a bar. Later he would harvest his produce and take it to Douala in Littoral Region about 350 km from Pinyin. These journeys became an integral

[13] Interview with Joe, Cape Town; 24/07/2011.

part of his activities as he would go to Douala in a truck every fortnight to sell vegetables. His business was thriving until he suffered two successive losses due to breakdowns of his rented truck on the highway for a week and the perishable nature of his goods. He lost everything and as a result of a lack of capital, he relocated to Yaoundé; the capital of Cameroon, in 1999 at the invitation of a friend. Nixon's life in Yaoundé was a daily struggle for survival. He moved between jobs and later (2001) he became self-employed, selling vegetables in the market from Pinyin, Santa, Babanki, Dschang and Yaoundé. Having made a name in the trade as an honest person, Nixon was constantly supplied with vegetables by farmers and he would sell on commission and send the agreed amount of money to the farmers electronically or by the same bus service that brought the vegetables to the capital. They cultivated a network of mutual trust and benefit with the farmers sending their produce on the overnight bus as unaccompanied luggage and Nixon collecting it the following morning with a list of the items sent. Good road networks linking the North West Region where most of the vegetable came from and the introduction of cell-phone communication were added advantages, ensuring that the produce arrived the next morning by 6:00. This is a far cry from when he started transporting vegetables from his village to Douala with a lack of knowledge about the needs of the market at the time. As we are reminded by Castells (2004), a network society functions when there is access to information and knowledge. This was lacking when Nixon started out, in contrast to the period when he moved to Yaoundé and when the phone was already omnipresent. According to Nixon, the cell phone was his organizer and he topped up air time for FCFA 5000 on a daily basis in order to reach his suppliers and update them on the market and what vegetables to send and/or not send. Invariably, the cell phone was the contact between him and his clients, who would call to find out about the availability of goods or he calls to tell them about new arrivals. Nixon's friend migrated to South Africa and a year later convinced Nixon to join him, opening up another avenue for further migration. Within three months, Nixon got a passport and left for South Africa where he was introduced to pitching (hawking) over long distances, trekking each day, and on Sundays he goes to the flea market to sell by the beach.[14]

Joe and Nixon's stories complement each other and show not only the extent of mobility but the level of migration that is engrained in their daily activities. Their livelihoods revolve around mobility and migration. While Joe's story epitomizes the challenges faced by earlier migrants in the mid-1990s, it also gives insight into when the cell phone was not very popular, seen from the length of time it took him to get to Cape Town (16 months), and was completely out of touch with his family, learning about his mother's death only a year after it hap-

---

[14] Conversations with Nixon, Cape Town; 19/05, 26/05 and 30/05/2011.

pened. Nixon, by contrast, reported that for some migrants, even before his migration to South Africa, the phone was already shaping their daily lives. Both stories show at one level the challenges migrants had to put up with due to limitations in mobile connectivity, and, at another, it shows how mobile communication has facilitated and organized people's daily lives and activities, with mobility centred on the mobile phone. Besides, the insights they gather from their sojourns in various places along their mobile trajectory come in handy at their final destination. At another level, one is able to glean information from the stepwise/chain migration that is characteristic of most Pinyin and Mankon migrants. Whether national or international, there is often a sentiment that transnational links by those who have gone ahead can be seen in the decisions taken by Nixon's friend to invite him whenever he relocates. He moved from the village (Pinyin) to Yaoundé and from Yaoundé to Cape Town and in each instance he invited his friend along and Joe has, in turn, brought most of his siblings too.

While migration has evolved from what it used to be in the past, the new trends put women in the spotlight as well and introduce a gender dimension to the mobility of Pinyin and Mankon women. Unlike when Pinyin and Mankon women were seen to be sedentary and taking care of the household and their children, they have increasingly become mobile in their own right and also to join their husbands or going into marriage. However, in relation to the number of men who migrate on an individual basis, migration among women is still hampered by factors such as doubts about finding a spouse in the host country and kinship obedience (Castles & Miler 2003: 36-37; Boyd 1989). Although the majority of Pinyin and Mankon women in Cape Town came because of spousal reunions and/or to study, there are still a handful who made the journey on their own in search of greener pastures. The small number of women is further upheld by the idea of machismo: to be a man one needs to import a spouse from home; a woman who will be subservient. This notwithstanding, migration to South Africa and the Netherlands is characterized by young generational movement. Those who engage in it are often in their late twenties.

The new mobile culture means that migrants can oscillate between return and transnational migration. In Cape Town, for instance, there are individuals whose families (spouse and children) have been left at home and have come simply to work, save money and go back home to invest. For these groups of migrants, their return is imminent, while the second group is more transnational and have opted for a foothold in the host country and another in the home country, like Joe and Nixon. Such a dualistic lifestyle could be led because they have increasingly appropriated the new communication technologies to increase their virtual presence although they are far away.They are bridging the distance between host and home country by constant telephone communication and other social network sites at their disposal. This dualistic lifestyle is echoed by a willingness to do so and to be seen as responsible, hence garnering respect and cultural capital.

Such connections, either among migrant networks or back in the home country, equally give rise to social spaces – the 'in-betweens' who are often propelled by communication technologies. Though the activities that take place among the in-betweens give insight into the real and daily lives of the 'mobiles', such as their economic activities, managing social relationships and ICTs in coping with the in-betweens as well as participation in association, being actively engaged invariably guarantees a 'posthumous' (Castles 2002: 1154) form of migration back to their country of origin. One thing both communities under study agree upon is the taboo of burying one of theirs in South Africa or the Netherlands. As a result, many migrants belong to their cultural association as a form of insurance that makes provision for a posthumous return home in the eventuality of death. Returning home either posthumously or for family visits requires the necessary documentation, with passports/visas giving one the right of entry into one's country or another country.

Passport mobility

Mobility, as we have conceptualized it is not simply the movement of people, it is an encompassing of the movement of things, ideas and fashions. The general trend is to travel with a legal passport, as is required by the regulations on international travel. Until 1914, people were largely free to move and live anywhere they wanted to until the post-war period when people were required to have travel documents (Taylor 1965 in O'Bryne 2001: 401). As O'Bryne, (2001: 400-403) put it: '(a) passport carries a significant political meaning and a subtle cultural or symbolic one' and he goes on to say that this is the onset of encroachment into the everyday lives of people. This therefore underpins the fact that passports have a multifaceted role: proof of citizenship, identity, the right of border crossing, of admission into another country and the protection of the issuing sovereign state. The passport thus could also be seen as a means of policing borders, controlling people's access to resources and monitoring international flows of people (Bakewell 2007). The question that this section seeks to address is whether there are sentiments attached to these identification documents by those who carry them. Do they attach the same importance to their passport as the issuing sovereign state does or are they simply a means to an end? The answer could perhaps be found in Bakewell's (2007: 14) article on identity paper in North-West Zambia in which he states that 'people may make statements with their mouths and sign paper to gain a new nationality, but they will retain their attachment to the nation from which they came'. In other words, people claim membership in more than one place. This suggests that migrants/refugees take up other nationalities in

> I used two passports; a Swazi and a Cameroon passport because a Cameroon passport is visa-free to go to Singapore. Now you need a visa from Singapore to Korea. So in the Cameroon passport, I had a visa from Singapore to Korea. I had to enter Korea and then apply for a visa to enter Japan (using the Swazi passport). I had two options – either to go to Japan or Canada. When I entered South Korea, they said my passport was a transplant. That's where the problem started and I was repatriated.[18]

The notion of identity is questioned through 'handheld' and 'heartfelt' nationality. In the age of globalization, identity, it is argued, must be seen as one of hybridity with migrants taking on a multiplicity of identities that are combinations of home and host countries. But as can be seen in Charles's case, taking on Swazi identity was to ensure entry into Japan as Swazi nationals at that time did not need a visa for Korea, from where he would apply for a Japanese visa. Flexibility of identity only helps to strengthen the argument for handheld and heartfelt identity. Identities are taken on as long as they serve the person's interest at the time, and a multiplicity of identities is taken on if necessary given the circumstances. Much as passports and valid visas are there as speed bumps against unauthorized migrants, migrants have also trivialized and contested this claim, while circumventing the speed bumps put to check them. This has made passports goods for commodification (Bakewell 2007) and they are therefore regarded by migrants more as a means to an end.

Migration to South Africa and The Netherlands

> In a single decade, between 1990 and 2000, South Africa became the new migration hub at the southern-most tip of the continent, drawing hundreds of thousands of migrants from Central, East, and West Africa and as well as Bangladesh, China, Pakistan and Eastern Europe. For some of these migrants, South Africa is just a temporary haven en route to Europe or North America. For others and unintentionally for many, it has become a final destination (Segatti 2011: 9).

While the quote above succinctly and aptly captures the new migration trends into South Africa, Europe is often still the preferred destination for most economic migrants, especially those from Sub-Saharan Africa. Cameroonian migrants are no exception (Alpes 2011; Ngwa & Ngwa 2006; Nyamnjoh & Page 2002). South Africa has had a long history of 'regional contract migrants' (Shindondola 2007; Crush & McDonald 2000) destined for the mines and the collapse of apartheid and the transition to an open and democratic South Africa opened the floodgates to migrants who hoped they would be able to enjoy a share in its riches.This has also meant that the traditional forms of migration that are being reconstituted are giving rise to new migrant networks of students, asylum seekers, traders/hawkers and professionals. The collapse of apartheid, the changing dynamics of migration and mobility, and the building of 'Fortress Europe' have

[18] See Footnote 16.

all helped to catapult South Africa into the spotlight. Pinyins constitute the largest number of migrants in South Africa compared with other Cameroonian groups. South Africa was considered by many to be a springboard from where they could launch migration to their country of first choice and as a place of transit. As Peter puts it; 'Even though South Africa was my last resort then, I want to tell you now that I wouldn't have been what I am today elsewhere. A lot of people do not know much about South Africa.'[19] Although this notion of South Africa as a transit zone has been not completely abandoned, most migrants have come to terms with South Africa as their final port of call. For instance, Charles and Mark from Mankon and Pinyin left Cameroon for South Africa with the intention of going to Japan and England respectively. While Charles was twice deported from South Korea en route to Japan and arrested for being in possession of fake documents (see above), the agent that asked Mark to move to South Africa in the hope that it would be easier for him to get a visa, cut off all communication with him. Jerry, however, after having lived in Cape Town for ten years finally migrated to France via Germany. From France he sent an email to friends announcing that he had 'finally achieved his dream of getting to Europe'. Jerry's story confirms that mobility is a conscious lifestyle choice and that time does not necessarily make one forget one's initial plans regarding mobility, it may simply put plans on hold. This also implies that migrants are not especially attached to any particular location and that there is often a tendency to look further to satisfy their quest for the Utopia they have imagined at a future destination. Accordingly, Charles shared some of his dream:

> … when I came here, my intention was not to come and remain here. You know that when we were in Cameroon, we always had the belief that South Africa was a stepping stone or a bridge to Japan and other places. I was a technician and thought that if I went to Japan I would easily 'make it', more easily than here. You see … so that was my intention, to come here, work out my line and go to Japan.[20]

Stories corroborate the finding of Alpes (2011) about Cameroonian migrants and their eagerness to fall bush at any cost or to go home using someone else's passport. However, the idea of using South Africa as a transit country is beginning to disappear as the first groups of Pinyin migrants are seen to be succeeding and engaging in development projects back home, unlike most of the earlier migrants who left for Europe. Europe to some parents in Pinyin is synonymous with the 'place where people get lost' and are not heard from again. As such, much as some migrants in Europe are held in high esteem, most are considered to be passive towards home and development projects. As a result of seeing how those

---

19 Interview with Paul, Cape Town; 01/03/2011.

20 Interview with Charles, Cape Town; 23/08/2010.

from South Africa respond to developmental projects and family subsistence, parents are now choosing to send their children to South Africa.

Migration to South Africa has also been assisted by the continuing wide-scale availability of networks and information flows that assist migrants and facilitate settlement in a host country. The huge migration network built up by the Pinyin community can be seen as the driving force that propels migration to South Africa more than to Europe. Having circumvented visa and entry requirements, this explains the large number of Pinyin migrants in South Africa. At the time when the South African Embassy for Central Africa was located in Gabon, two Pinyins in Cameroon set up a network to process documentation to South Africa for a fee. It is thanks to these networks that earlier migrants (those who went in the 1990s) made it to South Africa before the Mozambican route became operational.

Conversely, the group of Mankon migrants has not been as exponential as their Pinyin counterparts. This is partly due to what I term the loose kinship relationship that exists among the Mankon compared to their Pinyin counterparts. With Pinyin being a more highly polygamous society than Mankon, there is increased pressure on siblings from the same mother to assist each other and improve their 'mother's house'. Hence earlier migrants take along other siblings or help them to migrate elsewhere once the first member of that household migrates. The idea is that with more persons having left, the burden of assisting is lighter. Apart from spousal reunion, there are fewer than five Mankon migrants who have sponsored siblings. Contrary to the Pinyin community who see the need to bring in as many siblings into South Africa as their finances will permit, their Mankon counterparts are reticent about bringing in relations. For instance, Henry's nephew has repeatedly requested that Henry assist him in travelling to South Africa or Thailand, but he is not interested. Bringing in his (Henry) wife was the most he ever did. Despite this influx of migrants from Pinyin and Mankon respectively, the challenge they all face is the stifling government control policy aimed at controlling the country's borders.

For the Mankon and Pinyin respondents in the Netherlands, migration was more for studying for John, James and Ted, while Pat and Beatrice were joining their spouses. Those who came to study often overstayed their visas and subsequently had to seek other ways of regularizing their situation, for example by paying exorbitant sums to Dutch or other foreign women with Dutch citizenship who were prepared to agree to a marriage of convenience. John paid a Zambian lady € 15,000 to pose as his wife so he would be granted Dutch citizenship and could subsequently get a passport. For her part, Pat came to the Netherlands on a tourist visa at the invitation of her fiancé and a few days afterwards was advised to go to the asylum-seekers'centre with a well-framed (politically motivated) story to substantiate her case. Her story was believed and she was granted asylum. Migration to the Netherlands is difficult given the strict visa regulations, even for those who want to use education as a means for migrating. Nonetheless, migration industries have emerged to circumnavigate the tight border controls in Fortress Europe. Would-be migrants pay up to € 5000 to migrate or for a relation. EU countries such as Italy, Greece and some Eastern European states have become ports of entry for migrants because entry visas are relatively easy to obtain, perhaps by colluding with some officials. James, for instance, brought two of his sisters through Italy and had to drive from The Hague to Milan to pick them up. He used the migration industry to facilitate the process for him, although he thinks he was very lucky as some of these industries collect money and disappear, as was the case of Mark who was duped out of € 2000 and left stranded in South Africa. Like James, Pat brought in her brother who has successfully sought asylum in France. Although Emile from Pinyin came to France using one of the migration syndicates from Cameroon, he has relocated to The Hague but is continuing to seek asylum in France and spends a few days in France every month with his lawyers to follow up on his case. On the whole, migration to the Netherlands is much more difficult than to South Africa due to the various common treaties that have been put in place by the EU to restrict entry. While both South African and Dutch immigration policies are still overly centred on border controls and regulations, migrants have sought various means of challenging and circumventing them.

Migration governance or governance fragmentation?

> Aliens control stems from the basic right of a country to decide which non-citizens are welcome within its borders. (Lindiwe Sisilu, Deputy Minister of Home Affairs, South Africa, 10/04/1995. In Peberdy 2009: 1)

> Southern Africa's national policies … remain marooned in an approach rigidly based on border control and national sovereignty. (Segatti, 2011: 9)

These quotes emphasize the role of the state, particularly South Africa, in determining the fate of immigrants within its territorial space. Nineteen years after the pronouncement made by the Deputy Minister of Home Affairs, policies concerning border controls and closures have, if anything intensified. Fast forward to today and the current language and migration policies, the country has had an open-door policy to those that meet its skills requirements of extraordinary and/or exceptional skills, the guiding principles of exclusion remain fundamentally the same. Peberdy (2009: 147) has shown that 'since 1994 immigration policy has been mostly exclusionary, based on a strong national, protectionist and territorial vision', which is geared towards the exclusion of Sub-Saharan African migrants, especially those from Zimbabwe who have taken to tearing down border fences. The importance of this is the ability of the state to define the parameters of contemporary citizenship; as this is seen to be the basic fundamental of exclusion and inclusion, and ultimately belonging (Faist 1998).

The age of globalization that ushered in an accelerated flow of capital goods and presumably persons has, however, left developing countries questioning the embargo on the flow of persons. There is a call for the placing of immigration on a par with free trade and the free flow of capital (Freeman 2006; Ferguson 2004). But as Zolberg (2006) and Sack (1986) contend, it is borders of states that make international migration possible as a distinct social process and hence, in the absence of state borders, international migration would not be an abstract. It is in this light that Sack (1986; also Peberdy 2009) sees borders as powerful social constructs governing access to resources and signifying a powerful relationship between those who control and those who are controlled. Borders have become a mechanism for exclusion and inclusion. While they reveal the vested responsibility of nation-states, the power of choosing who to admit also reveals the concerns of these states regarding the 'contamination of non-members and citizens within its territories' (Peberdy 2009). This has led Basch *et al.* (1994) to caution that, in the wake of transnationalism, 'deterrtorialized states' are emerging with potentially destabilizing consequences for state security, national identity and international politics (see Maharaj & Moodley 2000). These controls invariably lead to contestations and transgressions of borders because of the restrictions imposed by nation-states that give rise to the margins and all forms of illegal activities, as shown by MacGaffey & Benzenguiny (2000) and Das & Poole (2004). The rise of the margins and contestations of state borders could be put down to the disjuncture between state policies and immigration realities in the sending and receiving countries (Ros *et al.* 2007). Migration has often been depicted by the receiving countries as a problem (Nyamnjoh 2010; Portes & de Wind 2007) and, increasingly, is the concerned of state sovereignty (Castles & Miller 2003). Except for a few cases of collaboration with other states (Nyamnjoh 2010), for example the wave of boat migration to the Canary Islands, states want to control immigration by introducing legislation in their respective countries (cf. Castles 2004). With the explosion of migration and mobility, governments of receiving countries have, in addition of being intolerant towards unskilled immigrants and refugees, taken a tough stance on unauthorized migrants. Yet, policies to curb this upsurge continue to yield few results, perhaps, due to what Castles & Miller (2003: 114; Castles 2004) describe as the direct consequences of the activities of the migration industry[21] or the 'liberal paradox' (Hollifield 2004) whereby receiving nations are prevented by human-rights conventions they are signatories

[21] This migration industry is in fact what is encouraging the high rate of Pinyin and Mankon migration to South Africa.

to from suppressing and effectively controlling unwanted migrants, especially those already within their borders.

States want to know who belongs and who does not, and questions of identity, belonging and citizenship are high on the agenda in this regard. Since 9/11, the European Union has sought global governance, increasingly building up Fortress Europe (Castles & Miller 2003: 118; Nyamnjoh 2010). Why then are some states overly concerned with border controls and do not see the benefits of migration like some Asian countries that have 'built national development strategies around migration' (Segatti 2011: 9) to enhance growth? Foucault's governmentality approach (Gordon 1991: 20) sheds light on the process of global governance, maintaining that countries seek to exert control over the 'territory' and the 'subject', which in effect are the targets of power. Central to understanding the approach of governmentality is the notion of disciplining the population; where the population is conceived as thinkable entity that is primordial for effective governmentality. Governing, according to Foucault, involves a form of surveillance and control similar to that of a family head over his household and dominion. He asks if it is possible 'to introduce this meticulous attention of the father towards his family into the management of the state' (Faubion 1984: 207) and goes on to exemplify that 'to govern a state will mean to apply … a form of surveillance and control as attentive as that of the head of a family over his household and his good'. To attain governmentalization, the state must adopt tactics that make 'possible the continual definition and redefinition of what is within the competence of the state and what is not' (*Ibid*.: 221). As such, those who do not respect the common good essentially have to be 'disposed' of. In this respect, migrants are not seen as part of the household and to be protected but rather as people who do not respect the common good of the state and who will deplete the resources of the household. Hence the household must be protected against them by adopting stringent laws that keep them out. This, as Merlingen (2006: 192) concurs, is about the 'suppression of otherness as a condition for the realisation of order(liness) imagined by authorities'. By the same token, such policies invariably lead to resistance and the reversibility of power relations (Merlingen 2006). Dealing with such mobile migrants calls for trackable security processes because, as Deleuze (1995) explains, societies that control with power are more fluid and de-centred. It is in this respect that governments seek to impose systems of reason upon the rapidly moving and the restless (Urry 2007). As such, migrants have become the target for both sending and receiving countries to exercise their power over them by employing various methods that have been put in place to restore borders, check their influx and bring it to an end as if it were a tap that can be turned on and off. For instance, when Charles was deported from Japan, there were 15 immigration officials and police on hand to 'welcome' him at the airport. He was then sent straight to jail (Postmor Prison). While South Africa is yet to exercise deportations to Cameroon, deportation is one of the favoured means of control adopted by the Dutch government, although some still manage to evade the long arm of the law.

Policies put in place by states could either be seen as 'policy failure' or 'securitization' (Boswell & Geddes 2011). These measures probably epitomize the failure to consider migration as a dynamic social process and not make the economic model the scapegoat. But again, the very essence of global governance of migration indicates the complexity of the subject and the difficulty of finding a comprehensive policy that will mitigate migration because, as Beck (2000: 11) notes, 'nation states are yet to comprehend, probably, the full scale of trans nationalism wherein sovereign national states are crisscrossed and undermined by transnational actors with varying prospects of orientations, identities and networks'. By the same token, states and policy makers have underestimated the agency of migrants to circumvent and circumnavigate border regimes and state legislature, showing the limitations of nation-states in addressing mobility (Nieswand 2011).

Following the end of apartheid, the new democratic South Africa was challenged with putting in place migration laws that would reflect its new status and preparedness for South Africa's arrival on the global stage, particularly on the African continent. Invariably, most other African countries have had to revise sections of their migration policies pertaining to South Africa. For instance, prior to the abolition of apartheid, passports issued in Cameroon carried the message that the 'holder is barred from entering the Republic of South Africa'.[22] The Migration Act (Crush & McDonald 2000; Peberdy 2009; Segatti & Landau 2011), has since undergone various revisions but the general trend still remains that of control and the policing of the borders. Policies have thus rarely achieved their intended goals and have conversely produced outcomes that, if anything, have created loopholes for migrants and the migration industry to exploit. Pinyin and Mankon migrants have taken advantage of these loopholes to regularize their migrant status, either through seeking asylum or marriage to South African women. Formerly, marriage to a South African citizen entitled the spouse to South African citizenship and when this law was enacted, a lot of foreigners took advantage of it and married South African women to benefit from it. And once they had citizenship, the marriage was annulled. However the high divorce rate caused the state to intervene and to introduce more stringent rules regarding marriage. For instance, the SABC talk show 'Morning Talk' had an official from the Ministry of Home Affairs as one of its guests on 11/08/2010 who confirmed that

[22] The first passport issued to this researcher carried this message.

> … the Ministry is drafting an amendment to the bill of marriage Act to send to Parliament for approval; whereby it would be punishable as a criminal act those who enter into a marriage of convenience with foreign nationals. They will revise the provision of Permanent Residence (PR) wherein once an owner of PR one can apply for citizenship, and also an amendment against a foreigner with a PR to bring in a parent. If the former is amended, one has to be a permanent resident for 5 years before s/he could qualify to apply for citizenship (it is now the case). As precautionary measures, those who contract such marriages will receive impromptu home visits from Home Affairs officials to check on them.[23]

Until this law came into effect, the following accounts from CJ show how some took advantage of it at the time:

> ... I opted to get married to a SA at the time we discovered it, which was something that is/was existing, but because there was no access to information, we didn't know that through a conjugal relationship, you could actually be given permanent residence and you could apply for a SA identity card, which meant that a resident permit was good enough coming from that background. So that's what I did and I applied for permanent residency and I was granted permanent resident (PR) status and was issued an ID card. Unfortunately my wife passed away in 2007. The pattern at that time was generally that there were a lot of marriages of convenience because the spouses that we took were those that one could not actually be proud of, so to say. I remember my wife, it was very evident that if I stood in front of a home affairs official and told the Home Affairs that this is my wife, they would probably look at me and laugh and say my friend clearly this is not your wife.[24]

Although visa requirements for South Africa are strict, Pinyin and Mankon migrants have found a way to circumnavigate this and now come into South Africa via Mozambique given the relative ease with which they can obtain a visa in this country, as was seen in the previous section on passport mobility. Others who are financially viable and want to have a genuine visa, apply for admission to a college or university, pay the admission fees and get a letter of admission that they can present at the embassy to obtain a visa. Once in South Africa, they temporarily or indefinitely suspend going to school in order to raise money for subsequent tuition fees and apply for refugee status, which gives them a rebate on their fees. This echoes the continuous attempts by migrants to mitigate border policing and controls in South African migration policies that determine those who are welcome to enter the country. But as foreseen by Basch *et al.* (1994), transnationalism was the start of fomenting border contestation and the authority of nation states. In the face of such contestation, how have South Africa and the Netherlands sought to establish their authority? A number of measures and laws have been put in place, including a reduction in the number of entry points, a strengthening of border and boundary controls, and the mobilizing of the armed forces to patrol their borders (Peberdy 2009). In South Africa, the various Immigration Acts of 1999, No.13 of 2002 and No.19 of 2004 aimed to curb and enforce influx control. While it gave legitimacy to government action, it also

---

23 SABC's 'Morning Talk', 11/08/2010.
24 Interview with CJ, Cape Town; 25/07/2010.

enhanced circumventing these acts. Policing borders has not always yielded complete success (Boswell & Geddes 2011). South Africa's borders proved futile when it came to ensuring proper control was enforced to counter migration syndicates that flouted and circumvented border checks and took advantage of every possible loophole in policy.

Europe is different due to the common European Union migration and asylum policy (Boswell & Geddes 2011) that helps member states comply with its aggressive policy of Fortress Europe, but the policy, as in South Africa, is primarily one of exclusion rather than inclusion. EU member states, including the Netherlands, have always relied on the numerous EU treaties signed to create a bulwark around it. First the 1993 Treaty of Maastricht, then the Treaty of Amsterdam in 1999 that included migration and asylum in the EU's main legal and political framework, and later the Lisbon Treaty signed in December 2009 that further consolidated the EU's migration and asylum policy (*Ibid.*: 7). As such, the EU is clearly strongly committed to the fight against illegal migration but also to a standardization of the rules in order to create a common EU asylum policy (*Ibid.*). This notwithstanding, it has its own share of policy failures, as their draconian measures to clamp down on illegal migration and on those overstaying their visas have been futile. In addition, migrants who come in as students rarely go home when they have completed their studies. Bogus documentation is on the rise due to the global migration syndicates. Cameroon is no exception here. Italy and Greece have become the loopholes for syndicates obtaining visas for their clients. For instance, James in The Hague had to drive to Milan to pick up his sister who he had brought to Europe using one of the syndicates.[25] Stringent border regulations in South Africa and Europe only cause further mutations in unauthorized migration processes, thus making a mockery of borders that are meant to control but have instead been subverted and domesticated by migrants.

These mutations may be due to the inability of immigration policies to include ICTs as an area for action or extended research in order to understand how this information flow fuels migration. According to Ros *et al.* (2007: 4), 'there is a complete lack of knowledge of the ways in which information flows shape the movements of people around the world'. Profound knowledge would help the formulation of policies that could take into account information flows.

Conclusion

The goal of this chapter was to set the stage by addressing some of the issues that frame the analysis of mobility and migration. Mobility and migration are dynamic social processes that, in the wake of globalization, have seen a change in their

[25] Interview with James, The Hague; 23/09/10, and follow-up cell-phone communication on several occasions.

character and nature; changes that are inextricably linked to advances in ICTs and in road and air transport. The repercussions of these rapid changes are the compression of space and time, with the past and the future all in the present. Although changes in migration and mobility have been phenomenal, they have also been informed by the migration patterns of the past, by bush trade and travel to the plantations in the south western coastal region of Cameroon. With an increasingly mobile society that wants to live a dual life with a foothold in its country of origin and in its host country, the coming of new technologies has provided opportunities for such a dualistic lifestyle. Implicitly, the ability to live a dual life hinges on the social networks migrants have created in their home country and among other migrant communities. Mobility, as Adey (2010) has rightly suggested, should thus be perceived as a relationship through which the world is lived and understood. Conversely, this dualistic tendency is eroding state sovereignty, national border security and circumventing national border policing. In a way, immigration policies have not optimally attained their objectives of borderland control and exclusion. This is because policies have often considered migration and mobility as a tap that can be turned on and off at will. Instead, mobility cannot be continuously perceived as stasis: it is a dynamic process that is informed by different variables (ICTs, networks, kinship relationship, double engagement and migration syndicates). Much as these variables have given mobility a new dynamism, it nevertheless hinges on internal embodied practices. In this regard, policies need to take into account the development strategies that seek to manage the migration flows in the sending country as well as that of the migrant in ways that serve national interests and aim to find a win-win-win solution.

By and large, mobility is not limited only to the flows of persons and goods. Cultural mobility delineates the ability of migrants to move with their cultures and, conversely, take on or assimilate that of the host country as well as those of other migrant communities in the host country. In the process, they also project their identity and belonging. Atypical to mobility is passport mobility. Together with the flows of persons, they are contesting states' sovereignty, disregarding borders and creating new social structures. As a result of states wanting to control who is within their borders, the question of belonging and identity emerges.

4

A new form of madness in the village: The arrival of information and communications technologies (ICTs)

Introduction

'I be used to see people them di waka tok for road them them
I check say them di craze, me too I don craze now weh I get ma own phone.'[1]

The arrival of mobile technology was received with awe by most and also seen as a must-have. But for others it was a new form of madness that had arrived in the village; with the few people who owned mobile phones running up hills, climbing trees or standing in the middle of the road where reception was good to take calls, according to informants in Pinyin. Owning a mobile phone greatly enhanced one's status in the village and such people became the link between their migrant kin and relatives in the village (de Bruijn *et al.* 2009; de Bruijn & van Dijk 2012; Nyamnjoh 2010). But for Mama Mangwi, getting a mobile phone a few months after her son had gone to South Africa meant that she joined the association of mad people '*wey them di stand for middle road di tok, wey motor fit jam them and they no go know*'[2] ('who, when receiving a call, are oblivious to where they stand because of the search for an area with network coverage and could be run down by a car').

Against this backdrop, the history of ICTs in Africa does not tell a single story. It is laced with different stages of evolution from embryonic to explosion. The Maitland Report published in 1984 grimly highlighted the embryonic stage and stated the discrepancies in telecommunication services between the developed

---

1 Interview with Mama Mangwi, Pinyin; 28/05/2010. 'I used to see people walking and talking all by themselves on phone on the road and I thought they were mad. Now that I own a mobile phone, I too have become mad.'

2 See previous footnote.

and developing worlds. From this report, one can glean what the communication situation was prior to the arrival of the mobile telephone. Juxtaposing that period with the current one of an explosion in mobile telephony, we can better understand the quest for appropriation of mobile technology as well as for being connected and not left behind on the information super highway. The report states that: 'More than half the world's population live in countries with fewer than 10 million telephones between them and most of these are in the main cities: two thirds of the worlds' population have no access to telephone services' (Maitland 1984: 13). In addition, the 2007 statistics on Internet use in Africa point to the continent lagging behind (Internet World Stats 2007; Cheneau-Loquay 2010). In the discourse on ICTs, Africa is generally seen to epitomize the digital divide and is still behind the developed world in the acquisition and application of ICTs and subsequently the benefits derived from them (Yumba 2004; Castells *et al.* 2007; Ling & Donner 2009). In both South Africa and Cameroon, access to the Internet is still limited to the cities and the urban areas. For instance, Pinyin has no Internet access, while Mankon got its first cybercafé in 2010.[3] The introduction of wireless fixed phones with access to the Internet has not produce the desired effects because Cameroon Telecommunication Service (Camtel) is yet to penetrate the most rural areas, and the cost of running the fixed wireless service is similar to that of the mobile phone. However, the former is limited in that it cannot be mobile as it is too cumbersome.[4] The statistics for Africa with respect to fixed and mobile telephones given by Castells *et al.* (2007: 22-23) corroborate Maitland's report, with South Africa being on the highest rung in terms of landlines and mobile-phone connection and usage, while Cameroon comes close to the bottom.

The era of analogue telecommunication appears to have been part of the teething problems that beset telephony (landlines). The switch from analogue to digital paved the way for the modernization and extension of landline services that subsequently witnessed the arrival of the first mobile-telephone service in Cameroon, which was provided by Orange Cameroon. The adoption of the mobile phone, according to Castells *et al.* (2007: 13), began to grow steadily from 1999 onwards and saw a sharp rise after 2001. Predictions that growth would slow down once the higher-income earners had been served proved not to be true (Horst & Miller 2005). On the contrary, the growth rate of mobile telephony has tripled, and penetration of the rural areas is on course as empirical evidence from this study illustrates. Today, Africa occupies a privileged position in terms of the highest growth rate worldwide (ITU 2004). Statistics on Africa's access to the

3 Fieldwork observation.

4 These are regular fixed phones but with the capacity to be carried around if need be for those with Internet access.

information super highway show a phenomenal leapfrogging and the gap described in Maitland's report with the rest of the world has narrowed. Close to three decades on since the report was published, the story of telephony in Africa, and Cameroon in particular, has changed tremendously with the mobile telephone making inroads into the nooks and crannies of villages such as Pinyin and Mankon and statistics for Africa showing it to be the fastest-growing continent when it comes to mobile connectivity (ITU 2009).

The ICT revolution has transformed Africa 'from (being) a world of landlines to a world of mobiles (that) provides us (Africans) with a unique chance to gain insight how a personal technology affects social organization, both for the better and for the worse' (Ling & Donner 2009: 4). While the 2009 ITU report predicted a continuous mobile revolution in which robust growth was envisaged, some villages, like Pinyin, have moved straight to the era of mobile telephony. However, Internet accessibility and landlines are still lagging behind.

Apart from condensing geographical spaces, ICTs are changing lifestyles by creating an informal economy and new job opportunities, especially in the Digital Versatile Disc (DVD) sector where migrants now champion the downloading and burning of movies for commercial purposes, with the Internet being a key factor in the promotion of this business in Cape Town (Chapter 6). In this respect, this chapter attempts to establish why there is a significant boom and anxiety to be connected, how the evolution of ICTs is changing the landscape of telephony and to what extent the change in communication technologies affected social relations. It also considers how far the appropriation of ICTs is an expression of identity and belonging.

Overview of current debates on ICTs

In an attempt to make sense of these technologies, we turn first to Fisher (1992; 266-268) whose early work on landline telephony in the US was trendsetting and unravelled the importance of telephony. Fisher concluded, on the one hand, that much as the telephone had intensified the rate of communication between families, it had not limited face-to-face communication, and on the other, it had bolstered and cemented kinship and social relations through calls. Yet while this study resonates with Fisher's conclusions in furthering our understanding and appreciation of the evolution of ICTs since its embryonic stage, this still does not explain the upsurge in new communication technologies. Moving away from landlines, Miller & Slater (2000) contend that the appropriation of technologies is a result of unfulfilled desires and the adoption of one over the other is because of the limited opportunities users could have with previous technologies. This useful insight may help us to understand the appropriation of mobile communication amongst the Mankon and Pinyin as well as its advantages to these societies and why de Bruijn *et al.* (2009) see this emergence as 'the new talking drums of everyday Africa'. Turning to Licoppe (2004) provides useful insight with his notion of 'connected presence'. Like Fisher, Licoppe takes us back to the era of fixed phones and distinguishes between landline calls (prior to the mobile phone) and mobile calls. He claims that the former are used for longer calls and can cover a wide range of topics, and often at set times. Conversely, the mobile-phone call excludes all meanderings and close-knit networks are at hand for quick calls. Explaining his approach, he opines that a mobile-phone call:

> … consists of short, frequent calls, the content of which is sometimes secondary to the fact of calling. The continuous flow of this interaction helps to maintain the feeling of a permanent connection, an impression that the ink can be activated any time, and that one can thus experience the other's engagement in the relationship at any time. (Licoppe 2004: 141; see also Ling & Donner 2009: 22)

Succinctly put by Ling & Donner (2009: 22), 'we are engaged in a series of shorter communications; an ongoing dialogue that lasts through the day'. While these insights explain both the use of fixed phones and mobile phones and are still very contemporary in describing the use of fixed phones, they also shed light on the need by the Pinyin and Mankon mobile communities to be connected through the use of sporadic *tuyaus*[5] that allow a 'connected presence' with home. Conversely, it shows how calculative agency is being remade and agency on the move. But the question is whether all migrants readily avail themselves for continuous connected presence. Why are some more connected than others? However, these questions do not pay adequate attention to the missing link or the 'disconnected presence'. With this, I allude to the 'mobiles' who have migrated and do not call home, yet the family is aware of their existence and receives information about them from returned migrants or secondary sources. In other words, this is the agency of the absent. However it is an agency in which 'distance is created; displacement is controlled; in which something is kept present whilst also being lost' (Callon & Law 2004: 10). Thus it becomes a community where actors are trying to anticipate encountering or losing themselves to other modes of ordering, and where agency is being reconfigured. These encounters include migrant associations and other migrant communities and, to belong, they have to accept the rules or seek to change or adapt them.

Horst & Miller (2006: 7) continue the discussion of agency through the concept of 'linking up', which resonates with the way we see the adoption and appropriation of new technologies in this study. In their study of low-income Jamaican families, they aimed to find out 'what Jamaicans have become in the light of their use of the cell phone and what the cell phone has become in the light of its use by Jamaicans', i.e. how the mobile phone shapes social relation-

[5] Different illegal means employed by Pinyin and Mankon migrants in Cape Town to call home.

ships and society, and how society is in turn shaped by the mobile phone. As regards the Pinyin and Mankon mobile communities, 'linking up' is less about having to store numbers for the sake of it than numbers that the migrant will draw social and cultural capital from at opportune moments. Similar to Horst & Miller's link-up, Scifo (2005 cited in Ling & Donner 2009: 18) propounds the notion of 'domestication'. According to Ling & Donner (2009), this approach's 'analysis follows the process of placing the object or the service into the broader context of our daily lives' and we 'redefine and adjust our uses of technology according to our needs'. However, these uses are not as elastic as one would imagine. Therefore, our understanding of the use/appropriation of the mobile phone will be greater if we heed Slater's call (2005 cited in Horst & Miller 2006: 10) to situate any particular technology within the context of 'communicative ecology' given that 'one needs first to appreciate its role as part of communication ecologies', where the principle of complementarity is supreme. This study shares this point of view but is not limited to the mobile phone alone. Although the focus on the mobile phone supersedes the other communication ecologies, it does not in any way preclude the Internet, Facebook or other forms of communication adopted by migrants. The different forms of communication reinforce our understanding of the scale of preference and also confirms the explosion and appropriation of mobile communication (Cheneau-Loquay 2010: 102; Etzo & Collender 2010). Looking at community members' use of social media, what is posted and the responses and comments that are elicited by the target audience enlightens us about the users of social media. In this regard, we get to understand the variations and tendencies of migrants' itineraries and their navigation and use of the various communication technologies through the different sociocultural and economic activities they engage in. The cell phone certainly occupies a privileged position in terms of usage given the rapidity with which it has met their needs (Manuel *et al.* 2007: 252) in favour of the Internet in the rural areas due to illiteracy rates and a lack of access to computers and Internet connectivity. It is probably for this reason that Jenkins (2006) considers the cell phone as the 'Swiss Army Knife', where the mobile phone sets up a conundrum of convergence and is the central actor in the 'convergence culture' where various media intermingle across multiple platforms and are consumed, re-cut, re-published and re-appropriated by active users in unpredictable ways. However, this is by no means to claim that the mobile phone has not got its own fair share of problems. Linking up means an availability of funds to buy air time, which is still a problem in most rural areas. However in the processes of linking up, migrants are in control of their agency because of their ability to determine their 'accessibility', 'availability' and 'reachability' (Licoppe 2004: 137-138) on their own terms. Whether it is the mobile phone or a fixed phone, Internet, Facebook and/or other social media, they lend themselves to the above three categories that are inextricably linked. This is because, as Latour (1999 cited in Urry 2004: 28) remarked, social relations are not fixed or located in a place but are constituted through various 'circulating entities'. Such circulating entities, it can be argued, result in multiple connections of immediate presence and/or intermittent absence. Yet, as is argued by Vertovec (2004: 219-220), global linkages have been facilitated less by the mobile phone than by the boom in cheap international calls that have been widely appropriated by migrants. He refers to this as 'the social glue connecting small-scale social formations across the globe'. Much as this maybe the case, we cannot credit the boom in cheap calls over the proliferation of mobile phones as these calls connect mobile phones and their users, hence I argue that the proliferation of cheap mobile phones have necessitated cheap calls. Cheap phones and cheap calls are not mutually exclusive but tend to reinforce each other. Equally, cheap calls do not necessarily translate to social glue as the unintended effects may be unreachability because of too many unwanted calls. For instance, it was only in 2011 that Joe connected to Facebook and Skype. At first he was elated by the fact that he could communicate for free, but soon became weary of both because they impinged a lot on his business hours and drastically reduced his connectivity. Even when available on Skype (on a mobile device), he does not answer all incoming calls.

In their multi-sited and longitudinal studies, Castells *et al.* (2007: 246) came up with some astute conclusions about wireless communication. Paramount to their findings is the fact that: 'Technology does not determine society: it is society, and can only be understood in social terms as a social practice'. This is a useful insight for this study given the way Pinyin and Mankon have adopted mobile-phone technology without first using fixed phones. In this regard, it is important to understand the context of how far these communities have come and to appreciate their appropriation of ICTs in rural areas before yielding to sweeping assumptions of bifurcation that are creating a 'Fourth World'. These inhabitants of the Fourth World cannot be described as 'structurally irrelevant' (Castells 1996, 1997; Donner 2008) as they determined the market trends of production with their paradoxical demands for increasingly cheap and/or sophisticated gadgets. They are not merely passive users but also active agents in defining the complexities of the telecommunications technoscape.

From another perspective, the proponents of transnational discourse see the importance of ICTs in linking families back in the home country and in other diasporic communities in virtual spaces, as well as the new opportunities opened up for mobile communities in relation to their double presence in their host and home countries. The ripple effect is giving them the opportunity to lead a dual life and expediency in remittances (de Bruijn *et al.* 2009; Tsagarousianou 2004;

Grillo & Mazzucato 2008). And Castells *et al.* (2007) concludes that if ICTs enhance transnationalism, it can therefore be presupposed that communication technologies reinforce kinship and social relationships as well as organizing the complex activities of everyday life. If that is the case, to what extent could we assume that ICTs, according to Vertovec (2004), are the 'social glue of migrant transnationalism' and binding to all migrants? Is it not time for us to check how far migrants are prepared to sustain their transnational lives? As I pointed out in the previous chapter, if mobility entails physical distance and ICTs either breach or bridge that distance, we cannot talk of transnational links in absolute terms. Being a transnational is a matter of choice, though it is sanctioned by cultural norms that the migrant is expected to comply with. Nonetheless, this is subjective and migrants could also choose not to be transmigrants. As will be seen below, much as there are those who desperately seek ways of connecting with home, there are equally some significant others who have temporarily or permanently disconnected from home. We have been talking about transnationalism but should not forget to mention disconnections too as the two are inter-related. Before adopting a transnational mode of expression, it is important to discover how these communities were faring before the proliferation of mobile technology.

Phoning before the proliferation of mobile phones (1980s-1999)

Prior to the advent of advances in ICT, telephony in Africa was a service reserved only for those in the cities. This period was marked by faulty landlines, acute shortages of materials for new subscribers, long waiting periods before lines could be installed and a service facing corruption.[6] As shown by Castells *et al.* (2007: 15), Latin America, like Cameroon, was also beset by these problems. Having a landline phone at the time was a mark of prestige[7] and phones, even though they often had technical problems, served as a link between families and relations abroad. Despite the truncated network, most phone owners were happy to have a phone as it gave the family an opportunity to communicate with relatives abroad, like my family then. There was also the issue of distribution and a lack of materials for repairs. These setbacks paved the way for its access to be limited mostly to the urban centres and the wealthy as, at times, they were called upon to provide cables and poles (or Indian bamboos) for installation and maintenance. In relation to the Maitland Commission, one begins to appreciate

---

6 Homes of service operators became telephone booths early in the morning before work and after working hours, and those who paid more were given free access to international connections for a day or all weekend.

7 I still remember when my parents had their first fixed telephone installed in 1986. This was seen as an achievement and instantly became the link between families in the neighbourhood and their relations abroad.

the depth of the report when it is viewed from the problems that beset the distribution of landlines in Cameroon.

One thing that marked this era was the fact that there was a clear notion of location, time and space. Calls had to come at predetermined time or were made to notify relations about when to wait for a call. Consequently this form of telephony impacted on people's lives as they became immobilized having to wait for calls from abroad for long periods of time at designated locations.

In Mankon, there was only one landline (that was plagued by cable theft and technical hitches) that belonged to the Fon of Mankon in the 1980s until the arrival of mobile technologies. The coming of the wireless fixed telephone made provision for users to have fixed phones at home but this type of telephone came with the possibility to pass for a mobile phone and with connectivity to the Internet. However, wireless fixed phones may yet take root in some rural areas, such as Pinyin that is still without fixed phones.

Pinyin never witnessed landline telephone communication due to the proliferation of mobile phones that radically changed modes of communication. Many families are now for the first time able to talk to their migrant children and actually 'feel' their presence. To date (2013), Pinyin still has no provision for landlines despite its proximity to Santa where they have been introduced. The arrival of mobile communication has broken down the barriers that isolated Pinyin from the rest of the world. Despite the gains made in terms of penetration of ICTs in Africa, it is nothing compared to what has been happening in Europe. Africa is still chronically trying to catch up (ITU 2009). It is against this backdrop that the early Pinyin migrants, as opposed to their Mankon[8] counterparts, sought alternative ways of communicating with family back home and at times going for long periods without news of their families before ICTs became widely available.

Maintaining transnational relationships and forming ethnic communities by migrants are not simply part of a quest to appropriate ICTs but, in the processes that ensue, they are a reminder for them of who they are, where they come from and also their reasons for preserving and fostering this identity and sense of belonging. It is also important not to lose sight of the fact that these communities do not form a collective identity in their place of settlement or with their homeland but share a common identity with members of the same ethnic communities in other countries (Tsagarousianou 2004).

8 Mankon migrants started coming to South Africa in the early 2000s when the mobile phone was already in use and did not experience the same problems when phoning home as their Pinyin counterparts did.

Conceptualizing ICTs in Pinyin and Mankon

The first mobile-telephone company in Cameroon was the French company Orange, a service provider whose services were limited to the cities of Douala and Yaoundé and that slowly moved to other small towns later. Although they seem to have covered a greater part of the territory, when I left Pinyin at the end of fieldwork in November 2011 and on a subsequent visit in May 2013, the Orange network still did not offer coverage in the research area. Similarly, the Cameroon Telecommunication service Camtel is also yet to provide services to villages, including Pinyin, although they are not considered far-flung or remote in the least. The arrival of the Mobile Telecommunication Network (MTN) in Cameroon in 1999 changed the country's telephony landscape as its goal was to reach the widest number of people as possible as was claimed by their slogan 'Everywhere you go', even though access to the network remains a serious challenge. Most of the villages that had never enjoyed fixed telephones bypassed this technology and directly adopted mobile telephones. The mobile phone has thus been hailed as a panacea for all communication problems in Africa in general (de Bruijn *et al.* 2009; Horst 2006; Horst & Miller 2005; Ling & Donner 2009; Ling 2004, Katz & Akhus 2002). As mentioned in the previous section, Pinyin has access only to MTN network coverage.

In Mankon, the story is different. The people there can access all three telephone service providers. When Orange extended its coverage to Bamenda, Mankon was also included, and with the inception of Camtel wireless fixed telephone, it was also instantly connected to this project. Although Mankon is still a village, what distinguishes it from Pinyin is the fact that it is only 10 km from Bamenda Town. With these various connections, it follows that there is Internet availability from all service providers available through a USB key modem. The arrival of mobile telephony thus changed the landscape of the village by linking it more closely to the urban town, but ownership of such a device was slow to catch on among peasant farmers. It was much later that people, especially those with children in South Africa, joined the information super highway, for example when they were sent mobile phones by their children or were given one prior to their departure to enable them to maintain communication. It was only in 2010 and 2011 that Akuma and Chi's mothers were sent phones from South Africa.[9] When Akuma left for Johannesburg, his mother did not have a phone and communication with her was via his brother's phone for some years. When the mother had an urgent message for him, she would ask the brother to beep him. What is significant here is the distinction he made between the mother's beep and that of his brother. The mother would beep twice while the brother beeped once. Ac-

9 Phones have often been sent to Cameroon via this researcher. On one trip I carried eight mobile phones destined for family members.

cording to the mother, this was to help him distinguish who wanted him to call back and, perhaps, the urgency to the call too. If it was his mother, the call was returned immediately or he called to tell her he would call later. With his brothers, he preferred an SMS to beeping and, depending on the urgency of the message, he would call back immediately or respond with another SMS. For her part, Chi's mother would receive calls via a neighbour's phone until she received one of her own. A good number of parents I met in Pinyin and Mankon showed off the mobile phones sent by their children. For some, it was their third or fourth because the previous ones had been stolen or taken by a relation who felt that children abroad think only of their mothers and immediate family. They were thus being challenged to send another for the mother. Achiri, also from Mankon, has been in Cape Town for about two years but within this period he has sent his mother four mobile phones (with this researcher taking the fourth). And each time he sends a phone, a sibling or relation takes it from her, especially as he always buys an attractive phone that is too complicated for her to operate. The arrival of cheaper mobile phones with the provision for dual SIM cards has ushered in a new wave of owners as FCFA 10,000 (approx. € 15) will buy such a phone. This means increased connectivity as people can be connected to both networks (Orange and MTN).

Much as these technologies are hailed in both Pinyin and Mankon, people there have equally expressed frustration at the intermittent connection despite the installation of an MTN antenna in Pinyin. Although the antenna's arrival has improved local reception considerably, the issue of network coverage has not been completely resolved,[10] probably due to the village's hilly terrain, with some quarters having full coverage and it being very limited in others. Quarters with patchy coverage have resorted to mapping out spots where reception is available. The enthusiastic manner with which they have embraced the mobile phone minimizes the challenges that they face and phones have metaphorically been compared to wives. As one informant said: '*dis woman dey so ee expensive for keepam, you pay for charge yi, you must give yi chop (air time), and na so ee di empty your pocket, but way for leaveam no dey, when you marry woman na for life noh!*'[11] This captures the fact that the arrival of the mobile phone is not a zero-sum game. While it has helped to bridge families, users have also had to adjust to its financial implications. People and phones are mutually constitutive and share an interconnected relationship whereby studying the phone in isolation is not possible as it has to be done with people *vis-à-vis* their interaction, which is embedded in their social and cultural practices. Despite the expense entailed in having a

[10] Conversations with Gabriel, a Pinyin who was instrumental in getting an MTN antenna in Pinyin; 21/05/2010.

[11] Interview with Papa Ndoh, Pinyin; 28/05/2010.

mobile phone, the rural population in Pinyin and Mankon have come to embrace it and have adapted to the changes it has required in their budgets. In the same vein, they hail the benefits of the phone; comparing it to a marriage contract. They have made a lifetime commitment to the mobile phone. One Informant in Mankon said that thanks to the mobile phone, her children no longer have to cover long distances to carry messages to their maternal aunt as she simply calls and delivers her message herself. At this juncture, Castells' relational approach to technology can guide our understanding of the carefully cultivated relationship that Pinyin and Mankon migrants and their families have developed with the burgeoning of ICTs, especially the mobile phone, given the historic link between migration flows and technology. This approach postulates interaction between society, history and technology in migration that is fundamental to understanding the current trends of migration in the information age (Ros *et al.* 2007). Although all the informants are unable to recall when the mobile phone took root in the village, the early users all tell one story: the phones they had then were as big as bricks and so were nicknamed '70-seaters' (after the 70-seater bus) and owning one was a sign of prestige. Equally, they all recalled the long distances they would cover to meet peers or relatives and regretted that the mobile phone had greatly reduced face-to-face contact in favour of virtual contact. Meeting with kin is now more occasioned by family events such as deaths, marriages, births or family meetings.

When I made my maiden visit to Pinyin in April 2010, only certain parts of the village were electrified so users had to go to other neighbourhoods and homes with electricity to have their phones charged, at a cost of between FCFA 100 and FCFA 200 for a full charge. I did the same as there were no lights where I was staying then. Having a fully charged phone is one thing and having network connection is still another. Often, people had to check out areas in the compound or on the road where they could get reception to make calls. In most compounds, areas where one could get network coverage were marked out and people would go there to make calls.[12]

At my host's compound, for instance, all the phones are kept in a small bag that hangs on the wall in the kitchen and when one rings, the owner takes it at breakneck speed to the spot where there is reception before pressing the answer button (if it does not stop ringing on the way). However not all neighbourhoods experience such problems.

---

[12] This was the first place I was shown by my hostess and I was given directions about which way to face as well as the position of my phone.

Photo 4.1 Pa Joshua making a call from the road where there is network reception in Pinyin.

Photo 4.2 (Taken 2010) Everyone's phones (including the researcher's) are inside this bag. The kitchen was the only place that had network connection but there was no electricity then.

Photo 4.3 (Taken 2012) Taken at the same spot when electricity had been installed.
The phones are still in the bag but on charge now.

The lack of network reception by extension makes it difficult for relatives in South Africa to stay in regular touch with family back home. Consequently, Jake in Cape Town asks the family to beep him whenever they are in an area with network coverage so he can call back. Otherwise he spends most of the day 'test calling' to see if they are in an area that has reception.

By the time of my third visit in September 2011, my host had installed electricity in the compound and nearly the whole village had been electrified, which had led to an increase in mobile-phone usage. Charging had become easy and communication amongst families was more regular as phone batteries no longer ran down or went uncharged for days. MTN had also upgraded the level of reception and network coverage had increased, and had reduced local call tariffs and the cost of international calls. Users were being lured with adverts announcing these reductions.

Photos 4.4 & 4.5 Mobile Telephone Network (MTN) adverts showing the proliferation of mobile technology

Photo 4.6 Advert for Orange service provider

The excitement about the high level of network coverage meant the number of persons with mobile phones increased, at least judging from the families I visited where one phone used to serve the entire family but now almost every household

member owned their own mobile phone. This is in sharp contrast to when the mobile phone first arrived in Pinyin and when I made my first trip to the village. According to Mama Mangwi, when the phone first arrived in the village, people saw it as a new form of madness; 'walking and talking to oneself on the street' she maintains that '*I be use to see people them di waka tok for road only them say na craze, and I be deny say I no fit join that craze, but now wey I don get ma own phone, me too I don join that craze people them noh*!'.[13] One can understand Mama Mangwi's pessimism about the mobile when it first came to Pinyin. But with a son in South Africa and another in Douala (Cameroon's economic capital), the need to communicate with them became apparent. Owning a phone changed her perception of them (cf. de Bruijn *et al.* 2013). In the early days, seeing people talk on the phone rather than with someone was a strange phenomenon. According to Papa Teneng,[14] one of the first users of the mobile phone in Pinyin explained that they also contributed to this discourse of associating mobile-phone communication with madness; they were very loud on the phone and complimented it with too much gesturing. It was strange to the people who had no other explanation than attributing it to madness and to its newness, which Nkwi (2010) refers to as *Kfaang* among the Kom people. These early perceptions of the mobile phone, however, have been put aside and the device has been appropriated by almost the entire village. Those unable to afford a phone have opted for a SIM card that is then inserted into any phone at their disposal and a quick call is made. MTN SIM cards cost FCFA 300 (approx. € 0.50). In other cases, families have chosen to buy mobile phones that take a dual SIM card in order for two persons in the same household to share a single phone distinguished by the numbers of the SIM card – SIM 1 and SIM 2. A case in point is Joan. Before she could afford her own phone, she shared a dual SIM phone with her mother-in-law and the phone was always in her keeping. When the calls were for the mother-in-law, she would hand the phone to her and take calls that were meant for her. This system worked well for them until she got a job and could buy herself a phone. The affordability of the mobile phone has compelled most people in the village to join the 'new form of madness' by entering into a marriage contract with the phone companies as papa Ndoh's remarks suggest (Footnote 11). The need to stay connected meant people devised ways of doing so with multiple users sharing one phone distinguished by SIM cards and/or having just a 'beeping credit' in their phones. What does increased connectivity mean to families at home and abroad? Much as it has eased phoning, especially for those in South Africa by making home close and at the press of a button, it has equally increased the rate of calls, beeps and demands from home, much to the displeas-

[13] Interview with Mama Mangwi, Pinyin; 28/05/2010.
[14] Interview with Papa Teneng, aged 71, Pinyin; 27/05/2010.

ure of some of my informants. Each time some of them receive a beep from home, the immediate reaction is '*Eish na weti again noh*!'[15] The consequence at times is partial or temporal disconnection from home.

Although the mobile phone amongst all the other forms of ICTs has gained enormous popularity, the Internet and other forms of social networking such as Facebook are gaining currency as well although their usage is limited to family and friends in the cities.

While studies on ICTs tend to be euphoric about the revolution in ICTs, especially in Africa, the snail's pace at which it was introduced and the still intermittent nature of mobile telephony in Pinyin, as well as the lack of fixed phone lines, are setbacks. Nevertheless, the coming of the mobile phone has given users the possibility to leapfrog.

The Internet and social media

The literature on the Internet has glowing praise for this new technology (Parham 2005; Edimo 2010) and the way it is rapidly penetrating the rural areas, linking the rural and the global. The Internet is seen as the essential communication medium through which contemporary societies communicate and it is increasingly linked to other media like broadcasting and print (Wilson III & Wong 2007). Despite this widely acknowledged claim and the availability of the Internet in developed countries, it is still to make the imprint in Africa that it has had in the rest of the world (Bratton 2013). The slow pace of the Internet compared to the mobile phone could be linked to low literacy rates and computer illiteracy, the high cost of electricity, the unavailability of the Internet and/or the high cost of computers and electrical equipment in general. Unlike Miller & Slater's (2000: 27) findings in Trinidad, the Internet is yet to make its mark on rural villages in Cameroon, especially Pinyin and Mankon, and connectivity in some urban areas is still below par.

The case of South Africa is unique and peculiar. Despite its isolation due to its apartheid policies, it was able to boast a fairly good, functioning Internet service compared to other states in the 1990s, although it was largely used by what Lewis (2007: 111) calls 'Internet geeks'. However, by 2004, the Internet had nearly reached its full potential in South Africa although current trends of Internet penetration in the country is on a rise (Lewis 2007: 107). This trend witnessed a steady climb with the introduction of the 40 km ultra-capacity optic-fibre Internet cables. International submarine cable connectivity will curb the country's overreliance on satellites and bring down the cost to users. Consequently, the Internet and social media are providing people with cyber spaces for social interaction

15 Reaction from Chris on receiving a beep from Cameroon in the course of our conversation.

and connectivity with the home country and with other diaspora communities and providing accessibility to a wider mobile community. Such network ties and connectivity have taken the form of 'virtual proximities' (Urry 2004: 33), with people switching from one form of technology to another. As Edimo (2010) showed in her study of Cameroonian migrants in France, the arrival of ICTs has not only brought them together in the host country but has equally given them the opportunity to become actively engaged in political debates back in Cameroon thanks to the Internet forum where ideas and debates can be shared. Migrants are also informed by issues at home thanks to the development of the cyber press and the online publications of private newspapers in Cameroon. A similar trend was noticed among the Pinyin and Mankon communities that are connected by their various Internet forums, such as the Pinyin News Forum (PNF) and the Mankon Forum (MF). This is what Parham (2005) refers to as Internet-mediated publics. With the addition of social-networking sites, the blurriness of boundaries is being further intensified, contradicting the notions of nearness and remoteness, or mobility and fixation. Bogard (2000: 40) points out that 'this blurring of boundaries between the monad and the dyad is an excellent image of the rapidly evolving symbiosis of bodies and computers, groups and communications networks, societies and cybernetic systems' that together enhance mobility, especially virtual mobility. This is the case with both communities' linkage with their Internet forums (Chapter 7) that promote discussion about the homeland, keep migrants abreast of happenings, discuss tribal politics and development projects, and share their views as well as uniting them virtually with the homeland. While this may be the case, for others the forum has become a place of virtual fighting. There are still a number of individuals who have no interest in being connected via the Internet, while others see it mainly as a conduit for sending photos of new-born babies back home as well as a way of receiving photos from home.

Although Facebook has taken the young generation by storm, some parents have joined it too, often encouraged by children abroad, or they have been registered by their children. They become actively engaged in communicating with their children through this medium and Yahoo messenger. Jessica in Mankon, for instance, is a retired municipal worker who has two children in the US, two in Russia and one in South Africa. As a result, she saw the need to ask one of the children to teach her the basics of the Internet in order to be able to connect with the rest of her children as Internet connections are less expensive than phone calls. She plans to buy a computer and access the Internet at home, with the possibility of then talking to them via Skype. With a computer at home, she could stay awake to talk to them in their own time zones, something that is difficult to do at a cybercafé. This is what Jessica said in this respect:

> My children taught me how to use the Internet and now I am able to communicate with them via the internet. This medium is good because It permits one to write at length, and say all what you want without being scared of airtime like with the phone. If I pay for 1 hour, I know I can send a long email and maybe chat with him if he is online. I usually chat with him on Facebook and although he has not sent us his pictures, I have seen him on Facebook. I was also surprised to see my picture on Facebook and the comments made by all of them; how they miss us.[16]

For the literate and for educated parents like Jessica and my mother, their enthusiasm about developing virtual connections with their children has prompted them to take a crash course on the use of the Internet and its related social-media sites. Jessica plans to go for another short course in computing in order to access all the social-networking sites available on the Internet so she can maximize her chances of communicating with her children.

Despite the popularity of Facebook amongst the youth, some see it as a gossip site and maintain that the 'thing with Facebook is that all relationships are mostly virtual and very few blossom to physical contact and most die a Facebook death'. To Fru, 'Facebook has increased the rate of promiscuity, and provides them an avenue to chat (flirt) with many girls simultaneously'.[17] But for Mark, it gave him the opportunity to have a long-term relationship with a partner he met on Facebook. This notwithstanding, migrants and family back home have increasingly taken on ICTs as a way of keeping track of each other and the various forms of technologies have been appropriated to meet their different demands and needs, as will be seen in the section that follows.

Appropriation of ICTs before and after the revolution of communication technologies

The first groups of migrants like Joe, CJ and Jake who went to SA in the mid-1990s and John who went to the Netherlands in 1999 will tell a different story of their experiences about connecting with home and the use of telephony. Their stories will be of the difficulties they encountered. This is remarkably different from the stories of those who arrived between 2001 and 2003 when mobile communication was gaining currency in South Africa and Cameroon simultaneously, when there were possibilities of connecting with home but calls were still expensive. And then there are the experiences of the third batch that migrated after 2004 and these are significantly different again, due to the proliferation of mobile phones in South Africa and Cameroon and considerably reduced rates for calls as can be seen from their stories. The stories are not homogeneous and represent the different periods of arrival of the migrants, and describe how they navigated the dynamic complexities of ICTs and mobility to maintain links with the home country. Telling a homogeneous story would be tantamount to telling a

---

16 Interview with Jessica and her husband, Mankon: 03/06/2010.

17 Conversation with Fru, Cape Town: 20/7/2011.

single story that is not inclusive: 'the single story creates stereotypes, and the problem with stereotypes is not that they are untrue, but that they are incomplete. They make one story become the single story' (Chimamanda 2009). This study eschews a single story.

For the first-comers, linking up with family and relations back home was challenging and a huge sacrifice in time, as will be seen in Joe's story. Joe, who we met in Chapters 1 and 3, was the first Pinyin man to go to South Africa and he has been there for 17years now. He is self-employed and works in property development for rentals. His accounts capture all the experiences of struggles with connecting home and the surge in connectivity, precisely because of his long stay in South Africa, having lived first in Johannesburg and now in Cape Town.

Joe's story

> Hey! It was tough, really difficult to get in touch with home. There was a Cameroonian who was doing import-export business and lived at Sea Point at the time (1998). Once I went to visit him, and as a Cameroonian I was very happy to learn that there is another in town. Towards evening I excused myself saying that I wanted to rush home to make a call to Cameroon, but he said there is no need for a rush that we could stay longer and I could use his fixed phone to call. Yoh! It was like a dream to see someone with a fixed phone at home that I had access to and, for the first time, the first dial went through. I talked for 2 minutes and it was as if I was at home. You see, such moments were so precious and a one-off. What usually happened was that there were some South Africans that worked for Telkom (Telecommunications Service of South Africa) and would work with locals at the location (townships). The locals would apply for a phone connection and more than one line would be installed at the house. Then they would tap into somebody's line and make calls. When this is done, word goes round amongst foreigners that there is a *tuyau* at, say, Khayelitsha, and we will go there. I used to leave my house at about 4 am just to secure a good position in the queue. When the queue was too long and meanders outside, those who were outside were sent off and asked not to loiter around as this could raise suspicion. The rest of the people were crammed into the room. Each person was entitled to 3 tries. When you make the first try and it doesn't go through, you give way for the next person to call. At the end of his call, you have your second try. If it's unsuccessful, you make way for the next person and after that you make the last try. If the operators were lenient, you were granted another chance. You may spend your whole day there and leave without reaching home. But if you are lucky to get through, you leave there a happy man and this steam keeps you going for the next few months before you reach home again. You see, but when you leave without calling, your whole week and those after are bad as you are in doubt about how the family is. This was the case with me at times. This was not the only way at our disposal to reach home. After this, there was another one that permits one to buy a number from Telkom workers and this number would be dialled in at the pay phones before one dials the phone number. This would appear as if it is a local call, and you can call home or wherever for unlimited hours. This number was very expensive to buy and so those who got it would use it to sell to those who wanted to make a call (as a call box). As usual, there was always a queue to call. This *tuyau*[18] lasted for a while and evolved to another *tuyau*. The next was that there were some people monitoring broken pay phones. Once one was identified, news quickly circulated and we

[18] The various illegal ways of calling home.

would use it till Telkom discovered it and repaired it. It required one to put a R 1 coin in the machine and when the number is dialled, only 50 cents was used and it ends there; permitting one to call for an unlimited duration. When the phone was repaired, we devised another way.[19]

Joe's story covers the period from 1996-2003 when calling home was difficult. Prior to advances in ICTs, calling home was not a simple link up, but required sacrifice and endurance given the long hours spent trying to make calls and the long distances covered to where there was a *tuyau*. It also highlights how the migrants' need to reach home caused local Telkom workers to circumvent ways of making cheap calls available to migrants. The culture at the time was to rely on clandestine methods of phoning due to the expensive rates for calls. *Tuyaus* were relatively cheaper but come at a price of having to spend long hours in a queue or of having to go to the 'location' (townships) at very odd hours to secure a good position in the queue. Whatever the case, migrants saw it as their duty to reach out to the family and the instinct to go looking for availability of where to call (for those who cared, as not all were subjected to this) speaks to the tendency of migrants' wanting to connect and be back in his home community although miles away. In this regard, we can also draw on the nature of social relationship that existed between them and the family – a close-knit relationship with an emphasis on the quest for home in their migratory trajectory. For instance, Jake, like earlier migrants, had his ears to the ground in anticipation of news about the availability of a *tuyau*. He came to South Africa in 1998 after gaining his first degree and worked first as a salesperson, then as a security guard at a car park and as a hawker to finance his studies, and is also assisting his family back home. He now teaches at one of the universities and is married with children. He corroborates Joe's story:

Because these phones were illegally connected ... eh ... it's not expensive because you pay like R 2.50 a minute. For someone coming today, it will be expensive. But for us who experienced the difficulty of phoning in those days, because we had to go around town asking, does anyone know if there's a *tuyau* phone. And if there's a *tuyau* phone in Stellenbosch, you take yourself to Stellenbosch (about 30 km from Cape Town) to phone and you wouldn't feel safe while you're phoning there. So today having these phones for like R 2 or R 2.50 feels like you're phoning for free.

Q: And were there mobile phones back at home then?

A: No ... or yeah, that's the whole thing ... there were a few cell phones ... I cannot even remember when cell phones suddenly appeared on the scene. But in those days, it was mostly home phones, a few home phones here and there. You phoned and gave a message or you phoned and they would go and call someone. I was lucky because in our house in Yaoundé, there was a phone. So I could phone often. But most people would make an arrangement with you that there was a phone in this neighbour's house or in this relation's house, and could you go wait there at this time?[20]

[19] Interview with Joe, Cape Town: 24/07/2011.

[20] Interview with Jake, Cape Town: 11/08/2010.

The availability of ICTs has had a more far-reaching effect than could have been imagined. It is not simply connecting to family and friends instantly, but the psychological relief and the satisfaction migrants get from talking to family back home. Calling home when the telephone was not common could be compared to a vaccine that keeps migrants going for a while and when the dose is finished, the need to call becomes imperative again. The long excerpts by Joe and Jake summarize what the early migrants in South Africa (1996-2003) had to endure to communicate with family and friends. This contrasts sharply with Joe's experiences with the wide availability of communication technologies from 2004-2007. According to him:

> The next were numbers sold by mobile-phone workers that would give unlimited access from a cell phone. Cell phones were very common then. This number is installed in a person's cell phone and the latter will use it as a call box for others to make calls at a reduced rate. Soon after, we discovered UK numbers. With this one, you convert the Cameroon number to a UK one via an Internet link. When you call home it is as if you are calling the UK and the cost is relatively cheap – R 1 per minute. This only came in much later when the cell phone was here but too costly. Today I can lie on my bed and phone home; even from here I just dial from my cell phone. It is quite cheap using MTN and even more so over the weekend. Calling the US on a fixed phone is 0.50 cents per minute. One does not have to go through everything we went through to make calls home.[21]

This second half of calling home is marked by the increased accessibility and availability of the cell phone in South Africa and Cameroon, the widespread availability of the public paid phone (phone booths) and the cost of phoning (R 5 per minute from public booths and R 15 per minute from a mobile phone), which was too expensive for migrants to make frequent and direct calls from their mobile phones. They therefore continued to rely on *tuyau*. But the difference here was that the Internet gave them the opportunity to access cheap calls through the Voice Over Internet Protocol (VOIP), and the chance to communicate by email. Advances in technology have meant the closing in on the *tuyau* networks and migrants have turned to VOIP calls. But this is not without its own challenges as migrants often complain of the poor level of audibility. However, with further reductions in rates since 2008, *tuyaus* have been rendered redundant as migrants can 'lie on their beds and make calls' to anywhere in the world. This is reflected in the fact that migrants have multiple SIM cards from various service providers to take advantage of the best deals offered at any one time (Thompson 2009: 369). Ultimately, this paves the way for migrants to have a foothold in both home and host countries.

The ripple effect is that the mobile phone offers migrants 'increasingly powerful information processing capabilities that was previously limited only to PCs' (Donner 2006: 3). By the same token, migrants and their families back home are

[21] Interview with Joe, Cape Town: 24/07/2011.

able to participate in events that are taking place either in South Africa or Pinyin and Mankon by watching DVDs of the events or seeing photos on Facebook. This gives an update of migrants' activities to the larger migrant community and friends back home. For instance, Chris in Cape Town posts his encounters with the traffic police on his Facebook page to tell friends about his day: '*chaii dis stupid traffic cop slam me a fine R800 JST BCOS I PACKED WRONG FOR JST 1MINUTE. THIS GUY DEM THINK SAY MAN DI FIX MONEY WE GO METUP FOR COURT MASSA*'.[22] Similarly, photos of the PIFAM party in 2011 were posted on Facebook and this evoked a series of comments from friends all over the globe.

Joe and Jake as well as some of the first migrants would make calls home but there were equally some early migrants who felt that the safe thing to do was to temporarily suspend communication with the family and focus on their daily survival. This was the case with CJ while in Johannesburg (before moving to Cape Town) and John in The Hague.

John is a bachelor in his mid-forties from Mankon who has lived in the Netherlands since 1999. He came to study and, like most Cameroonians, did not go back at the end of his course in 2002. Unable to get a decent job that equates with his qualifications (he has a Masters in Business Administration and a Masters in Management), he has settled for low-income jobs. Accordingly, he had prepared the family psychologically for his long periods of silence. He called to inform them about his busy schedule at school and at work, and told them that calls would be infrequent. The same goes for CJ who hails from Pinyin and was at Hillbrow (Johannesburg) when he temporarily severed ties with the family. The long spells of silence were due to their inability to assist the family financially. Calling home was a reminder of the challenges they had left behind and were unable to do anything about. Staying away and not wanting to get information was a way out for them. This brings us back to the issue of migrants' agency that was discussed earlier. Although ICTs are credited with connecting people, we should not lose sight of the fact that these technologies can also disconnect. What does this say about migrants and technologies?

It is worth noting how both John and CJ share similar concerns and these form the basis of their simultaneous disconnection from home. By virtue of sharing a similar culture from the Bamenda Grassfields, where migration has come to symbolize affluence and wealth on the part of the migrant (Nyamnjoh 2005; Nyamnjoh & Page 2002; Shindondola 2007), not being able to live up to this expectation means that migrants are not true 'bush fallers'. Absolving oneself from family ties in such circumstances seems the right thing to do. And connections

[22] Posted by Chris on his Facebook page on Wednesday 7/03/2012, accessed 08/03/2012.

are often re-established when the migrant is financially stable and can respond to the demands of the family, as CJ eventually did.

Underpinning migrants' disconnections from family is an element of fear at two levels that is explicitly and implicitly implied in the way they perceive the responsibilities and expectations imposed on them by relations. Although John can call home because of the fixed phone at his uncle's, he chooses not to do so for fear of receiving a message that will necessitate assistance that he is incapable of giving, thus increasing his worries. Jake, however, calls because of the fear of bad mouthing from relations who may accuse him of not calling them. The issue to be dealt with here is migrants' emotional and psychological distress and fear of the unknown, which challenge the rosy concept of migration or bush falling in Cameroon (Chapter 9). John's tone and the incoherent manner in which he speaks are suggestive of frustration, fear of not being in control and an inability to handle news of the death of someone back home.

The use of these technologies fits well in migrants' experiences of relating with home. Connecting plays a dual role as it helps to conserve their identity and in the process provides space for them to affirm their belonging, thus re-echoing the notion of home to them (Parham 2005: 373; Tsagarousianou 2004: 62).

While the boom in mobile phones has increased connectivity and been instrumental in sustaining relationships, it has also given users a means of personalizing their callers by attributing ringtones to specific contacts. Although such usage is particularly entertained by those with multi-stranded relationships (Horst & Miller 2006: 63-64; Miller 2010: 112), it also insulates migrants from demanding calls and/or offers them the possibility of accepting or declining a call. The ringtones in themselves provide a social clue as to the phone's owner, with Christians tending to choose religious tunes/songs or Bible verses and teenagers going for the latest songs of a certain pop star.

Le nouveau Beaujolais est arrivé: The gift! The marker! The mobile phone!

> In 1998 the first mobile phone I bought was not for me but for someone in Cameroon who asked me to buy one and send it via his friend and he would give the money to a designated relative at home. It was so expensive back then that I could not afford one and they were such huge things that you needed a bag to carry it along. I was not sure whether there was network coverage in Bamenda at that time and I was sure it was just for him to show off.[23]

Similarly, when I was in Cameroon for fieldwork in September 2011, I met the brother of one of my informants in Cape Town and his first question was 'did you tell my brother you were coming home so he could send me the latest Blackberry? Look at the old phone (showing me the two mobile phones he was hold-

[23] Interview with Joe, Cape Town: 23/07/2010.

ing) I am using, and this one, I borrowed from a friend.'[24] Joe's quote above and this example give an idea of how owning a mobile phone is fast becoming an identity marker.

Like mobility, the ubiquitous presence of the mobile phone has removed borders and linked relations in the village to migrant kin, and they are seen as a sign of globalization (Hahn & Kibora 2008). By the same token, the arrival of the mobile phone has remarkably changed the way people are perceived, depending on the kind of phone they own. The hype around the mobile phone could be attributed to that surrounding Beaujolais wine where the arrival of bottles from the new harvest receives wide publicity and people place orders in advance. The latest mobile phones, like Beaujolais wine, are received with excitement by both the haves and the have-nots (those who can only afford to window-shop), with the haves going for the latest model on the market and everyone else trying to extract the best from relations abroad or as gifts. Some buy Beaujolais simply to display it at home, while for others, it is meant for consumption only. In the same way, some people are fascinated by the latest mobile phones and apps with which they are not acquainted, while for others they convey status and the message that one has arrived, of belonging to that exclusive club of those who could afford the smartest phone and different apps, even if they do not know how they function (Smith 2006). It is in this light that Pinch & Bijker (1987 cited in Hahn & Kibora 2008: 91) note that the mobile phone is a 'socially constructed technological system'. Yet others clearly see it as a form of madness.

Although most people back home are interested in mobile phones as gifts from relatives abroad, the craving for 'bush things' (everything that comes from abroad) has equally been heightened by the rate at which migrants send mobile phones home as gifts, thus giving the impression they are inexpensive and easy to come by. Every time I travelled from Cape Town to Cameroon for fieldwork, I carried about ten mobile phones that were to be distributed to families in Pinyin and Mankon.

In cases where migrants do not send phones, every available tactic is deployed to get one from them. But for most, sending phones, especially to aging parents, is a way of getting in touch with them directly. For parents, owing a phone is an achievement in itself regardless of its make, and phones have become a symbol of the proximity and availability of their children in a single beep. Yet for others, the mobile phone echoes competing agencies at least regarding ownership of the sleekest smart phones. While the move to smart phones has meant that communications have been taken to a 'smart' level, status for most is being redefined to reflect the phone one owns.

---

[24] Conversation with CJ's brother, Buea: 08/09/2011.

Photo 4.7 Mobile phones and medication sent with this researcher to relatives in Pinyin and Mankon.

Navigating the conundrums of the mobile phone

The simultaneous availability of ICTs, especially the mobile phone, in both South Africa and Pinyin and Mankon could be, to borrow from Horst (2006), considered as the 'blessings and burdens of communication' to the migrant community and families back in Cameroon. While communication technologies help in strengthening kinship relations among some families, to others they extend the geographic and virtual distance due to the absence of calls or minimal calls, as migrants choose which to take depending on the caller (Horst, 2006). Calls may be seen as a burden, creating emotional stress due to demands for financial assistance or as the harbinger of bad news. It is in this respect that Siri Lamoureaux (2011: 40) notes that the cell phone has become 'both a tool for 'keeping in touch and for artfully avoiding that social obligation' (see also Powell 2012: 15-17). For some, ICTs have given them the chance to perfect social gaming with relations and friends, such as Penn. He like some other migrants oscillate between SIM cards, with one that is used only to call Cameroon and a regular SIM for everyday use in Cape Town (that is not known by those back at home). The SIM for Cameroon is inserted only when he wants to connect with home and, once he has called, the SIM is removed. This means that relatives and friends back home cannot get hold of him and calling home is at his discretion. Others go to such extremes as changing their numbers regularly or declining to call back after repeated beeps. In the same vein, some decline calls from home. Frustrated at his son not responding to his repeated beeps, Papa Joshua remarked that '*il ne peut pas contourner le vieux*'[25] (he cannot escape from me). As a result of migrants deciding their own 'accessibility', 'reachability' and 'availability' (Licoppe 2004: 137-138), the phone has become a curse for disgruntled families back home. Although Licoppe (2004) suggests that the failure of one interlocutor signifies unavailability, it is in this context that the 'mobile phone is used to reassure the other person and to reaffirm the link by means of very short call' (*Ibid.*: 145). Empirical findings, according to Liccope (2004), suggest that all societies have land lines but this is not the case in Pinyin and Mankon. Equally, mobile phones to the Pinyin and Mankon migrants may also provide an escape route from family, even though they can be used to reassure the family of continuous bonding. This presupposes that each call that comes in is taken, which may not necessarily be the case. The question that comes to mind is what happens when calls on mobile phones are not taken or the recipient is not willing to engage in communication. The mobile phone could be seen as an actor and actant at multiple layers, not only facilitating communication between home and the host country, conversely perpetuating the distance between migrants and relations back home in the situation where migrants have severed links with relations, thus eliciting a response from the respondent (actor).

The misconception about a mobile phone is often that a person has their contacts and/or family members with them everywhere. It would appear so but, contrary to popular belief and as Powel (2012: 15) points out, 'cell phones do not guarantee availability all the time', as 'people hide behind their phones'. At one level, it shows the embodiment of people and technology yet at another, it shows the disaggregate that co-exists. To these persons, the phone gives freedom and choice simultaneously; freedom from unwanted demands and the choice of taking or turning down incoming calls because they could be easily manipulated (Powel 2012; cf. Hanh & Kibora 2008: 90).

While some migrants in South Africa have adopted the hide-and-seek style of connection, others feel that the new communication technologies have confirmed their fears and dismay that they are considered by family back home more as 'moving wallets' and have been stripped of any humanity, and hardly would they received calls from home simply to check on them or to acknowledge receipt of remittances (Nyamnjoh 2005). Calls are geared towards demands and the norm is to beep, even when their money comes from the migrants. In Cameroon, there is an unwritten rule that migrants are expected to maintain links and call home. Mirielle's fiancé is in Cape Town and in the two years that she was in Cameroon before joining him, she never called him once. She is content to just send emails and texts because she believes he is the one who should call. Calls from home are

[25] Conversations with Papa Joshua, Pinyin; 12/11/2011.

often to announce a death, make financial demands or simply to beep. Jake, in Cape Town, expresses his frustration at such attitudes from his relations at home:

> You know ... for me it's not just about phoning and getting these complaints and demands, for me it's about phoning and speaking to people who really show emotions of joy and excitement that you have phoned, even over the phone because you can sense them ... And also the fact that perhaps you are seen as a commodity more than anything else, a producing machine.[26]

This excerpt from Jake aptly reflects the frustrations of most migrants in Cape Town towards their families but it also confirms the disconnection by John in Holland and CJ in Johannesburg towards their respective families. Communication between migrants and family becomes redundant when the migrant is not forthcoming with remittances. Much as they would like to maintain a 'continuous presence' within their family circle, the family's attitude towards them lends itself to materialism as opposed to filial sentiments and generates poignant feelings of exploitation and dehumanization. The resulting effect is phone fatigue on the part of the migrant, leaving some to resort to what Knorr & Bruegger refer to as 'response presence' (2002 cited in Urry 2004: 33). Although referring to traders' responses in the marketplace, it is suggestive of the attitude adopted by some Pinyin migrants to countering their frustrations regarding the lukewarm reception they get over the phone. Critical to understanding virtual ties is the fact that the tone as opposed to face-to-face contact is what matters. The tone cannot generate complete person-to-person contact but it has replaced eye-to-eye contact and it still gives away the mood of the receiver or caller. Although it is not as 'direct and pure' (Simmel 1997 in Prelipceanu 2008: 7) as eye-to-eye contact, migrants can sense how the recipient at the receiving end feels. In this regard, it could either create a sense of togetherness and belonging despite the distance or fragmentation and distance between caller and receiver.

Such an attitude is not exclusive to male migrants and most female informants share similar sentiments of dismay at the nonchalant reception they get over the phone from immediate siblings or children. However, such frustrations often go unnoticed by relations who are preoccupied more with their demands than any discomforts these demands may cause the migrant. This could be explained by the fact that in the Bamenda Grassfields, especially in Pinyin where polygamy is very common, wealth is still considered in terms of the number of children one has. They are considered primarily as long-term investments following the old adage that 'children are their parents' walking sticks'. Parents in Pinyin and Mankon make no secret of their expectations regarding their children. The experiences shared by Jake confirm Vertovec's assertion (2004: 223; Tazanu 2012) that telephone calls 'provide a kind of punctuated sociality that can heighten

---

26 Interview with Jake, Cape Town; 11/08/2010.

emotional strain as well as alleviate it'. From another perspective, it could be partly these excessive demands on migrants and the ingratitude expressed by relations that push mobiles to withdraw from the family fold, as expressed by Jake, or the fear of calling that John and CJ have. However there are still some parents like Jessica who call their children simply to find out how they are doing.

While there is excitement and euphoria about ICT, it comes with intricacies that cannot be disaggregated and quietly concealed and that may unfold following careful analysis within the context of everyday life. This brings us full circle to the notion of 'connected presence'. How do we, using this approach, understand the intricacies that underpin such communication patterns? Can we talk of a wholly 'connected presence'? While it helps to gain insight into the urge for a continuous and connected presence, this is just one aspect of the story, one that is homogenous and where everyone is seen to be connected. It is devoid of such conundrums as described above. Often, being connected is done on the individuals' own terms and migrants, being social agents, try to improvise. The phenomenon of having an all-encapsulating theory that explains behavioural patterns in the use of ICTs and the dynamism and complexities of these technologies is yet to see the light of day, hence the contentment of using approaches that enlighten our understanding to specific aspects and behavioural patterns (Katz & Aakhus 2002). Unlike Licoppe (2004), Katz & Aakhus (2002) catalogue the merits of the mobile phone as well as it shortcomings. But at what price does this embodiment of technology come? Would it not be an exaggeration to see it only in terms of its intrinsic worth? Is there such a revolution if we compare the levels of penetration in the developed world to those in Africa?

Despite the challenges migrants face from home, there are still parents who take pride in calling their children and even of having back-up numbers of their friends in South Africa in the event that their children are not reachable. Jessica, for example, said that: 'When I am unable to get through to him, I call his friend, a neighbour here in the village, Patrice … I keep Patrice's number as a back-up for when I can't get to him; I call the friend and leave a message.'[27]

Conclusion

To sum up, the evolution of ICTs has not only made its mark in the annals of technology but has also kept social scientists busy trying to understand the multifaceted ways in which it has been adopted into our daily reality. Its uses and appropriation have been phenomenal, going way beyond the realm of the manufacturers and helping to generate a strong sense of shared space and time that overlook the realities of geographic borders. The ripple effect is that most villag-

[27] Interview with Jessica, Mankon; 03/12/2012.

es, especially Pinyin, leapfrogged from zero techno-connection and no fixed phones straight into the mobile-phone era. However, prior to the wide availability of communication technologies, early migrants had an arduous task when connecting with families in their home country. Through a myriad of manoeuvres to connect with home, they experienced first-hand ingenious, yet clandestine, ways of linking up with family. To these migrants, it was more than simply linking up given the long process it took to call home. When the call went through, besides checking on the family, they wanted to virtually transport themselves back home.

Others, conversely, are gripped by the fear of calling and being informed of 'unpleasant' news that will upset them. For others, the fear of calling and being unable to solve problems that require financial assistance because they 'live by the day' and cannot afford to send money home leads to disconnection. This type of disconnect from the family may be temporary or intermittent and the migrant only reconnects when his/her situation has improved and they are ready to send remittances. From this perspective, we are given a clue to the mindset of families back home and the emotional duress migrants undergo as a result of excessive demands being placed on them. In the absence of remittances, relationships may in some cases be at a low ebb and this can impinge on migrants' emotions.

Despite these mitigating circumstances, the transformational impact of ICTs on society and especially the Pinyin and Mankon communities is undeniable. Through their interaction with ICTs, societies are maintaining connections at a distance, with complex recurrent patterns of presence and absence. In the case of the Pinyin and Mankon mobile communities, ICTs are structuring and reconfiguring social ties, a process that has either created emotions of pain and antagonism or nostalgia, which reinforces their dual nature. They are capable of both connections and disconnections. However, the arrival of ICTs has largely been characterized by the benefit they offer in revolutionizing the world, dramatically changing how people live their lives (dualistically) and how they perceive ICTs as deterritorializing spaces and compressing time. But the question is whether this change is affecting the social fabric of society or eroding its cultural values? To perceive ICTs only as connecting people, as suggested by Licoppe (2004), or only seeing the litany of its merits, as Katz & Aakhus (2002) do, obfuscates the fact that these technologies can be manipulated and people hide behind them (Powell 2012). At one level, the relationship is convivial but at another it is antagonistic. This notwithstanding, communication technologies, especially the mobile phone, have been used extensively to network among relations, even though they are also used to sever links. But this is normally only temporary and a situation of normalcy is usually established when migrants become financially stable. The mobile phone has been 'culturally appropriated' (Hahn & Kibora 2008) to strengthen family ties and solidarity.

The analysis presented in this chapter has focused on the evolution of ICTs from their embryonic stage to their explosive stage and has examined the teething problems that compounded the various stages. It also looked at the various periods when migrants appropriated mobile-phone technology in South Africa to connect with their home country and the different periods that coincide with their arrival in South Africa. Conversely, the arrival of mobile technology in Pinyin and Mankon was marked by both tensions and memories of nostalgia. Through ICTs, migrants want to be virtually connected to home and to participate in the daily activities and happenings in their home country.

5

Networks and shifting relations: Social and kinship networks and the formation of a network society

Introduction

The last two chapters showed how mobility and migration are not new and also how they are deeply engrained in human history. However what is new today is the form and dynamism that they have taken as a result of the advances in ICTs and transport networks. This has led to new mobile societies that are characterized by mobile flows and 'miniaturized mobilities' (Elliott & Urry 2010: 28). Secondly, ICT is what spins mobilities and reorganizes social structures, reconfigures existing relationships and builds new ones as well as maintaining existing ones. In the process, we see how everyday life is construed. Underpinning this is the fact that mobility and ICTs do not function in isolation but are inextricably linked to social relations of networks. In this respect, the formation of networking amongst mobile communities in the host country and with families in the home country is choreographed by ICTs (Thompson 2009; Horst 2006; Horst & Miller 2005). While face-to-face meetings are common and the preferred option in the host country given their daily mobility and hustling activities, they have come to be reliant on mobile-phone communication (including SMS & WhatsApp), Skype, cheap-rate calls, emails and Facebook for most daily connections and linking up. As regards the family back home, networking is by mobile communication and once in a while face-to-face meetings when the migrant visits home or parents are brought over to visit.

This chapter examines the plethora of social and kinship networks that together seam mobile communities and are characterized by negotiations and renegotiations. Until recently, studies on migration put social networks as the driving force behind migration, showing how social networks impact on migration at all stages in the process (Hagan 1998: 55).

This chapter focuses on how Pinyin and Mankon mobile communities are able to draw on the various forms of social networks such as family, friends, associations, and legal[1] and migration syndicates that are predisposed towards them in an attempt to establish how network ties work. It demonstrates how these networks constantly shift to accommodate new and relevant networks. In addition, the chapter explores whether mobilities have affected network ties and, if so, how. And if they do not, do these ties continue to adopt the norms that migrants have been brought up to respect, i.e. social cohesion? The chapter goes beyond cataloguing how these networks are being forged over time and focuses on how and under what conditions they are weakened. By the same token, I examine the role of weak ties in network flows and how they create bridges that link subsequent weak ties (Granovetter 1973). In summary, the chapter examines the multi-stranded networks that migrants accumulate, including strong and weak ties. Social networking, as such, is seen more as a dynamic process as a result of the emergence of flexible communities that reconfigure and are capable of producing and reproducing complex structures of communication to ascertain the unity of purpose and flexibility (Castells 2004). They are thus able to adapt to the host society. I attempt to understand not only the large presence of Pinyin in Cape Town in comparison with Mankon as well as the extent to which networks of family and friends and ethnic community ties play a role in the corporate sponsorship of migration. Given the dynamic nature of social networks, the gender structures of network are of interest given the differences between women's and men's everyday lives and the different types of networks they therefore leverage.

Networks, as was discussed earlier (see Introduction) are a set of interconnected nodes that process flows (Castells 2004). Critical to our understanding of these flows and the reconfigurations that involve the inclusion and exclusion of networks are their organizational form (Castells 2004) and bonding approach (Putnam & Feldstein 2000). The strength of the weak ties (Granovetter 1973) will also shed light on why there are more Pinyin migrants here than other communities. The mobile communities involved see their daily interaction as one that is informed by the multi-stranded networks they navigate, and this network approach also illuminates the social gaming that occurs. The bonding approach will explain why particular groups of people tend to bond more than others. Weak ties, as noted by Granovetter (*Ibid.*), are practical bridges that link various nodes and are stronger than strong ties. Drawing from the notion of weak ties will allow an understanding of why the Pinyin make up the dominant ethnic group and how

[1] A legal network refers to the lawyers that are used by migrants to enhance a secured refugee status or facilitate life-partner relationships at the Department of Refugee in the Ministry of Home Affairs.

migrants relate to their weak ties. Despite this octopus-like nature of networks, Castells (2004: 6) identifies three basic characteristics, including flexibility, scalability and survivability. He explains that

> Flexibility, networks can reconfigure according to changing environments, keeping their goal while changing their components. They go around blocking points in communication channels to find new connections. Scalability can expand or shrink the size with little disruption.
>
> Survivability, because they have no centre, and can operate in a wide range of configurations, networks can resist attacks on their nodes and codes because the codes of the networks are contained in multiple nodes that can reproduce the instructions and find new ways to perform.

These organizational forms can reconfigure networks and create a network society because of the ICTs that network societies have appropriated to reinvent themselves due to the powers embedded in information technology (Castells 2004). As mentioned in the Introduction, social interaction and linkages between migrants and family are not new but the newness is derived from the manner in which network societies have literally built their lives around ICTs that have caused the mutual shaping and reshaping of humans and ICTs.

What we should also note is that these networks work for people to be mobile because they are hierarchized, given that people do not have equal access to the various forms of capital and resources (economic, cultural and social capital). Although social capital is informed by history and capital and it is accumulated over time, it has become a kind of exclusive right of the elite class (Bourdieu 1986).

Social networks in this chapter, as the title suggests, include kinship networks and they are grouped as social and kinship networks because they are interwoven. Thompson (2009: 362) and Tilly (2005: 5) maintain that a social network 'is a chain of dyadic connections or interactions which is social and not cultural because practices rather than ideas define it'. To limit the working of social networks only to the social ignores the fact that social networks and kinship networks are often guided by both cultural practices and ideas. To say they are independent of each other, as Thompson *(Ibid.)* claims, is tantamount to throwing the baby out with the bath water.

Overview of network in the migration process

Social networks, besides their dynamic nature, are conceived as a social product in this study. This is not as a result of individual decisions but rather as an outcome of varying needs of interaction that bind migrants and non-migrants together in a complex web of social roles and interpersonal/group relationships. These networks are conduits of information/knowledge and social and financial assistance and shape migration outcomes ranging from no migration to immigration, return migration or the (dis)continuation of migration flows to the same destination or other destinations (Massey *et al.* 1987; Boyd 1989). The structures (kinship, political, economic and non-ethnic) that generate migration in both sending and receiving countries are channelled through social relationships and social roles that impact on individuals and groups.

It is worth noting that, prior to the rise and use of the term social capital (ties) in social sciences, Durkheim (1933 cited by Field 2003: 11) underscored the connections ties of society and asserts that society does not:

> … become (q) jumble of juxtaposed atoms ... Rather the members are united by ties which extend deeper and far beyond the short moments during which the exchange is made.

Drawing from Durkheim's statement, and in conjunction with Boyd (1989: 639), it can be seen that social networks provide the basis by which we gauge society's social cohesion. They also enable people to cooperate with one another. But networks are not exclusively enabling, they can sometimes serve to 'exclude and deny as well as include and enable' (Field, 2003: 3), just as the forms of capital available to people vary and determine to what extent people access social capital (Bourdieu 1986). Accessibility in this context hinges on the notion of interplay or social games of networks. The ability of interplay creates a shifting network, with actors and relations forging new networks to leverage advantages or using any capital that they can muster to access networks that others without such capital do not have access to. Similar to games, social life and social/kinship networks are strategic and call for constant improvisation and strategic creativity with every new challenge or situation. We become adept at navigating and negotiating often contradictory expectations by various other social agents involved in our game. This is internalized habitus that embodies 'the capacity to … improvise the next move, the next play, the next shot', as Calhoun (2003: 3) puts it. This echoes what Putnam & Feldstein (2000: 2) suggested earlier about social networking being a game of strategic interest. Ties may be strong, reinforced and weak or severed depending on what game is at stake. They go on to indentify two types of social capital – 'bonding social capital' that links people who are similar in many respects and tend to be inward looking, and 'bridging social capital' that encompasses different types of people and tends to be outward looking. (*Ibid.*: 2; see also Putnam 2000: 22-23). This bridging capital is synonymous with Granovetter's weak ties whereby bridging knows no boundaries but is guided by need and necessity.

To understand the shifting relations and the way networks bond and connect, these approaches will help us to understand mobile communities' social gaming. Bonding capital brings together people of similar ethnic background and acts as a kind of 'super glue' while the bridging capital bonds heterogeneously and provides a sociological 'WD-40' (Putnam & Feldstein 2000). In this regard, we can associate social capital with the 'infrastructural moorings' that support departure and anchor arrivals for migration to occur (Urry 2003; Cresswell 2010). Whether with family relationships, social relations or communities and much as social networks are espoused in real values, underlying strategic game-playing is embedded in it. The question here is about the role of ICTs in strategic gaming amongst networks within mobile communities. What does it mean for the social fabric of the society? How do migrants navigate these forms of network and under what conditions do networks weaken and/or disappear? What is the role of weak/strong ties in networking?

The role of networks has always been embedded in the migration process (Hagan 1998; Fawcett 1989; Gugler 1971, 1997). And current studies on migration show the importance of a network in the migration process from the decision to migrate to the continuous flow to a particular destination, transnational links and settlements patterns and the formation of home associations that provide organizational support for newcomers and additional settings for the flow of information and assistance (Nyamnjoh 2010; Mercer *et al.* 2008; Page *et al.* 2010). The most plausible explanation for migration is that persons follow the lead of others from their area and migration thus becomes self-perpetuating and every migrant creates the link for subsequent migrants to follow suit, as has been the case with Pinyin migrants in South Africa. Banerjee (1983: 185) explains that chain migration can broadly be subdivided into 'delayed family migration' and 'serial migration' (of co-villagers), with the former referring to the migration of members of the same family in stages after the first migrant. And the latter refers to interaction between individuals who are not family members but are known to migrants and/or their families. It is the expectation that when individuals migrate, they will seek ways of taking a sibling and the influx continues until most of the males in the family have moved, as was the case of Joe, Mike and Jake, whose brothers joined them in Cape Town. Priority is given to the male members of the family and the women are brought over when there is little option for them back home or if they go to study. In most cases, they migrate to join their spouses although some women may be left behind not to inhibit their chances of marriage. This may explain why the first two Pinyin women in Cape Town did not come as a result of spousal reunion but were widows who were brought over by their sisters who had been brought by their spouses. However, this does not in any way suggest that there is no female migration amongst the Pinyin and Mankon community, on the contrary.

Contemporary migration is redefining the role of network. Given the rise of information flows, ICTs have become the 'new talking drum of everyday Africa' (de Bruijn *et al.* 2009) that have propelled the network society. According to Castells (2003: 33), a network society is defined as a 'society made up of power, wealth, management and communication networks within the fabric of the social structure'. Concurring with Castells, Ros *et al.* (2007: 6) maintain that 'current international migration is the inevitable result of interconnection processes generated by the communication and information flows'. For the migrant communities in Cape Town, the high number of influx to the same destination is largely thanks to the information flows that link the home village and the host country, as well as the wide availability of information on the Internet and from returned migrants. Connectivity and access to networks become essential (Castells 2004). My contention here is that chain migration and the forms of network connections migrants leverage for personal interest and to seek asylum are informed by information flows and the availability of information in their own personal circles, especially those with significant numbers of weak ties.

Network intrigues

My understanding of social networks amongst these groups was enhanced by the long informal meetings I had with informants in Cameroon, South Africa and the Netherlands, and from interviews with leaders and members of both associations and other related ones. In addition, attendance and participation at associational meetings and cultural events were watershed moments in my comprehension of the intricacy involved in social bonding and bridging. For instance, with the sudden death of three PIFAM members, rather than rely on the Cameroon Association that would normally repatriate the bodies, the association decided to take full responsibility and received enormous support from other 'sister' associations. And at the end of this ordeal, they pledged their support and cooperation with these associations (see Chapter 9). Spending time with individuals at their places of business (for the sedentary ones) or hanging out with hawkers listening to their stories and seeing first-hand how they built and maintained ties was informative in categorizing the different networks. Clovis, for example, was issued with a 'you must go' by the Department of Refugee at the Ministry of Home Affairs and the days just prior to his departure were used to actively seek a reverse decision. It is at such moments that weak ties become extremely important to migrants. Acquaintances and ethnic enclaves in this situation provide information about distant parts of the social system. The weak ties between Clovis and his acquaintances formed a crucial bridge between the ethnic enclave and acquaintances, as each member of the enclave could draw on their specific ties. In the same vein, weak ties are quite important to businesses and the building up of a client base (Chapter 4).

Listening to their stories of kinship network and, when fortunate enough to be around when 'beeps' come in from home, I could follow the conversation between them and families. These are rare moments because through such conversations one is able to appreciate the intimacy of the network and thus understand the bonding. Back in the home village (Pinyin), my presence often provoked relations to beep their children in Cape Town to inform them about my visit. They were special moments as I had lengthy conversations with migrants who expected me to give them detailed information about the wellbeing of their family back home.

In Mankon and the Netherlands, one-to-one meetings with individuals were useful as they revealed how loose networks are here. James in Holland, for instance, successfully brought in two of his younger sisters to Europe via Italy thanks to the network amongst migration syndicates that could easily obtain an Italian visa for them. In an effort to hasten James's sisters' chances of integration in Holland, James contacted Cameroonian friends with Dutch passports to see who would be able to marry his sister. John in The Hague was contacted and an arrangement made to drive to Belgium to meet the sister but, John was quick to note that this was a calculated plan and dissociated himself from it. The repercussions were a rupture in his relationship with James, confirming the notion of social games by actors, as well as seeking to forge or intensify new or existing relationships to lever an advantage.

Although the rate of chain migration amongst Mankon migrants to Cape Town is nothing compared to that from the Pinyin community, there are, however, cases of chain migration among kin relations and assistance with information about travel to South Africa and settling in in Cape Town. Migration thus, 'builds upon a growing base of knowledge, experience, social contacts, and other forms of social and cultural capital in self-reinforcing fashion' (Massey *et al.* 1994). This has lead Wilson (1994) to synthesize the role of the network in migration as 'network-mediated migration' and she notes that migration networks must be considered as facilitating rather than encapsulating, as permeable, expanding and fluid rather than as correlating with a metaphor of a rigid bounded structure. This may be true but the tendency has often been to study social capital from the perspective of facilitation and building bridges, with very little on how social capital works (Portes 1998). Depending on the perspective adopted, it is necessary to bear in mind that social networks facilitate as well as encapsulate. And it is again safer not to classify them in straightjackets of 'either or' but 'more or less'.

Prospective migrants log onto the internet to learn more about South Africa and call friends to find out what skills they need to settle economically. Friends in Cape Town often encourage those at home to come, such as Nixon who was encouraged by his friend at every stage of his migration to join him (Chapter 3). Joyce was able to make all the necessary arrangements for her nephew to travel to Cape Town thanks to the network she had built up from the migration syndicates in Johannesburg and Maputo and, through the constant communication she maintains with him, she can give directives on making a passport and the actual migration process itself.

Equally, my trips back and forth between Pinyin/Mankon and Cape Town brought me into contact with these networks and how they are maintained/ forged, severed or re-established. On all the journeys I made, I took gifts of cell phones, money, medicines and Visa cards from migrants in Cape Town to relations to withdraw money directly from the migrant's account in Cameroon; and from Cameroon, I took back food parcels, *ashwabi*,[2] photos and jewellery. Jones, for instance, sent the sum of R 1000 with me with a detailed list of how the money should be distributed, and also a cell phone.

While Mankon is not as polygamous a society as Pinyin, kin relations also lean towards matrilineal lines and the immediate family. This bonding in both communities is heightened when the mother of the migrant is widowed. Major decisions cannot be taken in the family without consulting the migrant and his decision is often final. Families are keen to meet someone coming from the host country in a bid to get first-hand information about their children's wellbeing and the possibility of sending food parcels. But equally, some do not hide their disappointment when it turns out that there is no parcel for them. This is often met with the question '*you been see ... as you di come so*? '*Ee no send something.*' These are some of the questions I was confronted with in both Pinyin and Mankon by disappointed relatives and parents. How then do we study and understand these networks in ways that delineate the home and host countries and yet not side-line the agency of the migrants and, by the same token, extrapolate how migrants leverage advantages? Seeking to understand networks from the vantage point of who is linked to who, the content of the linkages and the pattern they form will help us comprehend how their networks enable them to broker information. Equally, by focusing not only on individuals *per se* but on the entire Pinyin and Mankon communities as represented by the PIFAM and MACUDA respectively, we can gain useful insight into the communities' social networking as social capital is both a 'private good' and a 'public good'. Beyond this, the network and bonding approaches, as shown above, will greatly enhance our analytical perspectives of social capital.

Studying networks

Social capital, according to Bourdieu (1986), Coleman (1988-9) and Putnam (2000), comprises personal and interpersonal interaction. These different interactions share common values and their durability is informed by the way they are harnessed. However all authors approach the topic from different angles. Bour-

2 Clothes (uniform) worn by family members and friends in the event of the death of a family member or at a wedding.

dieu's concept is based on Marxist thinking concerned with inequalities regarding access to resources. He maintains that 'capital does not exist and function except in relation to a field' and we are informed by history and capital, one has to work to earn social capital and it is accumulated over time (*Ibid.* 1986). Coleman (1988-9) focuses on the rational choice of the individual in pursuit of their own interests and social interaction is considered to be part of a form of exchange. While rational choice alone cannot explain why people indulge in social capital since people have the ability to pursue their objectives, they do not invest in social capital but delve straight into ways of achieving their interests. Putnam (2000) underscores the importance of civic society and the role of associations in its formation. It is fundamental to the smooth sailing of society and underlies the social capital theory as well as the value of social networks. In a nutshell, Putnam (2000: 19) conceives social capital as 'connections among individuals – social networks and the norms of reciprocity and trustworthiness that arise from them'. Although focusing on American society and lamenting its rapid decline, Putnam's simplistic notion of social capital and his overt celebration of the functions of social capital in society make him oblivious to the multi-layered and intricate interconnection of networks. Rather than conceiving social capital as Putnam does, where other persons benefit from someone else's time and energy, Bourdieu (1986) presents a different perspective and categorizes social capital in three forms, namely economic capital (material property), cultural capital (prestige) and social capital (family and network accumulation), with an emphasis on cultural capital and inclinations towards social hierarchy. In the process of hierarchy formation, inequalities are created and social capital becomes the exclusive right of the elite.

Even though Bourdieu's conclusions are one-sided in favour of competing for social capital, we cannot dismiss these assertions given that they fairly depict some contemporary associations. We are able to gain insight into how members of the Pinyin and Mankon mobile communities convert capital accrued from their fields to enhance and improve their standard of living. These capitals are accessed by the ways in which 'mobiles' internalize their habitus and navigate the various fields. Although much of the attention is on social capital, it does not exclude economic and cultural capital as these two forms of capital, give insight into how social capital is produced, transformed and retransformed. For instance, when a family member migrates to Cape Town, integration into the host society is dependent on the networks s/he creates. Similarly, the amount of economic capital that migrant can muster invariably leads him/her to accumulate cultural capital in the host society. In the home country, the provision of financial support and connectivity with family as well as participation in developmental projects in the village communities help migrants to accumulate cultural capital. One way of achieving economic capital for them is through the economic activities they have been introduced to by social and kin networks in the host country (Chapter 6). When Pinyin migrants arrive in Cape Town, they are introduced into pitching (hawking) and the income generated is converted to human/cultural capital by paying for tuition that gives them access to a university degree and sets them apart as their cultural capital increases as well as their social network. It will increase even further if they secure a white-collar job.

Networking is part to all societies and is a fundamental pattern of life. Monge & Contractor (2002 in Castells 2004: 3) opine that 'networks are created by flows of messages among communicators through time and space' but in the current context of ICTs, 'the ability of networks to introduce new actors and new contents in the process of social organization increased over time with technological change, and, more precisely, with the evolution of communication technologies' (*Ibid.*: 5). For example, Clovis was able to reverse his deportation order and got a two-year residence permit thanks to the networks that he could access with the help of phone calls and texts. These networks are defined by three distinctive traits; flexibility, scalability and survivability (*Ibid.*), thus giving migrants agency to reconfigure relationships and extend interaction to a plethora of human activities, the corollary of which is a 'networked society' (*Ibid.*: 33).

While social networks have the capability to include/enable and exclude/deny (Castells 2004), there is equally a tendency that with ever-widening complex mobilities systems, inequalities will be created. As stated by Elliott & Urry (2010: 19), mobilities have the capacity to 'produce substantial inequalities between places and between people in terms of their location and access to these mobilities systems', which is what they refer to as 'network capital'. The inability of some migrants to acquire network capital means that they have been unable to visit their home country since their arrival in South Africa because access to such capital can change a migrant's status from one of asylum permit holder to business status or temporary residence. As such, kinship ties play a significant role in perpetuating migration and weak ties such as local women and lawyers provide legal assistance to regularize their status, extending their stay in the country when they have been issued with a deportation order. These ties are reignited when it is time for the extension of an asylum permit, while new ones are also forged.

Much as I compare the formation of networks to octopus-like creatures in nature due to their ability to spread to all fronts that could possibly yield results, they do not have the same characteristics and functionability. While some are visibly weak ties, their interactive role is purposeful and acts as a bridge to other networks. These weak ties lubricate the various nodes and link them to one another. Using the metaphor of a bridge, Granovetter (1973: 1360-1365; see also de

Bruijn & van Dijk 2012) expounds on how weak ties are bridges because 'it is through these networks that small-scale interaction becomes translated into large-scale patterns, and that these, in turn, feed back into small groups'. It follows that actors (weak and strong ties) participate in social systems connecting migrants to other actors whose relationships have an important influence on one another's behaviour. Migration syndicates and legal and corporate[3] networks in this study are considered as weak ties because of the role they play in linking multiple nodes and, like a bridge, they are capable of linking networks that were even previously unknown to each other. These networks are the essential lubricant that propels migration. The essence of such ties is to get the job done, for example in the case of Henry regarding his passport mobility (Chapter 3). The success of his migration is credited to weak ties, including his brief stay in Johannesburg and subsequent move to Cape Town, confirming Granovetter's assertion that 'weak ties are an important resource in making possible mobility opportunity' (*Ibid.*: 1373). His assertion that 'weak ties play a role in effecting social cohesion' (*Ibid.*: 1373) comes across as simplistic given that not everyone is successful with weak ties. Mark, for instance, moved to South Africa with a promise of further migration to England. As soon as he moved to South Africa, the migration syndicate agent who asked him to relocate to South Africa stopped communicating with him. Mark had to warn others not to fall prey to this syndicate, thus breaking off all ties with them and any related syndicates. However, in cases that have worked, they may be seen to offer social cohesion.

The concept of fields and capitals are more useful than the structural explanations because of their flexibility and dynamism. The concept of structural relations maintain that social structures consist of 'regularities in the patterns of relations among concrete entities; it is not a harmony among abstract norms and values or a classification of concrete entities by their attributes' (White *et al.* 1976 cited in Knoke & Yang 2008: 4). Structural relations also exist only at specific times and places and may be static, with the tendency to wane after actors separate. While every society is structured, structural relations are crucial to sustaining cohesion and solidarity within a group. However they may also reinforce prejudices and fan conflicts with other groups (Knoke & Yang 2008). The structural relations approach perceives migration more as an individual movement, focusing on individuals and with less regard for migrants' agency. In the same vein, it denies migrants the chance to navigate within the various social fields and views society as one in which there is a permanent logical order in place behind society/culture. To have a fair understanding of network capital, it is important to look both at the structure and the action (relations and actors) that are

[3] By corporate networks I refer to the network syndicate that operates at the Department of Refugee at the Ministry of Home Affairs (officials, agents and security guards).

embedded in each other given they are dialectical and indispensable elements of any social network because actions reproduce structures (Bourdieu 1977; Knoke & Yang 2008). In this regard, social networks are considered within the various social fields in which actors and relations constantly adapt to circumstances that call for going beyond structural-relational settings. The notion of social fields here is tied to that of Levitt & Glick Schiller (2004: 1003) who define social fields as overlapping interactive spaces: 'multi-layered, multi-sited transnational social fields, encompassing those who move and those who stay behind'. Much as these networks function in the various social fields, there is also the possibility for them to function without collision, thus reiterating the flexibility of networks. This is also suggestive of the shifting and imprecise boundaries of the field but has multiple nodal points of entry.

The accumulation of the various forms of capital, especially social capital, is largely informed by the advances in new communications technologies (information flows), especially in the migration process. Information flows are the lubricant that propels the network to be physically and virtually present at every stage of the migration process, given that what is important is not so much the movement but how it enables connections that result in the network. It is in this light that the United Nations Global Commission on International Migration states that 'global communications networks provide people with the information they need to move from one place to another ... have made it easier for people to move to another and to adapt to a new society' (GCIM 2005: 5). In the sections that follow, I outline the working of social networks among the migrant communities in Cape Town and their families back in their home country, with particular emphasis on (matrilineal/patrilineal) kinship ties and how they inform social networking.

Networking amongst mobile communities

The Pinyin community is noted for its close-knit community in Cape Town and news of this has reached the home country. It could perhaps be explained by their non-integration in the host society, leading them to create and rely on their associations for a sense of common identity (Anderson 2006). This identity is rooted in their shared sense of remembering of the community's history and culture in creating the home they have left behind in Pinyin. The process of creating a home-away-from-home is guided by cultural norms and practices and the idea of looking after the family. Memory becomes a salient factor in the recreation of mobile communities. It is against this backdrop that these communities seek to network among themselves and with other migrant communities to maintain a social cohesion of network society that seeks to safeguard one another. While these communities' interests may be warm at the outset, with the focus more on unity and the creation of a mobile community, tensions can arise as a result of conflicting views, leading to frosty relationships among groups, between individuals and between kin.

The community may be woven together but there is also networking amongst clusters in both communities: the younger generation, pitcher men and newcomers congregate together and the first (successful) migrants have their own cohort. Despite this clustered networking, the groups exhibit a tendency to be flexible and permeable. In addition, both communities have clustered in separate neighbourhoods for easy interaction and accessibility. The first Pinyin migrants, as a result of their social mobility, have bought their own property and moved out of Parow, but a large number still congregate there and in Belleville. The community is essentially very inward-looking and tribalistic. I do not in any way mean to be scathing or spiteful because Pinyin migrants equally acknowledge these attributes and see them more as a strength than as a weakness. Thanks to this attribute, they have become a force to be reckoned with in South Africa and are seen as the strongest Cameroonian tribal group. When a new member arrives in Cape Town, he is often advised to join the PIFAM and is taken to the meeting and to the relevant sub-association for self-introduction depending on which quarter in the village he hails from. In addition, the newcomer is put into the care of an older pitcher man as an apprentice and the older man is charged with teaching the latter how to pitch, where to buy goods, what to buy and how to approach a client.

Although most of the Mankon community reside in Maitland, some are also found in other neighbourhoods usually due to work. The tendency is for newly arrived migrants to live with relations and friends either permanently or temporarily until they get their own accommodation, which is always in the same vicinity. In this way, village relationships assist by providing help to newly arrived migrants, as can be seen below.

Henry is convinced that 'what one becomes in South Africa depends largely on who receives you upon arrival in South Africa, as that determines what you become in South Africa'.[4] Some migrants leave home without knowing anyone who could receive them in the host country so the natural thing is to look for a fellow countryman in whose care they will feel safe. This was the case that Henry was faced with when he arrived in Johannesburg for the first time.

> When I first came, I stayed in Johannesburg for a while. Here I met some Mankon people who offered me accommodation and trained me in what they were doing for a living – buying and selling scrap metal for recycling.[5]

---

4 Conversation with Henry, Cape Town; 20/08/2010. A similar statement was made by Peter in a later interview; see Footnote 3.

5 See Footnote 1 of this chapter.

When Henry eventually moved to Cape Town, he sought residence near his cousin, and rented from a fellow Mankon man who has bought a property. It is interesting to note that the properties owned by Mankon migrants are almost entirely rented out to fellow Mankon. In contrast, when a Mankon youth arrived in Cape Town, he was taken in by a Nigerian and a Congolese (DRC) and was offered accommodation for a month. At the end of the month, he was asked to join them as a drug pusher to pay his share of the rent after the initial month of grace. In desperation, he phoned some Mankon people and begged for help because he did not want to become a drug pusher. Almost immediately, Peter[6] organized shared accommodation for him and he moved out. And Peter offered to pay his share of the rent for the first three months to give him time to organize an income for himself. In accordance with Henry's statement above, this migrant would have been lured into becoming a drug pusher had he not requested the help of the Mankon community as he was easy prey for his benefactors who took advantage of his newness to initiate him into the business by offering him free accommodation for the first month. While networks often have value, in some cases there are ulterior motives for forging them.

All the interviewees attest to having passed through a hawking stage during the early days after their arrival. While some found the job demeaning and would rather have gone back home, they were encouraged to stay on and others simply moved with the flow. It is interesting to note that before the social mobility of those who have decent jobs as well as those still aspiring to social mobility, the one common denominator linking them is pitching (hawking). Successful migrants stand out as those to be emulated by subsequent migrants, and they do indeed often provide advice to the latter.

Much as there is chain migration among kin relations, we equally have serial migration within these communities (Banerjee 1984). Serial migration refers to the assistance given by migrants to prospective migrants who have no blood or marriage ties with them. It is partly motivated by the visits of return migrants and the development projects they organize in the home village. This is facilitated by migration syndicates that assist prospective migrants with visa procedures and escort migrants to the airport. Most Pinyin and Mankon migrants who come to South Africa without the assistance of a relation based in South Africa have used one of these networks. In this regard, family members seek help and pre-migration information from them. For instance, Edmund and Alex travelled to South Africa without any prior contacts and their knowledge was limited to the high number of Pinyin based there. They contacted a co-villager who runs one such syndicate, provided him with the necessary documents and fees and some months later were informed of their travel date. On arrival in South Africa they

[6] Interview with Peter, Cape Town; 01/03/2011.

realized that most of their friends who had disappeared from home had relocated there so settling in was therefore not a problem for them. Similarly, Penn and his father were motivated by the success of their neighbours. The father was confident that sending Penn to South Africa would bring him financial relief as he had just retired (in 2008). Based on the information he had been given, he used most of his savings to send his son to South Africa in the hope that he would continue chain migration in the family. This shows the intricate interconnection of networks, how they operate across space and time, and what makes them happen as much the information and new technologies that transnational and translocal communities have appropriated. Equally the new age of unbounded virtual and corporeal mobility churns out information from the networks of their kin in Cape Town.

Kinship relations and social networks

As already mentioned, contemporary migration is embedded in the double presence between home and the host country, as well as with other migrants. Kinship relations and networks between migrants and family back in the home country are either transnational virtual ties or are face-to-face when migrants visit their home village once in a while. As noted by Massey (1992), 'network as a "place" is constructed out of a particular locus'. This suggests that migrants travel with the same social and emotional values to their newfound community irrespective of 'place', but that they are now bound together by a sense of the imagined community they have left behind and strive to replicate the community in their present locus. Kinship relations and networks and the large number of Pinyin migrants can be traced back to the first Pinyin migrant who went to South Africa in 1996. The network chart below shows a network of transnational kinship relationship. Since his arrival in 1996, Joe has brought in his immediate nuclear family (his wife and three sons) and has also brought in four siblings and some close friends and has provided information for others who may follow. In addition, he has assisted newcomers with settling in.

By kinship, we mean blood relations or relatives by marriage and the individual nuclear family, or close relations marked by community ties. This kinship network is also informed by the Pinyin adage that 'when a Pinyin man discovers a fertile ground he paves the way for many to follow'.[7] Earlier migrants became not only the communication outpost for those who remained behind but also created a bridge for others to follow. In this regard, kinship relations are seen more as a continuum than as a disjointed movement of individuals. This leads to chain

[7] Conversation with Papa Gabriel, Pinyin; 21/05/2010.

Diagram 5.1 Chart showing network connections and expansion (thanks to Joe)

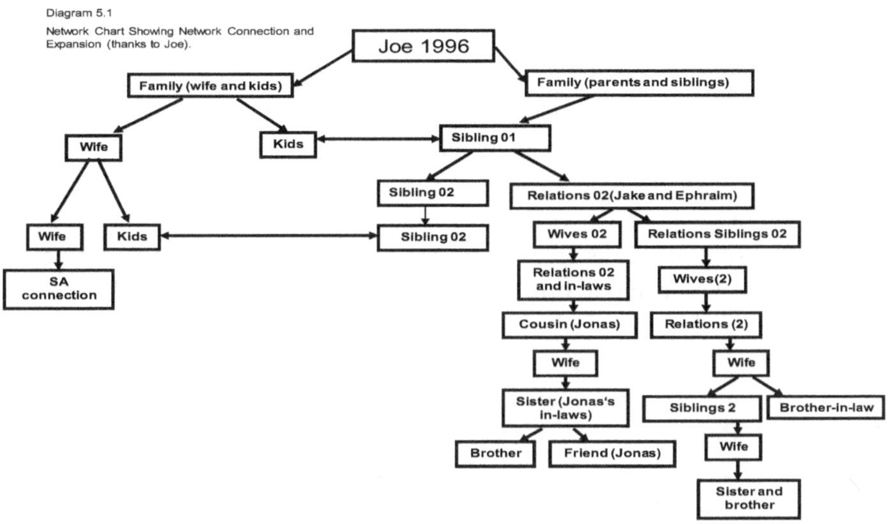

migration, something MacDonald & MacDonald (1964) defined as 'that movement in which prospective migrants learn of opportunities, are provided with transportation, and have initial accommodation and employment arranged by means of primary social relationships with previous migrants' (cited in Choldin 1973: 164-165). There is ample evidence from the literature on kinship relationship as well as empirical data from this study to substantiate the fact that these ties are the most important and well preserved. They are sought after in pre-migration, migration and post-migration periods, and are the most tapped from (Prelipceanu 2008; Choldin 1973; Pelican *et al.* 2008; Fleischer 2006). Relationships with kin and the home village are anchored in the notion of transculturalism and transnationalism (Levitt & Glick Schiller 2004; Vertovec 2002). Although migrants want to have a foothold in the host country, they equally want to maintain a network with family and kith back home. In the same light, these relationships are the primary ones that migrants have grown into and relating to them is manifesting internalized practices.

Communication back home and networking often intensify when a sibling is about to leave for South Africa and in the days immediately after his/her arrival. A case in point is that of Edwin and his younger sister Adele (who got married in Cape Town). She was instrumental in his coming to South Africa and settling in Johannesburg. But, like Henry, he was disappointed with the lifestyle of his host coupled with the constant police harassments. He phoned his sister from Johannesburg to inform her of the situation and was prepared to go back home than to sell stolen computers if this was the only chance open to him. She immediately made arrangements for him to move to Cape Town and after two weeks with her, she advised him to move to Parow where most of the hawkers live. Here, she had made contacts to arrange a mentor for him (Chapter 6). It is worth mentioning that all these plans were done behind his back.

Adele's spontaneous reaction (paying for a ticket for him to fly to Cape Town the next day) and her ability to draw assistance from the ethnic enclave networks show that kinship networks still remain the fundamental bonding ties. Mike, like other migrants, is seeking to bring in all of his siblings (he has already brought in his wife and two brothers, and plans to bring a further two as well) and has even outlined how they will eventually assist their mother with setting up a business that will provide an income for her and limit her dependence on her children. Thanks to the Pinyin networks he has in Cape Town, he has been able to secure a peer interest-free loan and admission to university for his spouse. Mike, like most migrants, juggles constantly within his wide range of social networks both at home and in the host country. Importantly, he has laid the foundation for subsequent family members to move there and this will greatly reduce the financial and psychological costs of migration by providing information on employment and material assistance and accommodation. It will also offer a supportive relationship during the adjustment period, as well as reducing the costs and strain on Mike of providing for so many people in the home country (see Banerjee 1983).

Frosty kin networks

Network relationships with kin require a lot of trust. Not all members in the family can be sufficiently trusted with the migrants' affairs back home. Information abounds about how kin have plundered migrants' resources at home and in the host country. Assisting siblings to join migrants in the host country is often considered a way of easing the financial burden on the first migrants back home. This, in some cases, turns out to be a burden. In extreme cases, as proven by empirical data, instances of sibling rivalry and feuds are not uncommon and this invariably threatens kinship ties back home. In other instances, the spill-over extends to the wife's parents-in-law. For instance, Terence single-handedly sponsored the migration of his younger brother from Bamenda to Cape Town and provided accommodation and financial assistance for him to start pitching. The expectation was that, after settling in, his brother would contribute to his rent and food. However, the latter turned a blind eye to these and never discussed his finances with his brother. And each time his parents or sister called from home, he would run outside to answer the call, away from any possible eavesdropping. His attitude soon became untenable, leading Terence to conclude that he was being ill-advised by his sister at home. Terence reported the matter to his father and threatened to kick him out of his house. The father called the brother and told him to calm the tension. A few months later, he moved out of the house. During this period, Terence's spouse, Joyce, was also making arrangements for her nephew to travel to Cape Town. In the midst of this feud, Joyce's nephew arrived and the news did not go down well with her mother-in-law who thought Joyce asked her spouse (Terence) to bring in her nephew so he could spend money on her family as well. The result was antagonism between Joyce and her parents-in-law and her brother-in-law and for a while communication between them ceased.

Another case in point is Mike. Prior to his departure for South Africa, he agreed with his father to assist his stepbrother and to have him join him once he had settled. When the opportunity came, he chose his real brother over his stepbrother. This immediately led to tensions between the co-wives with the stepmother accusing Mike's mother of having instructed him not to take her son. The stepbrothers immediately pooled their resources and sent their brother to Cape Town as well. Once in Cape Town, Mike opened a shop and had his brother manage it while he went to school. The brother let the shop become run down and left him with huge debts. Consequently, Mike fired him, then dropped out of school to work to pay off the debts he had incurred to start the business. The ramifications were felt back home and his mother tried to restore peace between them. He later decided to assist his sister by sending FCFA 300,000 (approx. € 500) so she could register for the police's entrance exam and as money for bribes. The money was appropriated by her spouse and this made Mike vow never to send money to her again. While the tendency for most studies is to show the linearity of kinship networks as encapsulating, facilitating, homogenous and compact, there is more to networks than meets the eye. They can also be impeding and generate simmering rivalries.

For those who have regular documents and can travel home, there is often excitement about seeing family and friends after long years of absence. This is mutual and the first weeks are spent performing rituals such as visiting the graves of deceased relatives and seeing relations. This was the case with Jonas[8] who maintains that for the first two weeks of his visit to the village, the euphoria and excitement about his homecoming superseded financial demands but from the third week when all was quiet, people began coming to him with their problems and in the fourth week he became edgy and was overwhelmed by the demands being made on him. He switched off his phone and went to live with his sister in Nso, Kumbo Division, after which he went to Douala to be with a friend for the last two days prior to his departure. Migrants are becoming increasingly weary of the insatiable demands of relations and have devised coping mechanisms to deal with

[8] Interview with Jonas, Cape Town; 13/08/2011.

them, as can be seen in Jonas's actions. Although they have a strong desire to go home and be connected with the family physically, this urge soon changes to eagerness to go back to South Africa, as if distance will mitigate the demands. Their lives are constantly being lived in the in-between. Nonetheless, frosty relationships are at times packed with emotional distress as the line between frosty and good relationships and emotions seems blurry and impacts on both of the parties concerned.

Networking through marriage

Marriage patterns amongst the Pinyin and Mankon show a preference for women from the home village and there is extensive networking involved in this. Often, these networks go beyond kin structures and can at times involve the whole neighbourhood when it comes to searching for a bride. The new communication technologies, especially the mobile phone, have given communities extensive networking over space and have redefined the traditional practice of bride search. This is similar to Andersson's (2001) conclusions in his study among the Buhera community in Harare, Zimbabwe. Apart from Emma and Esther (who we will meet below) who are South Africa spouses, the rest are from Pinyin and Mankon.

Marriage for Pinyin men resident in Cape Town usually depends on extensive networking to look for a bride in Pinyin. It begins with him sending word and his photo to family in the village to look for a suitable bride for him. They in turn consult their network to see whether there is a daughter in a close family's social network who could be eligible and sent to South Africa. In the absence of anyone within the family's social network, the search is extended to the quarter and beyond. When someone is found, her picture is sent to the prospective husband in Cape Town who in turn finds out about her from his friends. If he accepts her, he sends off money so the formal negotiations can begin and the bride price is paid. The prospective husband is represented by his father or brother throughout this process. This has come to be known in Pinyin circles as a 'picture marriage'. Doris in Cape Town and from Pinyin originally is one such woman. According to her, 'it was an arranged marriage over the Internet and (following) several phone calls. I barely knew him, and we had only met in passing prior to his departure for South Africa.'[9] She talked of how her father made several calls to different people in Cape Town and Sweden to find out more about this prospective husband before entrusting his daughter to him. She adds that on the actual day of payment of the bride price, it was her father-in-law who stood in for her husband and it was to him that she gave the cup of palm wine after drinking: '*I know na pa ya, na for ee I be drink mimbo give cup*'.[10] While the arrangements are going

[9] Interview with Doris, Cape Town; 24/07/2011.
[10] Interview with Doris, Cape Town; 31/07/2011.

on in the village, the groom has also appointed a member of the family or a friend to assist the prospective wife with making her passport and sending it to him either by post or with someone going to South Africa to commence visa procedures that will allow her to join him. According to Jones, this was the fashion in Cape Town, for a man to 'import a wife from home' because it was a moment of pride for the groom to be proceeding into manhood.

On the day of her arrival, members of the husband's quarter converge on the airport to welcome the bride, with the groom looking very distinguished in his suit and sometimes with a bouquet of flowers too. But equally, it is a moment of pride for whoever gives the bride a ride in his car and the men scramble for this honour. And there is a reception later in the evening to formerly welcome her.

The situation is no different for Mankon migrants but some would rather go home to get married and travel with their new wife, as Paul and Peter did.

Photo 5.1 Arrival of Pinyin members at the airport to welcome the wife of a member (the husband has a hat on)

Photo 5.2 Arrival of the wife

Photo 5.3 Husband and wife together

Photo 5.4 Reception at the home of a Pinyin elder

By contrast, only two informants in the Netherlands succeeded in having their wives join them, a process that took over two years. In Cape Town it takes a maximum of two months for the tuyau to be completed. However, not every wife is ready to be 'imported'. Nixon was not lucky and, after having gone through the process of bringing his wife over, she decided to return home because the situation in Cape Town did not match her Utopian expectations of 'bush'.

Gender social and kinship network relations

The literature on kinship networks has been too male-oriented, i.e. 'gender blind' (Gidengil & O'Neil 2006: 2), with little regard for gender balance. Whereas some studies, especially in the US, have shown that women tend to have more extensive general networks than men (Smith 1976; Choldin 1973), the same cannot be said of this study (see also Hagan 1998). Amongst the mobile community, female social networks are limited, with women having fewer networks than men to offer help and support. Women are often dependent on their family relations and the fact that most are married makes it difficult for them to navigate as easily as the men, a trend that is also raised by Hagan (1998) in her study of the Maya community that settled in Houston. Gendered social networks, especially in the economic domain, interact to produce outcomes that are more advantageous for men than for women. Women's contacts tend to be localized, often within the family and in the case of the Pinyin women, these are amongst each other, while men's ties more often extend beyond the association to a multitude of occupational and associational settings.

It may be difficult for women to establish their own networks outside Pinyin and Mankon circles because migration among women is primarily for spousal reunion and joining their spouses years after they first moved to Cape Town means that they are absorbed into the networks that their husband has already established. This shifts a woman's preoccupation from creating her own networks to studying her environment and engaging in business or studies. In addition, women's abilities to create new social networks faster is impaired by cultural norms whereby they are expected to relate more to their husbands' networks and family relations in the host country. Their networks are usually previously formed networks and their agency to create networks of their own is limited. Most have opted to adopt the ties that have been put in place for them by their spouses, as was the case for Judith and Jane from Mankon. They joined their spouses in Cape Town after many years and, rather than create new social networks, they chose to adopt those of their spouses, thus attesting to the flexibility and permeability of social and kinship networks.

In the same vein, Esther and Emma are South Africans married to Pinyin and Mankon men respectively. While both women have maintained kinship ties with their individual families, their social network relations have been largely reduced to those of the Pinyin and Mankon communities as well as the friends of their spouses. This could be partially explained by the fact that both Pinyin and Mankon, like most African societies (Adepoju 1994: 5), are still conservative patriarchal societies where women are expected to blend into their new surroundings either by volition or following the dictates of cultural practices. Emma's husband explains that is it the norm for woman to feel part of her new family because it is the man who pays the bride price (*lobola*) and so the relationship should not be reversed: the woman has to follow the man. True to this, Emma's husband Charles turned down swapping his Cameroon passport for a South African one because it would mean giving up his Cameroonian nationality. However this is not to say that they are unable to set up their own networks and new contacts are eventually created that help them settle in. Conversely, as Lowndes (2006: 224) notes with regard to women's social and kinship network, women are more likely than men to see relatives and speak on the phone with them. In this respect, she argues that women and men tend to have different social-capital profiles, with women's social capital espoused in 'neighbourhood-specific networks of informal sociability'. Acknowledging her findings, I could add that sociability is not limited to specific networks but instead spans home and the host country because of their mobility. While they seek to forge new social relationships (or none at all) beyond those already created by their spouses, they equally maintain strong kinship ties with their immediate families in the home village and to a lesser extent with their in-laws. This is well explained by Russell (1999: 219) who holds that 'the very building blocks of social networks are gendered' because 'women's continued responsibility for caring and domestic work tends to restrict the range of social activity they are involved in, but it does provide an opportunity to build up supportive social networks in the community'. The networks of Pinyin and Mankon women in Cape Town depend on the associational structures of the PIFAM and its sub-associations more than the men do and still cautiously keep to themselves. PIFAM women have made unsuccessful attempts to set up a women's group but some have gone beyond this to pursue networks within the church and those with children back home intensify kinship relationship, as Joyce has done (Chapter 9).

Some, like Bianca, consults a doctor in Cape Town on behalf of her mother to buy prescription drugs that she sends to the village. In the same vein, Jane's mother consulted her gynaecologist in Cameroon on behalf of her daughter during the latter's pregnancy and communicated information to her (Chapter 9). The kinship networks of the women, married or single, that I spoke to are almost entirely limited to their maternal and spousal relations. Even back in Pinyin and

Mankon, those who call their children abroad or beep them are usually the mothers.

Female migration has not changed the relationships women enjoy in their families. Instead, ICTs (and especially the mobile phone) are used to continuously maintain family networks and an awareness of what is happening in each other's lives. Equally, maintaining such networks is embedded in the cultural practices that see women primarily as care givers. This care is intensifying thanks to mobility and ICTs.

Negotiating *ngunda*: Social networks in asylum/refugee circles in Cape Town

Recent migration trends among Mankon and Pinyin migrants in South Africa are characterized by migrants entering the country as legal migrants and subsequently changing their status to that of asylum seeker. This is commonly referred to by migrants in South Africa as *ngunda*, or *adoro* by those in the Netherlands. Studies of the social networks of asylum seekers and how they have navigated and negotiated the process of seeking asylum are still relatively sparse (Koser 1997). The focus has been on the various processes of migration: from the decision-making, choice of country of destination and adaptation into the host society (Hagan 1998, 2003; Boyd 1989; Banarjee 1983), overlooking the fact that aspects of asylum seeking are part of the first step towards migration. Asylum seekers and refugees have often been viewed with contempt in their host countries, especially in South Africa. And this has recently led to a wave of xenophobic attacks against migrants and prompted debates on xenophobia (Segatti & Landau 2011; Vale 2002). This has not dissuaded Pinyin and Mankon migrants from emigrating to South Africa and the migration process of subsequent migrants to South Africa is often sponsored by settled family relations who facilitate the migrant's change of status from that of a regular migrant to an asylum seeker/refugee. It assumes that serious consideration has been given to this asylum route as a way forward and as a prerequisite to migration. Enough networking is galvanized and stories made up prior to the arrival of the migrant so that s/he can rehearse the story and then deliver it convincingly to the officer at the asylum centre. This is in contrast with Turner's (1995) view that 'organised state violence and torture as a means of destroying trust, and provoking community breakdown; compounds the processes of asylum seeking'. Instead, refugees and asylum seekers are not simply passive recipients of care. Asylum seeking is just a step to achieving the targets migrants have mapped out for themselves: it is a game of strategic interests. Migrants actively network for appropriate help on how to secure extended refugee permits to allow them either to complete their studies and move on to employment or to continue the migration pattern. Weak ties are particularly relevant here as migrants network with lawyers and workers from the Department of Refugee in the Ministry of Home Affairs to get the maximum extension (two years) or a temporary residence permit. These weak ties are dormant but reactivated for such purposes or when migrants are arrested and/or face imminent deportation.

In this respect, empirical data on the significance of social networking among asylum seekers corroborates with that of Crisp (1999) and Duke (1996) who maintain that social networks are of paramount importance to migrants as they enable the goal of self-determination to be achieved. Acknowledging Williams's (2006: 867) perception of refugees and migrants as 'tactical actors', I would add that migrants and refugees are equally strategic and shrewd in the way in which they go about seeking asylum. While seeking asylum is a means to an end, migrants also exploit loopholes in the system to access vital information and social benefits. The relative ease with which asylum is granted confirms claims from migrants that South Africa is yet to put in place adequate structures/policies to harness and rein in the number of migrants heading to South Africa.

Migrants have learnt the social game pertaining to the various social fields and how to strategically plan and map out their asylum case. I make the case that migrants' personal and associational networking are of vital importance and inform their daily interaction in asylum seeking, which transforms their status from irregular migrants to refugees and later to that of someone with exceptional skills upon the completion of their studies so they can find a job. Although I am looking at asylum networks within the Pinyin and Mankon communities, rather than grouping them together as a homogenous group, I treat them as individual cases, given that asylum seeking is negotiated by individuals. Inasmuch as the process is launched by family relations or other networks, each migrant has to activate the network relation to a specific end (MacGaffey & Bazenguissa-Ganga 2000). Although the community may assist in linking a migrant to a network, for the networking process to work it depends entirely on the individual and how well s/he is able to tell a convincing story or to pay enough kickback to the agent working with the refugee officer. It is important to mention here that, except for those with South African identity cards or permanent residence, almost all Pinyin and Mankon migrants have opted for asylum, including those with study permits. This is because migrants have come to realize that asylum permits allow one to pay lower tuition fees (as locals) in universities and it qualifies them for medical treatment in local hospitals at minimal cost.

Applying for and being awarded refugee status in Cape Town is relatively easy compared to in Johannesburg, as was reported by those who first lived in Johannesburg before relocating to Cape Town. Yet it is still a herculean task to obtain it in Cape Town given that migrants have to camp at the refugee centre overnight or go as early as 4:00 to secure a good position in the queue. The process of applying for asylum or extending one's refugee permit could take the whole day for some, days for others or just a few hours for yet others. It all depends on how willing migrants are to follow the '*tuyau* line' (Chapter 4) as there is a chain of connection that has been put in place to ease the process for the latter group. I accompanied some informants to the refugee centre when it was time for them to extend their permits to see for myself the reported daily experiences at the centre.

Photo 5.5 Migrants at the asylum centre seeking an extension to their refugee permit

This is an observational account of one informant that resonates with the rest of the community. Chris's permit had expired a month before he went to request an extension. His first dilemma was how to convince the officer that he was 'sick' and consequently had been unable to come for its renewal at the stipulated time. Chris contacted a fellow Pinyin who put him in touch with someone who could give him a fake medical certificate to show he was 'truly sick'. With this document, he proceeded to the immigration centre intending to fast track the process. While there, he phoned an agent who wasted no time in meeting him at the centre and the agent then phoned the officer. A few minutes later, the officer came out to see the agent and check Chris and any other migrant following this *tuyau*. Back in his office, the agent sent him an SMS with the name(s) of the person(s) he had. In the interim, the security guard was notified and was ordered to open the gate and let in the next group, so he slipped in with those who were waiting with the agent. It is such networking that I refer to as a corporate network, given that their relationship is more economistic and instantaneous. Once inside with their documents, extension is automatic between six months and two years depending on how much the migrant is willing to pay. About an hour later, Chris was called in, interviewed and his asylum permit was extended for nine months. Unlike most migrants who had been queuing at the centre from the early hours of the morning and still did not have any chance of extending their permits or starting their asylum case, Chris was able to buy an extension within hours of arriving at the centre, thanks to the connections he has. Underpinning his ability to network smoothly is the use of communication technology. During this time, he made more than five calls, all geared towards renewing his permit. This, it would appear, confirms what Smith (1976: 21) opined: 'it is not what you know – it is who you know' that matters.

By contrast, Tabufor from Mankon chose to use the religious network he created within the church where he worships to secure a residence permit. Religious networks like other networks play a significant role in assisting migrants to settle in the host country and also providing a family for them there. This is also a source of networking between the mobile communities in the Netherlands and their counterparts in Cape Town, although those in the Netherlands do not use theirs as a means of achieving refugee status. Nonetheless, religious networks help migrants overcome difficulties such as a death in the family, act as an intermediary in family conflicts among migrants in the host country and assist migrants in obtaining an extended temporary resident permit in Cape Town. Levitt (2006) has shown the significance of religious networks in the lives of migrants, how they help them to make a place for themselves in this case in Massachusetts and, at the same time, to belong to the home community by linking up with the religious community they left behind in the home country. However, a focus on religious networks limits transnational links between home and the host country.

This section hopes to fill this gap by looking at internal religious networks that are sourced by migrants in relation to asylum seeking and/or a temporary residence permit. It also shows how religious networks impact on the migrant's wellbeing and extends to relationships back home. Tabufor, for instance, worships at a Protestant church in his neighbourhood and has joined the choir there. He is also a handyman as he does carpentry, building and painting so whenever the church appeals for someone to assist in these fields he volunteers. Hearing about his status, a choir member advised him to talk to the pastor and request a letter stating that he is a volunteer with the church. He did as advised, presented the letter at the Department of Refugee and was given a two-year temporary residence permit. Similarly, each time Delphine has to go for an extension of her refugee permit, she fasts and prays with her pastor, and they also pray together on the phone in the middle of the night. When I met her on the eve of her visit to this department, she was at her business reading Bible passages that had been recommended to her by the pastor. According to Delphine, this has been her way of extending her permit as she is not ready to use a *tuyau* each time she goes for an extension and it has always worked for her. In this regard, asylum seekers have strategically formed and cultivated networks that attend to their specific interests and that can help them achieve their goals as asylum seekers and refugees and the benefits that accrue from this status (Baker 1990). While communication technology plays an important role in networking, face-to-face meetings are preferred when meeting asylum agents, while the phone plays a role in making possible the 'meetingness' (Urry 2003).

Conclusion

Advances in new technologies are redefining the way migrants are linking up with each other and with family and friends, and also how networks are forged and maintained. While face-to-face networking is still preferred, virtual networking is incontestably the norm. As a result of the new mobile culture, mobile communities have developed an octopus-like ability to spread their tentacles and create multiple networks. Whilst some are more significant than others, weak ties have the tendency to help them achieve their goals. These weak ties include the lawyers who assist migrants to fast-track the legalization of a 'convenient life partner' and bail them out of police custody. The weak ties also extend to the migration syndicate/industry and the corporate network (Home Affairs workers and agents) that assist in the renewal of refugee permits. Networking with weak ties is not permanent and links are constantly redefined, hence shifting. The purpose of weak ties is to act as a bridge, with their ability to shift constantly because these ties serve a particular purpose and if it is not met, alternative ties are sought. However it is important to note that weak ties connect better than strong ties. Such ties fuel migration and refugee processes as they depend on shortcuts that would have taken a strong tie much longer to accomplish. By the same token, earlier migrants have come to be the communication outpost, or 'gatekeepers' Williams (2006), for migrants and prospective migrants, handing out information that is of importance in either the subsequent migration of those in the home country or information that will assist newly arrived migrants to settle in. Equally, they provide information to navigate the contours of asylum seeking. The availability of communication technologies and the information that circulates leads to the formation of a network society. It is as a result of access to information and communication technology that the network society is able to constantly redefine itself 'transcending the historical limits of networks as forms of social organization and interaction' (Castells 2004: 6).

Kinship networks provide migrants with the possibility of double engagement as well as someone they can entrust their affairs to. These are usually friends but could be trusted family members. The multiplicities of networks have been described by van Hear (1998) as 'migratory cultural capital' that migrants leverage support from. But they also provide vital information about navigating the contours of asylum seeking. These networks are highly practical and lead to the formation of ethnic enclaves.

Much as the purpose of networks is to build bridges, they are not as cohesive as would appear to be the case. Given that they function in various social fields, they exclude and include, and in the process may lead to fragmented relationships.

Female migration has not changed the relationships women enjoy in their families. Rather, ICTs, especially the mobile phone, are being used to continuously maintain family networks. Although away from home, female networks and kinship relationships have followed traditional patterns and societal norms that see women primarily as care givers, and the care intensifies as a result of mobility. ICTs thus become the tool through which care is provided.

In summary, this chapter did not limit itself to cataloguing the different networks and how they are formed but went beyond a narration of network formation to show how they are intricately connected and their dynamic nature, as well as the simmering tensions that can lead to fragmented relationships.

6

'Going to the field': Pitching and migrants' economic activities

Introduction

Chris's story

> I sell pirated DVDs made by me and it is thanks to the Internet that I can do all of this. I have Internet at home on a USB, but I use it late at night when the rates are low. I often go to the cyber because it costs R3 an hour, where I download movies and their labels onto my USB key, and nobody sees what I am doing. Back at home, (as you can see, I have two computers and printers), I format the labels so that they will fit on the DVD, and save them onto the computers and I am ready for printing the movies. I print about 300 movies at night, so that by morning retailers will pass by and buy for their pitching sites. Others place orders and I print for them. During the day I take those left over to Parow to hustle a bit. I know what I am doing is illegal but I have no option because I have come here to make money. But what I am doing is nothing compared to the big companies who make all the monies, '*we di lick na oil for back bottle*'[1] (we are simply gathering the crumbs).

Chris has two phones with two service providers (MTN and Vodacom). While I was there, his phones rang constantly with customers calling to find out about particular movies, to place orders for movies or to see whether he was at home so they could drop round to collect movies. Chris's business reflects those of his generation who arrived in Cape Town at the time of ICTs and their affordability. They are in contrast with the rudimentary businesses carried out by earlier immigrants, such as Jake, CJ and Joe that entailed a lot of physical mobility and can be traced back to when they were still in Cameroon.

Chris's story heralds the role of ICTs in redefining contemporary migrants' businesses, unlike those of their parents and earlier migrants that were characterized by long-distance mobility, trade and head porters. Chris's business is thriving because of the networks that he can muster and he is known to produce better-quality movies than others in the business. And he is willing to accept

[1] Interview, Cape Town; 08/07/2012.

payment on credit, with his clients selling his products and then paying him afterwards. He has been able to circumvent the law and economic regulations on pirated DVDs. For Chris, like a host of other pitchermen, their trade activities are not in themselves marginal but the activities are carried out within the margins of the law. Accordingly, pirating movies is simply a form of business to Chris and there is nothing wrong with it. But he is aware of its illegality in the face of the law.

Following the way migrants view their form of business, I will leave out these dual categories on the dominant narratives of formal/informal and legal/illegal. Chris is very cautious about who he opens the door to. One has to phone from outside or use the local greetings (if it is a fellow Pinyin) of '*he-Y̌ëh*' before the door will be opened.[2]The phone serves as a security guard and migrants have become adept at evading the law. He has invested in computers, printers and empty DVDs, and makes about 2000 movies a week. The mobile phone and the computer are providing him with economic opportunities (de Bruijn *et al.* 2009).

Drawing from the above excerpts, this chapter focuses on migrants' economic activities, particularly pitching and other forms of income-generating activities. It will show how, with the rapid development of information technology, migrants are now able to do business differently; and have increased opportunities and improved and reduced mobility in some cases, with the mobile phone doing the mobility aspect for them. Pitching, in other words, is hawking but this is the terminology employed by those involved in it so we will adopt it in this study. It involves buying imitation designer items (sunglasses, football jerseys and belts) and cheap Chinese goods from China Town (a mall) and Somali shops, which they then resell in busy commercial areas and in neighbourhoods far from China Town. They have an extensive network thanks to ICTs, especially the mobile phone that connects them to a multiplicity of networks that play various roles towards achieving their economic goal and independence. This chapter questions how the introduction of ICTs has changed patterns of mobility and trade. To what extent do ICTs, mobility and networks propel the economic activities of the mobile community? What is the impact of ICTs in regulating profits? And what are the social dynamics that underpin the production of the informal economy? While this chapter focuses on the experiences of various entrepreneurs, it also discusses such themes as entrepreneurship, global products, networking and the implication of ICTs in the informal sector in the flow of goods and services, and the subsequent formation of status and hierarchization that emerges in the process due to the wealth accumulated.

---

[2] Such greetings opened the door for me in The Hague after two failed attempts to meet Pinyin informants.

Photo 6.1 Chris at work, pirating DVDs

Photo 6.2 DVDs on sale at Parow flea market

Informal economy

The reason for being involved in the informal economy for most migrants is above all to have an income-generating activity that will allow them to be financially independent and avoid subordinate positions. They also want to work on their own terms and be responsible for the income that accrues from it. By informal economy, I refer here to self-employed economic activities carried out by migrants that are unmeasured/measured or unrecorded/recorded, operating without formal licensing or outside regulatory control (see also Ilhaine & Sherry 2008: 246; Spring & McDade 1998: 10-11) on the one hand, and those that have been allocated recorded stalls or run formal businesses on the other. Pitching here is used synonymously with hawking but is the preferred term adopted by the migrants involved. These terms can be used interchangeably. Pitching to the 'mobiles' is not a survivalist trade but a means of wealth accumulation. For instance, after Peter finished his studies in Cape Town, he sought employment with a company, less for the income than to gain experience that he could use to operate his own business later, while those who are salaried workers enter the job market as people with 'exceptional skills' and they receive better remuneration and benefits. Central to understanding this chapter is communication, networking, especially the use of weak and personal ties, and the ability to strategize, be opportunistic and connect supply and demand (Portes *et al.* 2002). In this regard, the 'argument of embeddedness' (Granovetter 1985; Sheller 2004) addresses the role of social networks in economic coordination. And underlying this embeddedness is the question of trust and information, given that communicational technologies enable new kinds of mobile publics; one that is 'unimagined and unpredictable' (Sheller 2004: 41) but is also that of a mobile hawker and buyer.

These mobile communities attribute their economic activities to luck and the extensive networks they have been able to build. Much as the community seeks to maintain networks, it is a network that continually shifts to include or exclude others depending on what is at stake. In this connection, weak ties are of great significance as they act as a bridge due to their ability to connect several people to the same bridge, and they connect better than strong ties.The dominant narrative (the literature) on the informal economy tends to categorize such activities into formal/informal and legal/illegal, which are farfetched concepts for the community. To them, going to the field is seen as a battlefield where one must tread with care and each day brings its own struggle to outmanoeuvre the other, and to evade the law. The field to them is a place where all the strategizing and scheming of their economic activities is played out. Pitching is a game in which one's earnings depend on how well one plays the game, with the pitchermen doing whatever it takes to be smart. As a result of their mobility, it is difficult for the state to implement order.

The literature on the informal economy is vast (Hart 1973; Thomas 1992; Portes *et al.* 1991; Grosh & Somolekae 1996), while some authors have focused on the idea of economic dualism and marginality (Juma *et al.* 1993). And they consider the informal economy as a 'problem' sector with low productivity, seeing it as a drag on economic growth. Does this assessment reflect the informal economy? Cornia *et al.* (1992) and Hart (1973) see the potentials in this sector in terms of job provision and bringing unaffordable/cheap goods and services to the masses. To Chileshe (1992), the informal economy serves as a training ground for entrepreneurial skills development and management. MacGaffey (1987: 23, 1998) opts for the term 'second economy' to refer to the informal economy. But we differ in that she posits her study not only within the exigencies of the economy but also looks at the patrimonial state's indifference in regulating business, while I focus more on the dynamics of the informal economy, and the formation of status and hierarchization that emerges in the process. In this respect, I attempt to find out what the scale and extensiveness, organization and significance for status formation are. With any semblance of wealth accumulation comes the formation of status and hierarchy informed by success (Chapter 7). Status and hierarchy formation also echo the traditional tendencies in hierarchies in the businesses that migrants engage in, with owning shops and stalls at the top and pitching at the bottom. However, pitching is considered as a rite of passage for almost all male Pinyin migrants before moving on to stalls or owning shops. And this generates a different kind of status that is less focused on the formation of an elite through class, power and wealth accumulation than on new class and status formations informed by how wealth and profits from the vagary of businesses are used to improve the lives of relations (MacGaffey 1987: 14). This is because of the notion of 'wealth in people', wherein, the gains made have to be shared among the family (although this notion is shifting). Others invest in the acquisition of village titles through investing heavily in village development projects (Guyer 2004). Albeit and as noted by Portes (1983 cited in MacGaffey 1987: 23), 'the historical process of the ups and downs of class struggles unifies the phenomenon of "informal economy" in both advanced and peripheral economies'. This economy, I contend, thrives because migrants buy cheap imitation designer goods from China that are supposedly prohibited in Cape Town and sell them at a higher price. Such profits have been a means of capital accumulation for most migrants and this can lead to rivalry. The hallmark of the success of these migrants is their ability to contribute to the various *njangis* that are part of these communities.

Focusing on cross-border traders, street traders and those involved in small, medium and microenterprises (SMMEs) in South Africa, Peberdy & Rogerson (2000: 21-22) show that African migrants and immigrants are connected to strong formal and informal transnational networks of trade, entrepreneurship and migration, contrary to the xenophobic notion of 'stealing jobs', while they also have access to capital. The rise of informal trade, according to Holness *et al.* (1999), is due to the government's liberal and tolerant approach towards street trading. In spite of this liberalization, sporadic checks of contraband goods (especially pirated DVDs) are common.

Three basic tenets have been used in an attempt to understand the informal economy. These include dualist, structuralist and legalistic views. The dualists view the informal economy as being significantly distinctive from the formal sector, involving the marginal poor and providing income or a safety net for the poor (ILO 1972).They maintain that the informal sector exists (or persists) because economic growth or industrial development has failed to absorb those who work in the informal sector. While this approach is primarily concerned with the employment rate of the nation state, it fails to answer the question about those with a lack of human capital, given that employment is prioritized. The educated come first before moving down the scale to the less well-educated and the uneducated. The approach also fails to understand the global effects of an increasing labour force with diminishing employment opportunities. And given the benefits in the informal economy, some people have found a niche for themselves and would rarely choose to move into the formal economy. What this suggests is that there is bound to be unemployment and these groups of persons (the unemployed) work their way into the informal economy. However, most informants involved in the informal economy are less keen to seek employment.

For their part, the structuralists see the informal economy as a subordinate to the formal economy. However much this may be the case, we should not lose sight of the fact that there are those who have found income-generating opportunities in the informal economy and are unwilling to move into the formal economy. They are given tax breaks as an incentive to stay in the informal economy forever. They maintain that the formal sector is more preoccupied in its vested interests and thus minimizes costs by reducing labour (Castells & Portes 1991). Lastly, the legalists view the informal sector as a rational response by micro-entrepreneurs to over-regulation by government bureaucracies (de Soto 1990). They argue that those who run businesses informally do so to reduce their costs and increase their wealth. This partly accounts for the involvement in the informal economy but does not satisfactorily explain why most migrants start in this sector before moving into the formal sector where they stay, sometimes permanently. This is true for Pinyin and Mankon migrants. The informal economy provides migrants with easy entry into SMMEs and access to finances. While some may prefer this to the formal economy because of bureaucratic speed breaks by governments and/or have found a niche in an income-generating activity, others simply do not have the resources to start on a larger scale like the Pinyin and Mankon migrants. Alternatively, the informal economy may have developed as a result of government restructuring policies that tend to acknowledge the informal sector and make provision for them as a way of easing unemployment (Holness *et al.* 1999). For instance, a lot of concessions have been given to the informal economic sector in South Africa to ease the rising unemployment rate. Although these approaches do not wholly explain the informal economy, the focus in this chapter is on the sections that shed light on the way it is perceived.

These groups of 'mobiles' are aware of the ways of operating in the informal sector, especially those who operate from stalls in areas designated by the municipality where most navigate between formal/informal and legal/illegal. Those in formal trade often employ people to front for them, thus oscillating between the formal and the informal, while those in the informal navigate between the informal and the illegal. However, they do not see their activities as illegal unless they have been caught by law enforcement officers. Illegality here includes the sale of pirated DVDs and fake designer items. Even those with stalls are involved in selling DVDs. This makes the line between formal/informal and legal/illegal rather blurred.

Encounters

Information gathered on this chapter yields to the call of multi-sited research by Marcus (1995) and was obtained through semi-structured interviews, participant observation, life histories as well as following pitchers in their different fields in Epping, Bellville and Mitchelsplein or sitting with them at their stalls and restaurants in Parow, Maitland and Mitchelsplein on repeated visits. Often I participated by selling and sitting in for traders when they went to buy additional stock. At such moments a life-history toolkit comes in handy. Some of their stories corroborated those of relations in the home country and the mobile lives they lived in the course of business. Ben, for instance, grew with his grandfather who was specialized in buying and selling goats, travelling to Wum, Nkambe and Mesaje to buy and selling in Bamenda and the West Region. Similarly when Ben first arrived in South Africa, he spent a few years in Johannesburg where he pitched before moving to Cape Town and continued to pitch and to study. Although he had a salaried job, he still supplemented it with income from trade and did TV and radio repairs. A longitudinal study helped me to focus on the different locations (Pinyin/Mankon, Cape Town and the Netherlands) and I explored not only their interconnections but their differences too. In this respect, participant observation and informal conversations were important as they focused on the functioning of their trading activities, how negotiations and multiple networks were carried out amongst migrants and how they went about their daily pitching. Following David around during his pitching activities, I saw how it worked. It was a particularly slow day and he had not made any reasonable sales so he decided to go by car to meet some of his clients at the fruit market and the filling station to supply them with goods on credit and collect money owed to him. They all admitted they had made a bumper profit during the World Cup. He lured them with free sunglasses for anyone who would take goods worth R 250 or more. The timing of this is significant as it was getting close to the end of the month and David knew it was only a matter of days and he would get his money back. And the strategy of giving away sunglasses worth R 25 paid off as he was able to move goods worth over R 1000. His actions and way of relating to clients was strategic, confirming the fact that the informal economy thrives on strategies and strategizing because 'strategies are the ongoing result of the interaction between the disposition of the habitus and the constraints and possibilities which are the reality of any given social field' (Jenkins 1992: 83). A field here, to add to the notion of Levitt & Glick Schiller (2004), would refer to 'a field of struggles, in which agents strategies are concerned with the preservation or improvement of their positions with respect to the defining capital of the field' (*Ibid.*: 85).

Collecting life histories was an integral part of the research method. This gave insight into migrants' itineraries and business dealings to ascertain the extent to which their present activities are informed by the past and their family history of mobility. For instance, Nixon and Joyce, from the accounts they gave, were involved in informal business back home for a long time so once in Cape Town it was logical for them to continue in a sector they had already mastered. In the sections below, I detail the various (trading) activities, networking and lifestyle that form part of the migrants' being with regard to the informal economy but first we will take a closer look at the gender demographics of the traders.

The role of women in the informal economy

In Pinyin and Mankon mobile communities, men tend to be engaged in pitching while the women operate stalls and are generally sedentary in their business activities.

With a few exceptions, Pinyin women in these communities study and manage their stalls, while their Mankon counterparts assist in the family business (usually a food shop) or run their own businesses (salons or restaurants). Those who have completed their studies may seek employment as sales assistants in supermarkets while continuing with their studies for a higher degree on a part-time basis. The activities in this sector are engendered in terms of the business activities engaged in, decision-making and their degree of networking. All but three Pinyin women are unmarried, while those who have stalls run their business with their spouses. The implication of this is that the major decision-making does not solely rest on the women but, nonetheless, they play an important role as their knowledge of supply and demand is vital in keeping the business afloat. It is this knowledge that spouses use to replenish stock or buy what is in demand. As a result of their sedentary business, the women's networks are more inward-looking as opposed to those of the men who have a wider network circle and maybe involved in multiple businesses. Not all the men are pitchers as some own stalls and shops and they employ saleswomen to work for them so they can maintain their mobility.

By the same token, some of the men who have well-paid jobs have focused on the education of their spouses as the way forward rather than engaging in petty trade. Joyce and Bianca were actively engaged in business back home and mobility was an integral part of the business as it entailed going to weekly village markets to buy foodstuffs to sell in the city.

There is also the cultural/ethnic enterprise business that caters for migrants who want to acquire and consume cultural goods from their home country. Similar to the study by Landolt *et al.* (1999: 296) that was carried out in an El Salvadoran migrant community in Los Angeles, ethnic enterprises are either small formal or informal retail shops that import foodstuffs from Cameroon and (mobile) restaurants that cook exclusively Cameroonian meals. These retail shops, (mobile) restaurants and pitching tend to replicate life back in Cameroon, with little attention to style.

Pitching: An overview of economic activities in Cape Town

Pinyin and Mankon migrants are involved in a variety of business activities that they have either learnt after arriving in South Africa or they continue in the same business activities they were engaged in back home. This is especially the case with Mankon migrants. As noted by Hansen (2004) in her study of the informal economy in Lusaka, access to and control over public space is often contested by traders and is at the centre of ensuing conflicts. Such conflicts and contestations are not exclusive to Lusaka, as it is evident in Cape Town. Given the categorization of their business according to the two ethnic groups, the Pinyin are largely involved in (business) pitching and/or have stalls and shops, while Mankon migrants have continued with the trade they used to do back home such as motor mechanics, welding and building. They have equally grouped themselves in distinct neighbourhoods to allow an easy flow of information and interaction (Chapter 5).

Although some Pinyin men operate from allocated spaces provided by the municipal council, others (pitchermen) do not and it is these groups of persons and pitchermen that are constantly evicted from selling points. As pitchermen, their lives are often precarious, given that they are at times faced with the seizure of their goods and are vulnerable to arrests when raids are made. Pitching (hawking) involves trading from one neighbourhood to another with goods stuffed in a huge backpack. Some have carved out a niche for themselves at busy business centres where they pack their cars with goods and are ready to take off if there is any danger of their goods being seized by the police.

Photo 6.3
Pitcher-man† ready to go to the field

Photo 6.4
A pitcher-man in the field

Photo 6.5
The car as a mobile shop

With neighbourhoods linked by rail, it is possible for traders to cover about three a day but this involves a lot of trekking around. Pinyin migrants usually move in groups, especially when they are going to areas considered unsafe or if they are hawking in a busy business centre like Epping in Cape Town. Migrants liken pitching to a game of cat-and-mouse with law enforcement officers. But it is also a game of strategy and strategizing, positioning oneself in the face of stiff competition from peers. Simplifying Bourdieu's game theory, Calhoun (2003: 14-20) explains the phenomenon of migrants by comparing social life to a game by serious, deeply committed athletes who have to reconcile the challenge of playing by the rules while at the same time being creative and innovative enough to win or at least to avoid defeat. 'Every field of social participation demand(s) for those who enter it a kind of preconscious adherence to its way of working' (*Ibid.*: 20) and 'participation in social games is not merely a conscious choice. It is something we do prereflectively' (*Ibid.*: 19). The Pinyins are adept at this and other ethnic groups have still to match their shrewdness. Newly arrived migrants are initiated into pitching by relations or by ethnic enclave relationships, and are forewarned of raids and seizures. An older migrant is assigned to show the newly arrived the market and teach them the intricacies of business. If the new arrival has relations and friends, they contribute money to give him/her a reasonable starting capital. Edwin attests to the help he was given by his sister and fellow Pinyin pitchermen who helped him with finances, buying items and teaching him how to attract clients:

> We went to Retreat, that was my first day. We got off the train and sat down and unzipped the bag, removed all the wallets, belts, everything and prepared them and he gave me some. He told me that that was his own side of the road and this was mine. I said what am I going to be doing? He said you'll just be moving and showing these things to customers that you are selling belts. You're selling car chargers. I say what I must say (sic), he said tell them … their name is 'boss' when you see any of them you say 'boss'. I say so I should call them boss he said yes. He said when you see them you say boss I've got car chargers for sale. I've got phone batteries for sale; I've got wallets and belts. I said then how would I know the prices. He told me that if you sell one of these items you will have 75% profit so from R 30 – R 70. It depends on the customer. He told me we bought this belt for R 10, chargers R 7, so you can sell the belt for R 30 – R 70. I said what about a wallet? He said the same R 30 – R 70. There are some of the customers if you even ask them R 120 they will be prepared to pay. He also told me that if they ask you, is it genuine leather, you should say yes because some of them don't even know the difference between genuine leather and non-genuine leather. So on my first day, I sold one belt for R70 and I sold a charger for R 30. I sold a wallet for R 30 and that was R 130. The next day again we went, I came back with R 80 … By the weekend, I had realized R 850.[3]

Such assistance and education from Pinyin migrants is often given only to one of their own, and this excludes non-members from their circles. Accounts by Pinyin migrants attest to this and if these migrants insist on going with the non-Pinyin hawkers, they are taken to places where they could be easily apprehended by the police. Hence with no knowledge and rules of the game, they back out after a couple of seizures. While most have come to South Africa as economic migrants, others came as students and double as students and pitchermen, as pitching is the main source of income-generating activity for them to pay for their tuition and accommodation. Through this active disengagement, the agencies of the traders come to the fore either as individual agency or collective agency (Lindell 2010), which emphasizes salient contestation and the circumventing

[3] Interview with Edwin, Cape Town: 27/03/2011.

and undermining of the authority of the state through collective support and initiatives (Scot 1985).

Even though some persons seem to have branched out into the formal economy, the tendency is still one of a seesaw. Such oscillation as MacGaffey & Bazenguissa-Ganga (2000), MacGaffey (1987, 1998), Thomas (1992) and Spring & McDade (1998) have shown involves criminal activities. These traits are also visible among these groups and the nature of their business makes it hard to discern between the formal and informal or the legal and illegal because of the constant seesawing. Yet another group of skilled migrants, particularly those who have acquired education and formal employment are equally involved in business and buying property and renting it out to Pinyin folk. This partly explains the ethnic concentration of these communities in one locality.The choice of residence of the mobile community in Cape Town is not accidental. As shown in Chapter 5, they have consciously chosen to live in a concentrated area, with Pinyin mostly found in Parow and Bellville, and Mankon in Maitland. The formation of ethnic enclaves in specific localities helps circulate information faster and they can offer each other immediate assistance. In the process of fending for themselves and expanding their businesses, they are reproducing class status amongst themselves because the need to achieve status and success is the motivating factor in their entrepreneurial expression.

Conversely, the Mankon community is still conservative in trade and most have continued with the various trade activities they were engaged in back home. Before settling into these activities, they did some pitching to raise funds to set up their businesses. Their early contact with the colonizers and later the Nigerian administrators gave the Mankon advantages over their Pinyin counterparts who were mainly traders or cattle herders. Some were able to go to school and/or learn their current trade while others have entered into new fields, such as collecting and selling recycled items to factories. The Mankon are not as geographically disadvantaged as the Pinyin and, through its association, support newly arrived members with starting capital. At the time of this study, only three Mankon migrants were involved in pitching and understandably so, because one has a Pinyin wife who introduced him to her network of Pinyin pitchermen, and he in turn brought in the two others. One of them had gone back home long ago as his goal had been to make money, go home and relaunch his ailing business.

The situation is quite different in the two communities in the Netherlands where some have sought semi-skilled and unskilled jobs in firms. This could also be attributed to their relationship that is marked by a high degree of individualism. Intimate relationships are mostly formed on the basis of kin relations or friendship. However one of them is involved in informal business and ships goods from the Netherlands to Cameroon for a friend to sell. The difficulty of establishing an informal business in the Netherlands has led some to criminal activities, with one of the Pinyin migrants resorting to extortion and money laundering. He was in fact serving a jail term between September and December 2010 when I was conducting research in the Netherlands.

The question then is what accounts for such distinctive business trends and why they manifest such diametrically opposed features. Why are the Pinyin predisposed to venturing into unchartered territories with respect to business activities while those from Mankon are not? And how do the mobile phone and ICTs affect these differences of scale?

Most migrants attest to the fact that they grew up seeing parents and family members making the rounds from one bush market to another so taking up pitching in Cape Town is not new, and going to Grahamstown for a week to sell things at the Art Festival (Indaba) as well as Sunday flea markets by the beach is seen as normal. Their adeptness at pitching stems from their Grassfields experience of bush trade. Current migrants have internalized the mental and physical agility required and have taken it to another level thanks to the new technologies that allow them to extend their network beyond the bush trade, just as in the past. The mobile phone is thus inextricably linked to the informal economy given the vagaries of the role it plays, either impacting positively by expanding networking possibilities and actual trade, or negatively when friends, due to rivalry, keep information to themselves in a bid to have a monopoly over certain goods. It should be noted that increased activities in pitching and street vending area are results of the South African government's tolerant and liberal approach to street vending (Holness *et al.* 1999: 285-86).

One correlation between the Cape Town pitchermen and their parents who practised long-distant trade is the contestation of borders and their lack of regard for institutions. Chris's remark that he is merely gathering the crumbs with regard to pirating DVDs in a way is contesting institutional and state hegemony and control that seeks to frustrate his attempts to accumulate wealth. Similarly, administrative records in Bamenda Province specifying the reason for the Pinyin trader being jailed state failure to pay tax on imported goods, while in Cape Town arrests are frequent for either selling stolen or contraband goods.

The availability of mobile technologies such as wireless/smart phones with Internet connectivity, computers and the Internet has dramatically helped to boost business and lessen mobility, while enhancing connectivity and efficiency (Ilahiane & Sherry 2008), and made information about market supply and demand faster. The contrast between earlier migrants (Joe, CJ & Jake) and the likes of Chris reflects the differences.

Whilst Chris & Jake's principal source of income is through the informal economy, what sets them apart is the introduction of communication technologies that determine the nature of Chris' business, and the absence of it in Jake's case. The latter's could be traced from a historical perspective characterized by physical mobility given communication technology was still in its infancy and not widely available to all. Jake sums up his business as follows:

> I started business way back in my secondary school days through to university. In the holidays I would buy potatoes in bags and take them to Douala to sell at Marché Sandaga. I was constantly moving between the village and Douala because back then there were no phones to communicate back home for them to send me more bags of potatoes, so I had to go back and get some more and spices and return to Douala. Even when I moved to Cape Town, it was the logical thing to do … pitching, but I did a lot of things in my first two years before settling on pitching full time ... working for Joe for seven months, working in a security company for about five months, doing car parking for a couple of months and then finishing the last part of my period before my studies doing pitching then while studying. I would go pitch in the afternoon and on Saturdays and sometimes on Sundays. Also, when someone was travelling to Cameroon, I would buy materials for men's trousers and suits and send it to my brother in Mutengene - Cameroon (who is now here); he would sell and send back the money. You see... this was a kind of foundation phase for me.[4]

Drawing from informants, one major characteristics of the informal economy is its diversity in activities and situations (Castells & Portes 1989).

The profits made from pitching are quite significant to migrants and families back home but migrants rarely tell the family the nature of the business they are engaged in. Often, the family knows they are in business, but what kind of business? While some have confided to particular individuals in the family, others have lied about what they do, such as Edmund who informs the family that he is a secretary in a company given that the family is aware of his limited computer knowledge. The informal economy is, however, a source of capital accumulation for these migrants.

By and large, migrants start from humble beginnings and move up into the formal economy and, as a result of financial success, gain status. For instance, Joe from Pinyin had humble beginnings but has moved from the informal economy into the formal economy, or may navigate between the formal and informal due to its blurry nature, and has employed workers who front for him. His business is transnational as he initiated his first visit to China and subsequently ordered goods from there. However, life at the start was a herculean task.

> Life was very difficult when I got here, at the refugee camp I had to save up to ... start business ... it took me three weeks to save R40 and I bought my first carton of chips which took me three days to sell. From there I understood the market and within two months I was able to buy a variety of flavours. By the third month, I went to train as a security guard. Upon completion I worked for four months and was able to bring my wife just from selling chips and what I had saved from my salary. I had also employed a sissy (sales lady) to stay put

---

4 Interview, Cape Town; 11/08/2010.

> while I hawk during the day and I was a guard at night. From there I moved to the parade ground where I had four stands.[5]

Accordingly, Joe left Cameroon as a 'result of frustration with the system'. As a university graduate and having completed his training as a lawyer, he was still unable to practice and get a job, and his business ran aground. Migration was the obvious thing for him as he puts it. However, being the first Pinyin migrant in Cape Town and a resident at the refugee centre meant that he had to be self-reliant. Joe's entry into the field of snack hawking was not without its repercussions. Going into this field without any prior knowledge meant that it took longer to sell his first box of chips than it would normally have had he mastered the rules of this trade. But it was a learning process and his initiation into this institution. He quickly learnt that he needed a variety in order to make meaningful sales.

Mobility and social networking in the informal economy

Historical/archival sources and studies on mobility and trade (MacGaffey & Bazenguissa-Ganga 2000) show how traders make enormous use of networks as well as kinship ties at every step of their trade. In Cape Town, ethnic and social relationships facilitate this process for newly arrived migrants. The latter rely on established migrants to help them through the informal economy trade, as seen through Edwin. Some of these relationships date from childhood, with others are recently forged due to ethnic enclaves, while others are due to interacting with migrant population or migrant cohorts (those who travelled together or came in about the same period often form strong bonds and share a room). Over and above some of these ties/networks are re-established kin and family relationship. Nonetheless, success is heavily dependent on how much social capital a person has accumulated, given one's level of social capital is informed by the level of accumulated social network (Bourdieu 1986; see also Chapter 5). In this respect and because of the context in which migrants find themselves, I will also focus on how their social network is a springboard to understanding the way migrants' businesses are premised on forging new relationships and maintaining existing ones. In line with Smith's (2001: 141) argument, these relationships are 'transgressive and affiliative, freely formed and yet socially produced', he maintains that it is by looking at networks in this light that we will understand how migrants organize their activities be they economic, or socio-cultural as this will give an insight into how they relate with others. These personal ties and social and ethnic networks operate on the unwritten ethics of the reciprocal exchange of favours/gifts that obliges receivers of favours to return them although they do not

5 Interview with Joe, Cape Town: 24/07/2011.

necessarily need to be returned exactly (MacGaffey & Bazenguissa-Ganga 2000: 108-9). As explained by Mitchell (1969: 20), 'actors and observer in any social situation are able to understand the behaviour of those involved because it is accorded a meaning in terms of the norms, beliefs and values which they associate with this behaviour'.

Such institutional ethics lay the basis for norms and codes of conduct that impact on relationships. Part of this ethos is the shared solidarity that ethnic enclaves enjoy and, as we have seen, it is paramount to economic success (see also MacGaffey & Bazenguissa-Ganga 2000). Ben's case sheds light on this. When he arrived in Cape Town in 2002, he was robbed and a fellow Pinyin offered him accommodation and some starting capital. As he was beginning to settle in, his mother passed away and, devastated, he decided to go back home. The community rallied around him and talked him out of going permanently. They contributed money and gave him to send home to assist in the funeral. These gestures stayed with Ben forever and when he got married five years later, he asked his wife to prepare a meal so they could go and formally recognize the assistance he was given by the friend who helped him when he was robbed. Equally, he put his energies into seeing that the PIFAM became a great association under his leadership. Such measures constitute a framework for people to be continuously indebted to each other, and to an extent form the basis of social cohesion. Such long-term indebtedness and obligation has, according to Lomnitz (1971: 94), to be 'kept in a savings account of services to be reciprocated'. Personal ties are thus hinged on the expectations of reciprocity and trust, and are defined by the institutional context in which these relationships are embedded. In addition, amongst the pitchermen and those operating stalls, when a member buys an item that is in high demand they phone those in their close circle to pass on the information.

With regard to trade transactions amongst the Pinyin and the Mankon migrants, personal, ethnic and kinship ties are by far the most tapped into.Although these networks are to a large extent the oil that lubricates the circulation of such mobile trade, and trade in general, they are however constantly shifting and redefining themselves and 'mobilizing networks according to circumstances, as the trade changes direction to respond to new market opportunities' (Macgaffey & Bazenguissa-Ganga 2000: 16). Such ties are not limited *per se* to kinship but stretch across various frontiers such as religion, ethnicity and friendship in the diaspora community. The Pinyin community are prone to such constellations of shifting networks because of the tendency to flock to a new business each time that one of them ventures into a business and it proves successful, before thinking it through. At one point, it was to own a shop at Mitchelsplein, and most of them rushed there and opened shops. Some succeeded but others, like Mike, failed dismally. Next was the taxi business and there were a lot of taxis on the road, and it soon became clear that it is a difficult sector to work in as some of them had their vehicles stolen and drivers went for weeks without paying in the agreed amount. Currently the new trend is either to have a tuck shop or a furniture shop in the township, and again a host of them have branched into this area. These networks thus flourish according to the needs at hand, prompting Smith (2007: 28) to conclude that they are dynamic in relation to migrants' 'priorities, obligation and change of interest' (see also de Weerdt 2002). It is critical to understand that these networks are seamed and kept alive by the new communication technologies that migrants have appropriated to mitigate the challenges of mobility, thus enhancing virtual mobility over physical mobility. The case of Chris and his cousin Nelson will elucidate migrants' networks, mobility and the use of communications technology.

Photo 6.6 Phoning to inform fellow traders about the arrival of grey knitted caps at Parow

Nelson, until his brief relocation to Cape Town, was resident in Johannesburg where he owns a small computer shop. In order to focus on his education, he moved to Cape Town. Prior to his arrival, Chris phoned him and asked if he could bring printers and a CPU that duplicates 7 DVDs at a go in order to save time. Once in Cape Town, he moved in with Chris and both of them embarked on pirating. While Chris is out distributing and selling, Nelson is at home pirating more. Nelson soon realized that the business was negatively impacting on his studies, he phoned a maternal cousin to find out whether he could move in with him. This he did and allowed Chris to carry on with the business.[6]

Nelson's story has touched on chain migration, kinship relationships and ethnic enclaves that together form the catalyst of migrants' insertion into the community. His story provides a case study that reflects on how migrants negotiate and navigate into the informal economy. By the same token, it shows how the enterprise is sustained and maintained through networks and the dependency on others. Similarly, Bianca and Rachael run small shops with tropical food (Cameroonian) respectively. They both independently buy their stuff from Johannesburg without necessarily travelling there. For Bianca, her cousin is resident in Johannesburg and was offered accommodation by her when arriving in South Africa before relocating to Cape Town. Hence she has a contact that is useful to her business. They communicate frequently. She phones with a list of orders and pays the money into the cousin's bank account via ATM banking, and the latter will buy the things and put them on the train or bus depending on the quantity, and phones her with details about collecting the items. In the same vein, Rachael, with no family network in Johannesburg, made a maiden trip there and all subsequent supplies have been coordinated via mobile phone and emails. On other occasions, her spouse makes the trip to Johannesburg. To Rachael, her phone is 'the one thing that she cannot be without'. All her financial transactions with regard to payments are done via the cell phone or Internet banking. They are equally called from Johannesburg with information about the arrival of goods. Besides, amongst the pitchermen and those operating stalls, when a member buys an item that is in high demand s/he phones others to pass on the information.

While others do well in pitching, some have sought an alternative business, like Ron†[7] who ran a small grocery shop (see Chapter 7). Although his business was very community oriented as he was in close touch with the community in the area, it was also one of those that was more than meets the eye. He explained how he ran his business:

> I have a shop next to the road; I meet so many people here a day ... I've been here for years and I've been surviving mainly because of this shop and having some capital that you can buy certain goods and sell; someone could just come and is selling a phone, I'll buy it for say R 300 and sell it for R 600 and I get another R 300 on it, or somebody can come with a laptop. I've been buying things like laptops also, that's how I've been surviving day in, day out. I've sat here, and some children come (sic) here with dollars, they don't know what money is that, they tell me, 'hey brother do you know this money'. I say yeah I know we use it in my country, they say we need something for this money, and what can you give us for this mon-

---

[6] I followed their activities for three successive days: 8-10/07/2011, and other days when I visited Parow. It was Chris that I accompanied to the refugee centre on 08//7/2011. See Chapter 3.

[7] Ron died in a ghastly accident in Cape Town in September 2011, ten weeks after I interviewed him.

> ey? Then me too I look at it and say OK this money I can give you like R 2000, but I went to the bank to have R 15,000 for that dollar (sic). So that's how one has been surviving.[8]

Ron's shop, we could ascertain, was a façade of an informal economy. This sheds light on similar businesses run by his peers that oscillate between the legal and the illegal, but the line between them is blurred. His networks extended beyond ethnic enclaves to that of the host society. And it is host-society networks that make his business thrive.

The significance of communication with regard to business transactions has been well articulated by Mallard (2004) while hawking in South Africa in the past with all the hurdles involved has also been well documented by Holness *et al.* (1999) and Rogerson (1989, 1990), but they stopped short of showing how a personal network weaves together this form of business; something Mitchel (1969 cited in MacGaffey & Bazenguissa-Ganga 2000: 12) has usefully expounded on. Mitchel sees a personal network as a 'set of linkages which exist simultaneously on the basis of specific interests and persist beyond the duration of a particular transaction' (*Ibid.*: 12). Similarly Castells & Portes (1991: 11) refer to such networks as 'horizontal networks'. Networks as I have shown above resonate with Mitchel and Castells & Portes's descriptions. Underpinning this relationship is one of indebtedness, obligation and mutual dependency that is guided by information flows, one that is linked by solidarity and/or jealousy and in-house fighting. These similarities and differences, however, still manage to keep the group together and may be the fundamentals needed in the smooth running of such an informal business that most operate in Cape Town. Trade networks are individually constructed but not exclusively limited to individuals and are activated to a specific end, although they involve individuals from a common ethnic background.

Much as the availability and affordability of communication technological gadgets have opened business avenues for migrants, it has led to new forms of circumventing hegemony and control of the rules that have increased sales of particular goods. Often, the ability to circumnavigate the law rests on the networks that have been set up in these sales areas. Following Scot's (1985) portrayal of non-violent resistance amongst the peasants and the poor through their use of 'weapons of the weak', we can see how pitchermen are able to resist the long arm of the law. Their resistance is re-enforced by their use of mobile phones that send information faster and alerts others, especially when the law enforcement team has come to raid those selling pirated DVDs. Such extensive reliance on networks is inherently present in the bush trade practised by Pinyin traders during and after the colonial period and shows how bush traders would travel in

[8] Interview, Cape Town; 12/07/2011, Ron died ten weeks after this interview.

groups using scouts,[9] and also making use of kinship relations and ethnic enclaves on the journey. Today communication technologies have changed all that the current migrants experience good communication patterns and can alert each other when necessary.

The success of being a pitcherman depends by and large on the hawkers' mobility and their dependence on trains for mobility, which can at times slow down their business. At the start of this research (2010), very few of my informants had cars but all acknowledged the importance of owning one for their business. By August 2011, almost all the pitchermen in Epping had them and they could therefore supply customers with goods immediately, depending on the amount of profit there was to be made. Alternatively, they are able to get scarce goods once they receive word of their availability in the market. As seen in the photo, their cars have become mobile shops and this allows them to store the bulk of their goods while carrying only a few with them in case of police seizures. They still have the bulk of their goods, unlike in the past when they carried items in backpacks that were easily confiscated.

Friedman (1990: 327) argues that 'goods constitute self-hood: and the practice of identity encompasses the practice of consumption' and this study concurs with this statement. While DVDs are generally aimed at the South African community, the migrant population is more interested in Nigerian movies. Imported food and restaurants catering primarily for the Cameroonian public, and West Africans in general, attest to the fact that consumption is cultural in character, as can be seen by the goods and services provided by the Pinyin and Mankon migrants in large part. Having seen the need to import food from home, Paul is now established as an importer of local food from Cameroon. Bianca and Rachael buy in Johannesburg and sell it in Cape Town where they have established networks so they do not need to travel themselves. Others have taken to operating small mobile restaurants. Prisca from Mankon moved from Gabon to Cape Town and continued with the same business she was in in Gabon and Cameroon prior to migrating to Gabon, namely a restaurant; where she prepares Cameroonian dishes. At weekends she serves special meals, such as *eru*, *koki* (black-eyed-bean pudding) and *achu* (the Pinyin and Mankon staple meal made of mashed cocoyam and yellow sauce). When business is slow, she phones her customers to tell them about the menu of the day. Similarly, Patricia from Pinyin is involved in multiple businesses as a seamstress and does most of the sewing for her community. She also operates a weekend restaurant; selling *achu* on Sundays to the Pinyin community in Parow and during the week, people call to make reservations for the weekend.

---

9 File no: cb, 1955. Annual Report Bamenda Division: Customs, (1955: 4).

However, not all the goods and services provided reflect this cultural dimension. Those involved in selling drugs, according to one of my informants, are drawn into it due to their circumstances. He claimed to have become involved as a result of financial hardship and the huge profits involved he is now finding it hard to extricate himself.

In spite of all the diverse businesses, one thing is certain: the ultimate goal in these communities is to be successful and to be seen back home and among their peers as the 'authentic bush faller'who has accumulated material wealth and readily shares it with kin and kith as the notion of sharing is very important.

The notion of success and material wealth

While the notion of success is measured in a multitude of ways by migrants, material wealth and accumulation are the most defining forms of success (Rowlands 1988). The entities that define wealth are in themselves meaningless if looked at in isolation, but instead we should view them in relation to society or in relation to other things (Rowlands 2005; Miller 2010). Expanding on this point further, Miller (2010: 53) notes that 'by learning to interact with a whole slew of different material cultures, an individual grows up assuming the norms that we call culture'. The societal norms that dictate what could be regarded as success are in turn the things that the migrant community live and work for in order to be part of the category of successful migrants. It is from this premise that this section seeks to understand success and its significance in relation to material wealth accumulation. Human capital (knowledge wealth) is often considered in relationship to material wealth and seen as a prerequisite to material wealth given that most of those in Cape Town are engaged in the informal economy as a means of generating income to pay for their studies. Informed by social ranking, success is equally defined by the need for persons to continually demonstrate, cogently so, who they are in relation to others (Weiner 1985), thus the constant quest to accumulate to prove oneself or to spend to prove oneself. By the same token, they inform the system of order intrinsic to the inculcation of habitus (Bourdieu 1977) and, quite often, there is a link between migrants and flashy and huge houses built in the home village.

Successful migrants among both communities in Cape Town are considered to have succeeded in business, own houses in Cape Town and back home and have human capital. The pitchermen are rated according to how much they contribute towards *njangi* and how disciplined they are with respect to being frugal and savings.While success is attributed to having *ifinti* (Rowlands 1992: 128) or the consumption of western goods and technologies (Rowlands 1988), migrants strongly believe their success is the result of blessings from parents prior to departure or from ancestral shrines and tombs by those representing death (Rowlands 1992:

122-123). Parents or successors often smear would-be migrants with camwood and drink from ancestral cups made from cow horns. In the process, the performer invokes the ancestors through incantation to look after his/her child and ensure that whatever s/he sets out to do will meet with success. In conjunction with this, libation and the transmission of breath (the breath of men) to the migrant is known in Mankon as *fogho-njzwi*.Their success is accentuated by hard work and the self-abnegation of worldly pleasures for fear of failure. Mike, for instance, attributes his success to the blessings he received from his father prior to his departure (Pelican *et al.* 2008: 121).

Mike's story

> Mike arrived in Cape Town in 2003, and has not been back to Cameroon since, but has regular communication with his family. Prior to his departure for South Africa, his father called him into his bedroom to give him his blessings and advice. Camwood was rubbed into his feet, hands and forehead; all signifying that the road will be opened for him as he sets out into the unknown, whatever he does to earn a leaving will do well and he will succeed in his studies and in whatever he plans to do. His father added that 'those who have been there have succeeded, you too have to succeed. Know that you are going to the farm and whatever it takes to earn clean money like the others, you have to do as well, and know that the rest of the family is counting on you and do not let us down'. True to this blessing, nine months after arriving in Cape Town he was able to bring his brother as well, and in September 2012 he brought in his spouse and younger brother, and plans to bring in two more siblings before the end of the year. His father's blessings have accompanied him. Upon completion of his undergraduate course in electrical engineering, he got a job and is now pursuing an MA programme as well.[10]

Ron† prided himself on having received many blessings from parents and extended family because of his generosity. Until his death, he sent money to his mother every year to buy two boxes of soap, two bags of salt and two gallons of palm oil and distribute it to her maternal and paternal aunts and the spouses of uncles. It is by such small gestures of giving that Ron believed he would be rewarded with success because money is not only regarded as a goal for securing wealth but also as a medium for pursuing a relational life (MacGaffey 2005). Going home in December, he believed would bring him more blessings.

> Maybe that year when I come back here (from Cameroon) things will happen just like that because they will give me a lot of blessings. They always say your parents are your God on Earth. So if I go home and spend that New Year with them, the blessing that they will give me when I come back here that year will make it a year of success; things will happen.[11]

Although Ron prided himself of receiving blessings from home due to his generosity, amongst his people, there was a lot of discord with regard to his form of business (he was a drug pusher) and this constantly provoked negative com-

---

10 Interview with Mike, Cape Town: 22/08/2010 and a conversation with his father; Pa Joshua in Pinyin: 10/10/2011.

11 See Footnote 7 of this chapter.

ments from them. And some of them deliberately refused to have his phone number and the only thing they have in common with him is the PIFAM. Similarly, Joe has brought most of his siblings to Cape Town and pays for the education of the others at home. He has houses in Cape Town and recently bought a house back in Cameroon. The duty of looking after his siblings is what he believes his mother would have wanted him to do and he does it for her and she in turn watches over him. If migrants are unsuccessful, it is believed that '*kontri fashion don catch them'* and they are advised by peers to go '*kontri make they wash you*' (go home and perform certain rites to get an ancestral blessing).[12] The case of John in the Netherlands is worth mentioning.

> John came to the Netherlands in 1999 to study. When he finished his studies in 2003, he decided to stay on in the hope of getting a job. In May 2013 he had still not managed to find any work at the level he hoped for (MBA). He still earned the minimum wage working in a linen factory. Although he has a Dutch passport, it has not helped him in any way. For 14 years he has not been able to bring any family member into the Netherlands and is barely able to assist the family back home. He was widowed in 2008 and remains single in his late forties. He has not started any development project back home. John acknowledges that the decision to stay on after his studies has cost him a lot and he has not been as successful as his contemporaries. In June 2013 he finally went back home to start anew. In the eyes of his friends he is an example of a failed bushfaller that has ignored repeated calls to go home and perform *kontri fashion.*[13]

However successful or unsuccessful migrants may be, their origins can be traced back the home village. Increasingly, with the notion of bushfaller (Nyamnjoh 2005, 2011; Pelican *et al.* 2008; Alpes 2011), migrants' success is measured by the level of investments and goods (cars and containers with second-hand goods) that they are able to ship home. Building houses and the acquisition of building plots and/or business investments back home are their top priority, as is participating in village development. Joe from Pinyin epitomizes this success and within the PIFAM and back home he is regarded as a member of the elite. Summing up, Alpes (2011: 12) posits that 'bush has become an important site of economic accumulation and socio-economic mobility'. The quest for success, it can be concluded, is a leitmotif of the avaricious desire for accumulation.

Success, however, when limited to the individual is not success enough because it has to be distributed to the wider kin. Thus we could metaphorically view success as the inlet and outlet of a lake where water comes in and is distributed to the various outlets. Being successful is linked to one's ability to share one's wealth with kin.

The importance of such investments is the value attached to them by both the migrants and those back home as it is believed that 'houses and land do not rot'

---

[12] This is the case of two informants in Cape Town and one in the Netherlands respectively whose peers have suggested that they go home because they are considered to have failed.

[13] Interview with John, The Hague: 29/10/2010 & 22/11/2010.

and they are seen to be the best way of saving money given low levels of trust in the banking system (Niger-Thomas 1995). But contributing substantial amounts to *njangis* and having this money wired in Cameroon for investment as well as investing in business back home is a benchmark for measuring success in the migrant community in Cape Town. By contrast, success in the Netherlands is not as obvious as in Cape Town given the level of individuality, but it is known by close friends and family.

Success in the home country is equally measured in terms of the investments made by migrants. Papa Muluh proudly showed me around his house that is under construction and is largely being funded by his son in Johannesburg. By contrast, Papa Khan commented that his neighbour's children had built a nice house for their father and bought cows to increase his livestock, while lamenting that his son had severed links with the family. Apart from such concrete material investment, going home on a regular basis is viewed by most as a sign of accomplishment and the ability to navigate between home and host country to bring the spoils of hunting is also seen as a sign of success. Ron† summed it up as follows:

> Am going to Cameroon and the first thing I am going to do is to buy a house … if I succeed in going to Cameroon at the end of the year because I am planning to be in Cameroon by New Year and to celebrate New Year with my parents. Then I will know that's an achievement of a lifetime.[14]

Going home is proof of having graduated from being an asylum seeker (*chop adoro or ngunda*) to a documented migrant. Those who have gone hunting and failed to come back are referred to as *nyam mu-mba* (has become an animal in the bush) or '*ee don loss for bush*' (has severed links with the home country). Such persons are regarded as failures. The bush thus has a dual meaning, as a place one can come back from with hunted game or a place where one gets lost and is difficult to trace. Bushfallers often go for the first option, namely they bring back game.

Olsen (2003: 87) calls on us to 'remember things' because understanding consumption patterns that account for success require that we 'ascribe action, goals and power to many more agent than the human actor' (see also Rowlands 2005). It is by so doing that we, I contend, can understand and appreciate the value attached to certain actions whose end result leads to success. One is to marry a wife from home and bring her to Cape Town, i.e. importing a woman, or spousal reunion. The following SMS substantiates this fact. 'Hi makon members, the wife of Pa Xxxx Xxxx is expected to arrive cpt (airport) Saturday the 27th and will be welcome at Pa Xxxx resident at 6:30pm. Pay ur contribution to the chief wipe (sic), attendance is compulsory'. Amongst the Pinyin community, there was a period when they had a boom of reunions with men bringing in wives from

---

[14] See Footnote 7 of this chapter.

home. This act of getting married is viewed not only as a coming of age but as the ability to move beyond fending for oneself and the readiness to start a family and be responsible. And this is a great moment of pride and success for a man, as sub-association members and friends converge on the airport to welcome the wife. In this regard, men are unwilling to marry a woman already based in Cape Town because they want to be seen as 'responsible', having to prepare her documents to travel, spend money on her ticket and have a welcoming reception in her honour.

The accumulation and consumption of goods, as noted by Rowlands (1995), is essential in establishing the creditworthiness of the entrepreneur[15] amongst his peers especially for those who are documented and legal migrants in Cape Town and are thus eligible for bank loans. Charles prides himself on being a successful migrant and, to back up his claim, he lists the achievements that would be expected of him a successful bushfaller according to societal norms. Charles hails from Mankon, is married to a South African and has children. His eldest daughter is in Cameroon with his mother. He is a mechanic (a panel beater), owns a garage and works with his younger brother and another Mankon migrant who he trained while in Cameroon. He tells his story:

Photo 6.7 Charles' house in the Mankon village

[15] There are others who seek loans from the credit unions back home and whose guarantor is their savings.

> I am happy with what I've got. First, I believe that what I've had in this country, most of those who go to that Japan has (sic) not even achieved up to half of what I've achieved here; only the house that I've built in Cameroon, I have age mates and even my elders who don't have that kind of house in Cameroon. So at the end of the day what I've I done in life? I've built my house, and completely furnished it. Also, I have two houses here financed by the banks. They send messages (SMS) that I can take a loan that I am doing well ... they (the banks) say I am doing well, I am not a debtor, I pay my debts, fulfil all my promises, if not I would not have been able to get properties in this country, and their approval because the first house that I bought was R 410,000. I deposited R 100,000 physical cash (sic). And then I spent R 50,000 to transfer sales and renovate etc. so that time I spent about R 150,000 physical cash (sic). The garage is directly in my name, all documentation carries my name. So I've already had a lot of achievements. If I go somewhere and I want this or that, it'll be given to me. They won't judge.[16]

Charles's story illustrates how the acquisition and building of houses has a double meaning. It shows wealth but also a degree of power as houses 'do things on people' (Miller 2010). While they bring status and fame to the owners, they equally open doors for further bank loans.

Successful Pinyin migrants, like Joe, Jake and CJ, are often called upon in meetings to advise members on the need to aspire to grow from humble beginnings (through pitching), given that they were once pitchermen too. Jake and CJ started off as pitcher men and now have white-collar jobs. In other words, the lives of the majority are sustained by hopes of this kind and most are ready to involve themselves in petty trade or the informal economy ranging in scale from the most trivial activities to major business deals. In the midst of all this, what sustains migrants' enterprises and economic activities are the new relationships they forge as well as their existing ones, and the role played by ICTs in facilitating communication and diffusing knowledge and vital information. Such relationships are usually used as leverage for economic gain.

This is by no means to suggest that there are no unsuccessful members within these communities. Of course there are and they are subjected to constant bad mouthing and ridicule by other members. They are often considered not to be creditworthy and cannot be given a loan from the *njangi* and, if they are, the criteria are stiffer than for others. And those who migrate and still expect financial assistance from home do not represent the image of a bushfaller. Like John in The Hague, Nsoh, who migrated to South Africa in 2010, has never sent any remittances home and his father was forced to sell a pig to send him money. They are weary of his excuses and would rather he came home. Unable to hide his disappointment, he retorts that '*pekin them di go bush di send money back and do*

[16] Interview, Cape Town: 23/08/2010.

things them for their parents, ma own go na me di send money, dat ye own bush na which kana one" He goes on to say '*if na so ee go dey make ee come back*'.[17]

Unable to find work and having taken to drinking, his brother in Moscow sent him US$ 1000 to buy a ticket and return home for them to meet and plan another migration exit. Unfortunately, he failed to buy the ticket and did not go home. According to their mother,'*ee take the money chopam, fool we say ticket dear*'[18] (he squandered the money and deceived us by saying that the tickets are expensive). Ndoh earns R 500 a week as a cybercafé assistant, can barely eke out a living and is unable to offer any assistance back home. He is ashamed to go back empty-handed.

Within the PIFAM, there are also those who the community has little regard for, as described here. 'He's (Jones) a good person. He's really good, he's appreciative. He works extremely hard but throws his money away; by having multiple relationship with sissy. And we've all spoken to him until we've given up. Even the family back home I'm sure would have given up by now.' The family back home are deeply disappointed in him. And this led his sister to be critical of the photos he sent back home and remark that 'people are out there toiling and making sense of their lives, all he does is take photos by the beach and send them back home, what time does he have for such work?'[19] This has led some of the elite to distance themselves from him and not have telephone contact with him. A good character is not all that counts in this community. Above all, it is the ability to be frugal and to be seen to be saving a good proportion of one's money. These are the traits that are equated with a good character.

Commodification of relationship

Many authors have documented the economic, social and kin relationships amongst Cameroonian migrants in the host communities and home countries (Nyamnjoh 2005, 2011; Alpes 2011; Ngwa & Ngwa 2006; Fleischer 2007, 2009; Pelican 2008). This section focuses instead on relationships amongst migrants and host communities in the host country to examine the extent to which they are used as leverage for economic and social gain. We should guard against the over-glorification of personal and kinship ties as well as social capital that boosts migrants' business opportunities. Although we have seen how effective these ties can be for business purposes, it does not negate the fact that they may come with strings attached. Both permanent and temporary relationship with others are nowadays crocheted by the wide availability of modern communication technol-

17 Interview with Papa Che, Bamenda: 03/01/2011.

18 Conversation with Jessica (Ndoh's mother), Mankon; 17/06/2012.

19 Interview with Jones's brother-in-law, Buea; 28/06/2010.

ogies that have made it possible for migrants to maintain relationships either at a superficial or a more profound level, i.e. weak and strong ties.

Relations are forged at times according to migrants' needs and momentary gains, as mentioned earlier. But multiple relationships are also forged, sustained and nurtured or deactivated at different periods for economic and social gain. Such commodification does not imply a reduction or a way of cheapening the relationship, as has been pointed out by Comaroff & Comaroff (2009). Instead, according to most migrants, it is the reassurance they want that comes from knowing they have someone to turn to at any given moment or that they have numbers stored in their phones to call if the need arises (Horst & Miller 2006). Nonetheless, some relationships appear to be sustained by monetary values and migrants often view relationships from a vantage point depending on what they stand to gain from them. Usually, it is based on priorities, obligation and a change of interest. This, it would appear, has affected relationships and the way they are used to display wealth could deface the meaning of such a relationship (Nyamnjoh 2011).

To focus on his drug business, Ron† brought in a fellow Pinyin to work at his shop in exchange for him (Ron) contributing to the latter's weekly *njangi*. Money became the interface in their relationship. Bringing in this person to assist in the shop while he was away did not mean they were best friends, but as Ron put it, 'I realized he is unable to contribute his *njangi* and I asked him to assist me in the shop so I could be more mobile and distribute the "stuff" (drugs) to those who call asking for it'.[20]

Migrants and families in the home country have developed ways of reaching one another and following up on relations in the host country, which has prompted these to be reduced to extractive ones. By having Joyce's children with them in the village, her parents-in-law have used this as a bridge, often calling to ask for financial support for the children. Sometimes the children are put on the phone to inform their parents of their needs to the point that it becomes difficult not to yield to their demands.

Having a relationship with South African women is frowned on because of the perceived notion of their extravagant taste and their reckless consumption, like the Beti people in Cameroon. Some Pinyin and Mankon migrants do, however, marry or enter into contractual relationships with South African women. Whatever their reasons for this, it would appear that economic gain seems to be the motivating factor and also their desire to acquire valid residence documents. For instance, when earlier migrants like CJ learnt that by marrying they would be eligible for permanent residence, a lot of them chose this option, not necessarily

---

[20] Interview with Ron, Cape Town; 12/07/2011.

out of love but as a means of legalizing their position in South Africa. As Jake explained:

> ... there are people who are married to South African women, perhaps in most cases not out of love but simply because they want to acquire documents and the women are aware of this ... So they put them through hell because your continued existence in South Africa is dependent on them. So there's no level of mutual respect or whatever because they know you are trapped and they put you through all kinds of emotional, physical and financial traumas. I know people who at this stage cannot even qualify for any kind of credit ... But because the women know that you have a right of entitlement, they will go and take credit and in most cases people sign these marriages in community of property and once a woman takes credit of R 100,000 and she cannot pay, you're compelled to because if she cannot pay and she's blacklisted, it directly affects you and you have to pay or you cannot get any credit. So it's really a terrible thing ... And truly I also have friends who have gotten married to South African women legitimately, out of love, and it's working out very well because it seems it's something that women actually pick up that this person got married to me because of this; and they really try to pay you back.[21]

The commodification of relationships is not exclusive to migrants but both migrants and hosts (women in terms of marriage) exploit the relationship for mutual gain.Clovis's case ties in with Jake's story. Clovis lived on a refugee permit in Cape Town for five years until it was revoked and he was issued with a 'you must go' deportation order. In the middle of this, he made frantic calls to seek advice and was linked up with a Cameroonian who lived in the township. Together they arranged for a lady to stand in as a life partner for him. She was paid R 1500, and was given a cell phone. She was also assisted with air time and her groceries will be paid for if necessary. The importance of the cell phone is twofold: she could reach him whenever the Home Affairs officials conducted random checks to verify whether they were actually cohabiting and he could also reach her to reassure her she had not been forgotten. She could also call for assistance in the event of a major crisis. Thanks to this arrangement, Clovis was given a two-year residence permit. The corollary is that he can focus on his business and even engage in long-term projects as well as the possibility of going home on a visit. Encouraged by the outcome, Clovis and his spouse have agreed to look for another male partner who could sign a similar life-partnership agreement with her.

Joe and Charles from Pinyin and Mankon respectively are legally married to South African women, both of whom have travelled to Cameroon. Charles has chosen for his children to have South African nationality because of the advantages they will have when they enrol at university. But he has rejected it for himself because he feels that, according to African custom, he married his wife and she should be the one to eventually change her nationality.

---

[21] Interview with Jake, Cape Town: 11/08/2010.

Although the commodification of relationships has been given another twist with the arrival of new technologies, the same cannot be said for earlier migrants for who, back then, mobile phones, the Internet and cell-phone banking were non-existent. Whilst such relationships were favoured, people were required to move to access services, such as the transfer of money from the home country to the host country. Migrants thus formed a network of monetary transaction syndicates similar to those set up by the Mouride Muslim Brotherhood in New York (Babou 2002) to send money through business persons travelling to and from the home country, or via other means. For example, parents who had to pay fees for their children studying in Cape Town were asked to pay the money to a designated relation in Cameroon by Jake and the equivalent was paid out in Rand, thus avoiding formal banking charges.

By the same token, calls and SMSs from those who run shops and restaurants are geared towards inviting clients to come whenever new stocks of food arrive from home. It is in the owners' interests to have as many telephone numbers as possible in their phones. These numbers are also useful at times of sales and for announcing the arrival of the latest consignments of foodstuffs. Succinct SMS are sent out to customers in such cases and restaurant owner phone customers when business is slow and they can come and collect food on credit. One such message read as follows: 'Dear costumers (sic), this is to inform you that there is green plantain (*unripe*) at cybella African delicacies. 17 essex st maitland (sic). R 35/kg'.[22]

Discussion and conclusion

Although most of the literature contends that the informal sector is disadvantaged due to a lack of capital to expand businesses, evidence from this study attests that migrants are supported either by ethnic, kinship and/or personal networks with loans to begin or grow their business. Migrants have found a thriving niche in the informal economy and some have chosen to stay in this area because of the profits that can be made. The dominant narrative on the informal economy is one of duality: formal/informal and legal/illegal. While it is easy in some cases to track the graduation from the informal to the formal economy (Castells & Portes 1989), it is difficult to give precise details about such changes regarding the Cape Town community. Having followed some migrants and been a part of their respective communities, it became apparent that this is not a survivalist business but one that they master in the informal sector where the 'current-account balance' is measured by how much one contributes to one's *njangi*. This is in sharp contrast with the premise put forward by Hart (1973: 86) that favours the formal

[22] Message sent on 23/08/2012.

sector over the informal economy in terms of a 'current-account balance of payment' with respect to the Frafras rural-urban migrants in Ghana. While this may be the case for the Frafras, such conceptualization obscures our understanding of the gains that are accrued from the informal economy. Activities seesaw making it difficult to distinguish between the legal and the illegal and to analyze them within the framework of dualism. This is also because even those involved in trade activities consider their actions as a means to material wealth accumulation. Illegality to them is only an issue when they come into contact with the police; otherwise it is business as usual. The issue of duality as such does not fit in the context of these communities because of the blurring of their businesses. This chapter posits migrants as a self-determined group that shrewdly and wittingly circumvent the legal authorities (Nieswand 2011) to further their business activities. In the same vein, it shows the complexities of agency and the determination needed to succeed. Migrants have become astute and adept at games of cat-and-mouse that they play with law enforcement agencies. But also, through these narrations, the theme of 'life as a struggle' crops up as well. Recounting all the challenges that informants go through is more, in my view, to portray them as self-made people rather than as villains. They have 'succeeded' despite all the odds. As Powles (2004: 13) insightfully put it, 'such elaborate emphasis placed on suffering and challenges was not meant to provoke pity but respect, for sacrifice and suffering are morally ennobling for migrants', especially those considered to be successful. Success is partly measured by the migrants' endurance and achievements, but also by the social and cultural things that define this society, such as material wealth accumulation (houses, land) and huge savings from contributing to *njangis*.

From historical analysis of mobility patterns (Chapter 3) and due to their proximity to the Bamileke from the West Region of Cameroon whose entrepreneurial prowess is widely acknowledged[23] (Spring & McDade 1998), it would appear that the Pinyin have acquired some business skills from their Bamileke neighbours.[24] The way Pinyin migrants do business in Cape Town seems to be informed by the skills they have inherited and internalized from the past and a long history of mercantile trading. The ethnic clustering of national groups around particular activities, on the one hand, illustrates the essential characteristics of the people from the ethnic group (especially the Pinyin) but, on the other, reflects the 'skills, knowledge base, education, capitalization and awareness of new market opportunities, as well as the transnational networks of these migrant entrepreneurs' (Peberdy & Rogerson 2000: 28-29).

---

[23] Discussion and interview with Papa Nkeng, Bamenda; 11/01/2011 & 07/10/2011 respectively.

[24] The Pinyin people are commonly referred to as the Bamilekes of the North West.

As I set out to show for Pinyin and Mankon migrants, their involvement in the informal economy in Cape Town has to a large extent been informed by historical factors, particularly by those who witnessed or accompanied parents to weekly village markets as well as the stories told by parents. Enhanced by the velocity and efficiency of new communication technologies and no longer shackled by the weight of conservative trade, current migrants are able to better organize their affairs, and mobility has become targeted or they are mobile while sitting in the same spot. The informal economy and all the actors involved are, in a way, a melting pot of 'unauthentic' designer goods as well as a point of convergence for the global and local with marketing insights and discovery. Looking more closely, the informal sector is used to support various levels of activities (Hart 1973), with legality and illegality at variance. Conscious of what is at stake in pitching, migrants strategically position themselves to evade the police at every stage in their daily business activities. But in their social interaction too, they strategically place themselves in a position where they can vie for different forms of capital.

While the informal and formal economy is, by and large, a way for migrants to adapt in the host country, profit heralds their success both at home and in the host country, with most striving to own a house in one place or the other. Houses, the 'elephant of all things', signify wealth and power. Although owning a car is seen to offer increased mobility of transaction, having a car for most hawkers increases the speed with which they can operate and their mobility as they can pick up goods that are in demand and deliver them within a stipulated timeframe. In addition, it ascribes a new status of wealth and social mobility. In line with the societal norms and pressure from peers to be a successful migrant, migrants are guided by hard work and the fear of failure, and want to be counted among those worthy of the name 'successful bushfallers'. This does not presuppose, however, that all migrants are successful. Some have performed dismally and are encouraged to go home for traditional cleansing.

While personal networks and ethnic enclaves play an important role in relation to migrants' success, they are dynamic and graded in the scale of their importance, reachability and availability. But equally, networks are created with respect to the circumstances and changes in the direction of trade in order to maximize opportunities. These networks are interlinked by new communication technologies that facilitate the transmission of information and knowledge about economic activities and, especially the weak ties, play a vital role in linking and informing migrants about the state of the market. As a result of their connections and the network society that has emerged, they are able to contest and evade state control and the rules of trade regulations. Through these networks, they become each other's keeper by alerting others when necessary, for example at times of raids. By the same token, these networks have been commodified and some relations are maintained because of the benefits to members.

Conversely, this cannot be said of the migrants in the Netherlands who do not have similar opportunities to operate in the informal economy because of the strict rules regarding the setting up of businesses. Those with educational qualifications tend, therefore, to seek formal employment.

Subsistence in the informal economy in South Africa is due to the government's creation of enabling policies that support this sector and allow room for self-employment. The stringent policies in the Netherlands constitute what Foucault described as state power (Faubion 1984). The state is thus able to assume its powers through state bureaucracy that controls what happens within the confines of its border. As such, migrants who cannot find a place in the legal formal/informal economy go underground into the illegal economy. By contrast, the increase of unregulated and informal activities in South Africa could suggest the state is unable to exercise its authority over certain forms of trade. In a bid to exercise control, the South African government has resorted to violent means, such as seizing goods and arrests, as a legitimate means of state control. Nevertheless, the migrant community in Cape Town is able to thrive because of the social embeddedness in which they have recreated a home-away-from-home. Associations in this regard play a vital role in the lives of migrants and have come to be recognized as their 'parents' in the host country, as will be seen in the next chapter.

7

'Your mami and papa for this country na meeting': PIFAM and MACUDA as agency in a transnational world

Introduction

> The ubiquitous 'home town' associations are another example of the efficacy of ideologies of homes in ordering communal responses to modernization. These associations bind together individuals from across socioeconomic stations and in spaces far away from their ancestral communities. They serve as powerful vehicles for the mobilization of resources, through levies on their members or prying of access to state resources, in order to promote the 'development' of their home villages. (Eyoh 1999: 291)

The Pinyin Family Meeting (PIFAM) and the Mankon Cultural and Development Association (MACUDA) are home-village associations that represent the Pinyin and Mankon migrants in Cape Town. Similarly, the defunct[1] Ngemba[2] association in the Netherlands was created to cater for the Ngemba Grassfielders there. The formation of village associations can be traced back to the period of inter-village weekly trading and they were further developed by bush trade as migrants often travel in clusters. These associations bring together groups of persons who not only share a common place of origin but are also attracted by the aims and objectives of the association. The opening quote by Eyoh forms the backdrop through which I seek to understand the PIFAM and MACUDA home-village associations. The term 'home village' is chosen here instead of the term 'home town' that is used in most studies on associations (Mercer *et al.* 2008). Home village is in line with references made by the members of these associations who

---

[1] Its collapse, according to informants, was due to its admission of non-Ngemba members, which led to financial misappropriation.

[2] Those who hail from Mezam Division and have a common language such as Mankon, Akum Bafut, Chomba, Nsongwa, Mbatu, Mendankwe and Nkwen.

feel that they are representing their village, and home to them is Pinyin and Mankon and not the regional capital Bamenda. However, within this defined locus of Pinyin and Mankon, there are imperceptible demarcations and members emphasize the quarter from which they hail. Apart from face-to-face associational meetings, transnational connections with members of the diaspora and the home-based members are linked in virtual spaces through Internet-mediated spaces, but with a common idea of development at home.

This chapter seeks to explore the emergence of home-village associations and their role in maintaining social cohesion within communities in the host country, and serves as a platform to project their identity. I examine the ways in which the Pinyin and Mankon migrants are 'crafting lives and communities' (Reynolds 2009), how they maintain ties with other associations and the formation of virtual associational ties with diaspora groups and home-based elites as well as with other Internet-mediated associations.

It is posited here that, although being a member of these associations is seen as a logical conclusion to proving one's belonging in the home village, belonging, to migrants, is seen as an insurance policy. In addition, due to the internal rules and regulations of the associations, the latter have taken up a parental-style role towards members through assistance, sanctions and disciplinary measures to curb unruly behaviour. In this regard, I question the role of the home-village association in the sociocultural life of its members. What is the role of ICTs in associational formation and cohesion? Central to the formation of communities is the availability of communication that permeates and creates a 'gel-like character of contemporary communicational settings' (Sheller 2004: 47), thus creating 'coupling and decoupling'. This emphasizes the notion of belonging that acts as a pull factor in both the host and home countries concurrently and connects migrants as well as the spatially dispersed. Despite this dispersal, the common factor that binds them is their sense of place. And in consonance with the findings of Mercer *et al.* (2008: 6; 2009: 143), 'place' is not bounded but is instead fluid with focus on the home village, while meeting venues are momentarily considered as the home village (see also Feldman-Savelsberg & Ndonko 2010). For instance, MACUDA regularly meets in the home of one of its member, a house that has come to be known as '*nda-ah Makoŋ*'. Loosely translated, this means the 'house or seat of the Mankon people' and emphasizes the notion of place with reference to home. What all researchers seem to agree on is the importance of place, as place is fluid and not bounded. Whereas I make a case that although migrants operate within multiple spatialities, place to them in this study is the home village because of their common shared identity and culture as is reiterated by this elite: 'I love Pinyin where my nivel (sic) was buried'.[3] A similar claim is made

[3] Email posted on Pinyin News Forum, 15/02/2012.

by van den Bersselaar (2005). Such firm rootedness could be explained by the fact that most migrants live their lives in the future and procrastinate in anticipation of a better life in the home village, the place where all their development is geared. They see their (often long-term) stay in Cape Town as only 'temporary'. The overt focus on home – Pinyin and Mankon – reflects their efforts to engage with the 'ideology of home' (Eyoh 1999) as well as the notion of belonging through the revitalization of ethnicity. In other words, these migrant communities, besides a continuous regard for the place of origin as home, lend critical financial support to home-based associations as concerns development projects because they need to be a 'place, a home, for oneself, … a place one can come to when necessary … a place to be buried when one dies' (Trager 1998: 373). This ideology of a burial place is firmly rooted in the migrant's repertoire and is at the centre of his/her relationship with the home village. This also underscores the reason why, despite the hurdles encountered in repatriating the bodies of some Pinyin migrants, an elder in the association had to appeal directly to the corpses (Chapter 8) about the importance of going back home. The following section starts by describing the associations, showing them as agentic structures that operate within multiple social arenas as well as in space and time.

Overview and characteristics of PIFAM and MACUDA

PIFAM

The general assembly of PIFAM is the body that represents all registered Pinyins in Cape Town and non-Pinyin who choose to become members of the group. If non-Pinyin belongs to the association, it means temporarily giving up their own ethnic identity in favour of that of Pinyin in order to be accorded full rights like any other Pinyin member. In the same vein, the new member has the same obligations towards the association as other Pinyin members. Members weigh up their options prior to becoming members, which resonates with the claim by Feldman-Savelsberg & Ndongo (2010: 374) that belonging to ethnic associations draws on a combination of primordial, performative, strategic and reactive identities. This is because, as they are relegated to the margins of the host society with less room for integration than tolerance (Nyamnjoh, 2011), village associations are useful in upholding the values of the home country. However, as of 2013, PIFAM's constitution was revised and the admission of non-Pinyin members was restricted to ensure their Pinyin identity, which is fast eroding as the language that gives them the Pinyin identity is dying out, as meeting sessions are now held in Pidgin English. Meetings are now held in the Pinyin dialect and in English for the benefit of non-Pinyin members. In addition, the president, the vice-president and the secretary are now elected on their ability to speak the dialect. These measures will ensure that the language that gives them their Pinyin identity will be strengthened. By contrast, migrants in the Netherlands, rather than projecting their Pinyin-ness or Mankon-ness like their counterparts in Cape Town, have chosen for Cameroonian identity and see themselves as members of a nation. This leads us to question why there is so much emphasis on an ethnic association in Cape Town and on the 'nation' in the Netherlands.

The aim of these associations is to unite members from the respective villages, provide welfare and economic assistance, and promote the cultural values of their home village. Integration into South African society is less of a concern. This, it seems, has been encouraged by the wave of xenophobic attacks on migrants and the absence of a well-defined government policy on integration. Contrary to Mercer & Page's (2010: 124) notion of multicultural identities with 'Britishness' and 'other cultural identities' upheld by Cameroonians and Tanzanians in their study, the Pinyin and Mankon have, primordially, maintained their cultural identities of Pinyin-ness/Mankon-ness, Grassfielder (*graphi*) and, lastly, Cameroonian.

PIFAM was set up in 1999 by six founding members and celebrated its tenth anniversary in 2009. The group is currently expanding and has 115 registered members due to chain migration and kinship relations. As a result of its growth, sub-associations of members from various quarters in Pinyin have been formed, with PIFAM being the 'mother' association. There are five sub-associations, which means that there is hardly a quarter in the village that is not represented in Cape Town. In addition, there is a weekly rotating savings and credit association (ROSCA) that is commonly referred to as a *njangi* (Rowlands 1995; Niger-Thomas 1995). While the PIFAM (General Assembly – GA) meets bi-monthly, ROSCA meetings are held on Mondays at 19:00 and are open only to registered PIFAM members and not to those only registered in the quarter associations. Weekly *njangi* contributions start from R 250. Meeting sessions are preoccupied with issues relating to the home village as well as the wellbeing of its members and serve as a platform to reinforce the expected moral ethos and to advise on how to go about obtaining legal documentation (Mercer & Page 2010). The binding factor is a sense of belonging and security and is underscored in this statement by an informant: '*your mami or papa for this country na meeting, na the only place wey you fit secure yourself*'.[4] Though used figuratively, this phrase highlights the fact that associations are given parental qualities and this heralds its significance among the mobile community. And true to parental nature, sub-associations organize end-of-year excursions for its members and the members contribute by purchasing a cow that is slaughtered and shared over the festive Christmas season.

4 Interview with Patricia, Cape Town; 26/08/2010.

Although PIFAM is seen as a family group, it has put serious development on the association's agenda as well as on that of the various sub-associations. PIFAM is a subsidiary of the Pinyin Development Association (PDO) and this umbrella organization oversees development activities in the home village.

For their part and in addition to socializing, the sub-associations are taking on functions discarded from the social calendar of the GA due to the increased Pinyin population in South Africa. These include born house (celebrations marking the birth of a child, Chapter 8), birthdays, weddings and developmental projects pertaining to the various quarters. The removal of these social events may not have gone down well with some members, as can be seen from the following excerpt.

> For me ... perhaps (I) am yearning for something lost that can never be recaptured because we were quite small; 6 of us to 10 maybe to 15 to 20 at some stage and these people were like a family ... These things are long lost ... because they are too large now anyway ... to recapture the past or to bring back this attitude ... I think it's difficult which is why I think we're trying to transfer those into the quarter meetings because the meeting is too large for people to cooperate and to do things in unity and to live the kind of cordial life that we could in the past.[5]

It has equally provided reasons for sub-associations to be more relevant and work collaboratively with elites and parents in their respective quarters to provide funding for projects that enhance the quality of life of those back home. For instance, each quarter is responsible for the provision of potable water as well as electrifying its quarter after the main high-tension cable was installed. According to Reynolds (2009: 210-11; Evans 2010; Mercer *et al.* 2008), such efforts create new groups and maintain existing networks that allow for these deterritorialized communities to live spatially, yet still connected to the home village. The home village thus becomes the locus of focus and concern despite being in Cape Town

Photo 7.1 PIFAM meeting in a rented hall

Photo 7.2 A PIFAM meeting at a member's home

5 Interview with Jake, Cape Town: 11/08/2010.

because they want to continuously internalize basic values and their orientation towards the home village, which continues to be their point of reference (Trager 1998). In summary, PIFAM and MACUDA serve a dual role in their respective communities: they are a safety net and have a parenting role for the newly arrived and older members. Serious disciplinary sanctions are applied whenever the need arises.

MACUDA

MACUDA is much smaller than PIFAM and was established on 6 June 2009, ten years after PIFAM and following the death of the spouse of a Mankon migrant. Faced with this event, the few Mankon migrants in the city at the time hurriedly gathered together to assist 'one of theirs'. After bringing the corpse home, the group met formerly, drafted its by-laws and MACUDA was born. Due to the relatively small number of Mankon in Cape Town, the GA decided to hold its meetings fortnightly to gain momentum before the schedule was reduced to monthly sessions. Its aims and objectives remain the same as those of PIFAM. Membership of this association is restricted to Mankon people and the non-native spouses of Mankon men or women. Exceptionally, they also admit members from the neighbouring Ngemba villages who were born and raised in Mankon town and can be vouched for by Mankon friends. This means that the group only admits those from Mezam Division, perpetuating the concept of autochthons and allochthons from home in their host country (Nyamnjoh & Rowlands 1998; Eyoh 1998; Konings 2001, 2008; Konings & Nyamnjoh 2000) as well as placing renewed emphasis on 'autochthony and belonging' (Geschiere & Gugler 1998: 313). They have turned down applications from other Grassfielders who do not hail from Mankon and its Ngemba surroundings because the group wants to avoid bringing in people whose background they cannot trace, as well as those they do not know very well. A recent example is someone who was asked to leave his home-village association and was hoping to join MACUDA but whose application was rejected. MACUDA has adopted Pidgin English as its medium of deliberations although many members are not in favour of this and feel that using the local dialect offers the perfect opportunity for them and their children to learn and preserve their language. This notwithstanding, some members still express their contributions in their local dialect at meetings because everyone understands the language. However, at the end of the formal session when food is served, members entertain themselves with riddles in the local dialect.

The small number of Mankon migrants in Cape Town could be attributed to the near absence of chain migration compared to that among Pinyin migrants. It could also be attributed to the polygamous nature of Pinyin migrants' kinship relations, with each migrant being responsible for the welfare of his immediate family and earlier migrants encouraging others to join them. About five members have brought in to South Africa relations or their spouses, while most came individually and have gravitated towards the group. However, evidence of kinship support abounds amongst Mankon migrants as some have come thanks to the information they got from relations in South Africa. The ripple effect of this low rate of chain migration is seen in their limited numbers and, on a positive note, they pay attention to all the social events that PIFAM has abandoned, as mentioned above, and this marks a major difference between the groups. It is at such joyous moments that the group's potential is noted by all those present. Nevertheless, the social networks of members of both associations are, first and foremost, among the respective members and are then extended to those with other special ties

Photo 7.3 A MACUDA meeting at the 'meeting house'

Photo 7.4 A new member introducing himself.

Language of operationalization

At the start, PIFAM meetings were held in Pidgin English, which is the *lingua franca* of all Anglophone Cameroonians. When serious decisions are to be taken, anecdotes and proverbial sentences in Pinyin are often used to drive home the importance of the message. The use of proverbs indicates underlying meaning and is a critical aspect of any culture. It is the medium used to transmit messages, attitudes, habits and knowledge but they do not necessarily mean that everyone acts in accordance or that everyone does what is expected of them in the group (Evans 2010). However, through mechanisms of social pressure, including ostracism, PIFAM and MACUDA have proved to be very effective in ensuring compliance. As of 2013, PIFAM has reverted to holding meetings in the local dialect in an effort to maintain their identity, which they feel is fast eroding. The use of Pidgin English before at meetings is contrastive to interpersonal communication amongst members as one would rarely hear them speaking in Pidgin. The tendency is to speak in the vernacular once they meet (Pinyin), which perpetuates assumptions about their inward-looking nature and affirms their identity. The language, according to most, is their strength and is what binds them together and enhances their belonging.

This is in contrast to their MACUDA counterparts whose Mankon dialect is rarely used at meetings or amongst members. In spite of the role of both associations in uniting their people and giving them a sense of belonging, language seems to be somehow neglected amongst their children as they are not taught or spoken to in the local dialect. Parents have chosen English as the medium of communication with their children.

By-laws

Both associations have internal rules and regulations spelled out in their constitutions. Although social cohesion and moral conviviality are the underlying reasons for the formation of associations, they do not hesitate to punish those found guilty of deviant behaviour or of bringing the association's name into disrepute. In extreme cases, members may be expelled from the association. These rules are subject to amendment and change as the need arises, especially in the wake of unforeseen events. The death of Ron and subsequently of three others illustrates how PIFAM amended its constitution to deal with such problems in the future. When Ron was killed in Cape Town in September 2011, he had three children from two relationships with South African partners (two children by the first partner who he had legally married) but neither the children nor Ron had documentation stating his paternity and, when he died, Ron also did not have legal documents that allowed him to live in the country. It was therefore difficult for the association to send the children home with his body, as they would have wished to, due to a lack of documentation. After this incident, PIFAM included an article in its constitution demanding that all members present photocopies of their asylum permits and a member of the association was elected to be in charge of ensuring these were kept on file. Members are now notified when their documents need renewing and have to present updated copies to the group. Admission of new members can only take place upon the presentation of a valid document that permits the person to stay in the country legally. The association also advises members on how to obtain various documents. Again, after the deaths of three members in a car crash, the group put a ceiling on how much members would have to contribute towards repatriating a corpse. Should the contribution amount to less than what the funeral parlour demands, then the difference would be given to the family. Members tend to be overburdened with the financial responsibility of repatriating a corpse and this relieves the family of some of their financial responsibilities.

As of June 2012, MACUDA reverted to having monthly meetings and included a clause in its constitution to ensure good attendance. Five consecutive absences by members now results in them being expelled. For PIFAM, a member is expelled after two absences. This does however question the whole notion of unity and belonging that both associations uphold. If the idea of forming associations in the host country is to create a home-away-from-home, then expelling members from its fold only highlights the power that these structures can exert over their members.

Major decisions are taken by consensus but when this provokes heated debate in meetings, there is a vote and the outcome is respected. However, the topic may later be put back on the agenda and there is a tendency for it to be rejected the second time around, such as the proposal to provide an ambulance in the district hospital in Pinyin, which was rightly rejected by PIFAM.

Discussions on finances take up a great deal of time in both associations, but more so in PIFAM given its reputation for being a business with members' savings that are given out as loans two days after their meetings. What is innovative here is that most of these transactions are done by mobile phone – calls and SMSs – that leave no trace and the problem of unpaid loans. Members call the financial secretary and provide him with their bank details and the money is transferred into their accounts, excluding the interest. Although they acknowledge the need for a loan form, mobile communication is not ruled out because it is accepted on the basis of trust – which binds the group together – and so members who prefer such methods are strongly urged to keep their promises.

Conflict resolution

Conflict resolution is very much an integral part of the associations' means of putting checks on wayward behaviour and unpaid debts, and is organized by a committee of elected members from both groups. The committee, known as the Conflict Resolution Committee (CRC), is comprised of respected associational members (the elite), elders and some women. The idea of a CRC in PIFAM imaginatively reflects the role of the Village Traditional Council (VTC) in Pinyin where members involved in disputes are expected to attend convened sessions with a keg of palm wine (a crate of beer in Cape Town) that the committee enjoys in the course of its deliberations. Most of the cases tabled before this committee by PIFAM members relate to unpaid loans or personal conflicts between individuals. With respect to unpaid loans, those who are seen to be recalcitrant are named in the GA in the hope that bringing them to the attention of the entire group will embarrass them and speed up payments.

The conflict in MACUDA often revolves around leadership. At the end of conflict resolution meetings, sanctions in the form of financial fines and/or drinks for the group may be meted out to defaulters. Other sanctions include providing a stipulated number of chairs for the association (MACUDA) or preparing the meeting venue and cleaning it up at the end when PIFAM hires a hall.

Gendered associations

Most home-village associations are male-dominated and provide little space for women. Unlike associations that include a women's wing and/or activities organized by women, PIFAM and MACUDA fall short in this respect. Women have often been given only minor roles and are not entrusted with such challenging tasks as their male counterparts. Instead, marginal decision-making processes, especially those that are considered to be traditionally female-oriented, are given to them. However and as argued by Chilver (1988; see also Awasom 2005), this does not resonate with the critical and historical role women have played in the village and in local development more generally, as well as in the political sphere. Both associations organize regular sports competitions but these comprise only men's football and no activities for the women, although PIFAM has attempted to include women's activities but with little success to date. During such competitions, women are encouraged to come as supporters and cheer on the men, while nursing students are given the responsibility of providing first-aid assistance.

Differences

What can be distilled with reference to both associations is that Mankon people are not keen on chain migration as they tend to control who comes, but this has led to their limited numbers in Cape Town. On the other hand, Pinyin people are constantly attracted from the village by settled Pinyin migrants because the latter act as nodal points of contact for would-be migrants. The mobile phone thus becomes an important vehicle in organizing and attracting new migrants from Pinyin thanks to the information sent home. But Mankon are less engaged in such networks and chain migration, which is why the Pinyin are more engaged in development projects back home than the Mankon.

In addition, there are substantial differences with regard to associational rituals, with PIFAM limiting its activities to death and MACUDA participating in elaborate celebrations of activities such as born house and welcoming brides to South Africa. PIFAM pays more attention to deaths than any other ceremony and has relegated other ceremonies to the sub-associations. In this vein, we are confronted with a degree of individualism vs. collectivism. PIFAM could be equated with centrifugal individualism and MACUDA with centripetal sharing. Although both communities are involved in business, Pinyin are more preoccupied in the rational sense of success and are too busy for additional social activities that are very business driven and profit oriented, whereas the Mankon take pride in indulging in them and are more oriented towards sharing.

This collectivism, as opposed to individualism, is also seen in the treatment of newcomers. Peter explains MACUDA's role:

> Recently we formed a committee to look into the situation of newcomers and initiated a project to raise money to make funds available for them to start their life here. Not like giving the funds to them as a gift but in the form of a loan. When you come, the committee sits with you, discusses with you, looking at what you were doing back home before coming here, to try to understand what you have in mind to do here and make any proposals to you if necessary … we try to initiate them into business. We give them R 500 and we give them a year to pay the money and R 50 as interest. The intention is to get them to be responsible and to assist another new member with that R 50.[6]

Conversely, such assistance for a new PIFAM member is given by relations and friends since the group has no formal structure to assist newcomers. Similarly, because of the nature of their business (Chapter 6), in the event of arrest, PIFAM bails a member out only if s/he is considered not to be at fault but was arrested for selling contraband goods. The member can request help from friends. With MACUDA, irrespective of the crime, the member is bailed out by the association and has to reimburse the group the amount paid as bail.

The actions of the association lead one to conclude that they act as agencies in themselves and are the structures that create social spaces and new social relations informed by status, hierarchy, inequalities and power (Karp 1986: 134; Nyamnjoh 2002, 2005a, 2005b; van Dijk *et al.* 2007: 5). For these individuals to feel at home away from home, this offers the recreation of Cameroonian-ness through the hegemonic powers these associations have taken upon themselves and how their structure affects the production of action. The notion of agency and structure presupposes the relationality between actor (agent) and structure, thus creating an agency space. The agency space is considered as the 'freedom that actors take, in their interaction, to manoeuvre between the stipulations set by structure. Agency then becomes not so much the enactment but the denial (*negotiation*), the compensation, the improvisation beyond structure' (van Binsbergen 2007: 17). The terms actor and agent are used interchangeably here. Following Karp (1986: 137), the actor refers to a 'person engaged in action that is framed, as in all social action' and the agent 'refers to a persons engaged in the exercise of power in its primary sense of the "bringing about effects", that is, engaged in action that is constitutive'. How do they navigate, negotiate and circumvent the circumstances, structures and spaces of their daily lives? And how do we understand the continuous mediation and negation of social relationships within the parameters of agency? This plays out in new power relations (kinship, elite and

6 Interview with Peter, Cape Town: 01/03/2011.

social) within the various associations and transnational relations in which power relations are continually produced and reproduced in context and related to the 'invocation of rules and mobilization of resources' (*Ibid.*: 136). This is because structure is virtual and intangible, and in a perpetual state of being in a plethora of mediated spaces because of some dominant agency over another. In this regard, I propose examining the above associations as agency that is mediated through the appropriation of social space and structures.

Overview of trends

The study of (elite) associations and the theories developed around them, especially those conceived of in the colonial period and its aftermath (Nkwi 2011; Delancey 1988; Rowlands 1992) and in the late 1990s, are based on the existence of an elite in combination with a group of persons – associations – with social power and a religious, political, ideological or cultural value that is transcendental to the social organization (Naville 1963). The associations were also associated with mobility between regions and the home village. Such associations tend to originate from those outside the village, given its importance to those living in the cities and wanting to reconnect with the village (Gugler 1971, 1997). But a relationship like this is often marked by one of intimacy and inequality at different levels. Associations have thus become the mechanism that those in the cities and the old in the village use to safeguard their hegemonic powers and to get hold of and have control over others, thereby creating an elite class within these associations. The status quo is being challenged today, as can also be seen in the growth of Christianity, which introduced egalitarianism and new elites connected to the modern state and economy thanks to education (Rowlands 1993: 92-93), mobility and easy connectivity. New-found wealth by migrants is challenging traditional elitism and the hierarchy through the 'reinvention of hierarchy' as a result of an entrepreneurial ethos that informs mobility. One thing that marked the literature on associations was more their preoccupation with the formation of the elite class than the development agenda, which was believed to be the reason for their formation. Today, the emergence of transnational studies has put migrants and home associations on a par and at the fore of home-village/rural-urban connections with a focus on development, such as the equipping of the local high school and the roofing of the church and church hall in Pinyin and Mankon respectively. A critical assessment of the literature will confirm the above mentioned narratives.

Dating back to the colonial period, Delancey (1988) noted the formation of the new elite to assist colonial rule and, in turn, sought to replace it with a focus on power. This period was replaced by the formation of alma maters whose focus was more on developmental issues pertaining to their schools than on political issues. Gugler (1971, 1991) in his (re)study of connections between urbanites in Enugu and those in the village shows that urban-rural connections have intensified, with a new mutually interdependent relationship reinforced by kinship ties being developed. The study by Ferguson (1999) among the Zambian Copperbelt miners reiterates this dualist habit that was well entrenched in society and amongst miners. However, the introduction of multi-partyism and the rise of oppositional politics in most African countries, and especially in Cameroon, altered the role and functions of elite associations. In line with this trend, Geschiere & Gugler (1998) and Nyamnjoh & Rowlands (1998) have shown how the elite leverage support from their home villages to garner political power from the state. They (*Ibid.*: 324-325) note that 'the tendency in recent years for elite associations to reinvent or rework ethnic loyalties as part of the political mobilization of regional support to gain advantage in a less organized and predictable arena of state politics has encouraged more explicit recognition of political ambitions of these social movements'. We should not lose sight of the fact that these elites do not exist in abstraction, as their power base is equally attributed to kinship relations to whom they owe allegiance and visit regularly. In addition, such visits and village development are prompted by the fear of witchcraft that has compelled both the rich and the young to go back to the village, with urbanites seeking to invest in the village and the villagers clamouring for urbanites to share their new-found wealth. Life and death are important rituals and events that necessitate rural connections given that urbanites want to be buried back in the home village (Geschiere & Nyamnjoh 1998). While most of these early studies dwelt on the notion of witchcraft, Trager (1998: 360) opines that 'home-town linkages are central to members of local and national elites, committed not only to their personal ties to the home-towns, but also to assisting in local development'. This begs the question as to why people go abroad and do not forget about home. While the salient themes of mobility, rural-urban linkage, power and elite formation that underscored the studies during the 1990s are still present today amongst migrant communities, the current tendency is an increased focus on development (Mercer *et al.* 2008; Mercer & Page 2010). Better communication technologies and networks have seen the pursuance of a development agenda to its logical conclusion. And connections back home are informed less by the fear of witchcraft than by 'moral conviviality' (Mercer & Page 2010), moral duty and prestige that is associated with village development and also the fear of bad mouthing.

Although this trend has continued, what seems to have altered is information and communication technology and road and air transport that have ushered in increased connectivity with home-village and other associational networks. But these groups of persons are imaginatively Pinyin and Mankon and reflect the socio-historical structures of the societies they represent (Trager 1998) and the strong sentiments and mental pictures according to which members organize themselves and undertake their cultural practices (although non-natives may be present too). These are, however, 'communities of memory' that reaffirm people's sense of place and attachment to their villages of origin (Barkan *et al.* 1991).

Associations as agents of development

Development is an integral feature that underlies the viability and credibility of these associations in the place they consider home (Trager 1998; Mercer *et al.* 2008, 2009; Mercer & Page 2010; Riccio 2011). In their view, home associations have the power to re-orientate belonging-based policies by mobilizing collective action and resources to mitigate the struggle for power, gender imbalance and inequalities through new by-laws. Although the projects are not on a massive scale compared to those carried by the government, they are quite significant given the impact they have on the lives of the beneficiaries. A case in point is the electrification project undertaken in Pinyin by the people at home and those in the diaspora coordinated by the Pinyin Development Association and the construction and equipping of the first high school under the supervision of home-based elites. Development projects are, however, not negligible and go a long way towards filling a void and bringing the migrant community alive in the village given the huge amounts they are expected to contribute. The question is about the setting of the development agenda. Unlike Trager (1998) who posits that the agenda for community projects is set by the elite who are not resident in the community, I maintain that development projects are set and supervised by national elites or 'translocal elites' (Comaroff & Comaroff 1993) that are not resident in the village. They make regular visits to the village, while the elites abroad play a key role regarding financial contributions. Nevertheless, a very critical and non-intuitive aspect in the strengthening of home-village associations is the role played by their relationship with home-based associations, as will be seen below.

PIFAM's development activities are initiated by the mother association, namely the Pinyin Development Association (PDO), to which all other associations are affiliated, and a communiqué is sent out via the PNF. Similarly, in an effort to synchronize development, associations often liaise with their mother association to find out what urgent project needs to be carried out or someone travelling home is mandated to meet the translocal elites and study a project's feasibility or suggest alternative ones. This, according to Trager (1998; Mercer *et al.* 2008, 2009), is prompting agency shifts between locally based home associations and Africans and international associations, as the centre of power lies with the former with respect to financial and managerial capabilities. However, given the differences that permeate most home-town associations, it is not possible to have similar conclusive functions between home-village associations. While the study shown by Mercer *et al.* (2008, 2009) upholds that the 'domestic diaspora' is more influential and has financial and managerial capacity, the same does not hold for Pinyin and Mankon migrants where the diaspora has the financial capacity and the home diaspora has the managerial capacity in terms of the time they invest in meetings and projects. Paying attention to their multiple spatialities, we will zoom in here on some of PIFAM and MACUDA's development projects. Distance is not the problem as communication flows and mechanisms are in place to mitigate it. I argue that distance is bridged by committed individuals – translocal elites who have chosen to be the bridge between home and international associations and can find compromises to avoid anything that could disrupt information flows and connectivity. For example, the following email sent by the translocal elite about pressing infrastructural needs generated a flurry of affirmative responses.

> The Government High School Kwang has grown into a large community with its incessant compelling problems. As at now, one easily identifies three main areas: an acute shortage of teachers, the damage of 34 computers and the pathetic situation of poor and needy students.
>
> 1. Acute shortage of teachers
>
> At least four senior temporary teachers are urgently needed in the following subjects for the examination classes: Geology, Mathematics, Physics and History. …
>
> 2. Damaged computers
>
> On Wednesday 14 December 2011, a total of 34 computers were damaged as a result of an electric surge … You may be willing to offer one all alone or join with someone. You may like to contact Mr X based in Bamenda, a dealer on all kinds of computers. Certainly the best offer is an instrument to prevent future surges.
>
> 3. Poor and needy students
>
> You bless yourself when you bless others. Everyone has something to offer. Never say you cannot help someone in need because you are not rich enough ... The act of giving is cultivated. Cultivate it. Be a blessing to others and see them as opportunities and not adversaries. …[7]

In response to the call, a good number of Pinyin elites supported one of the three projects that were announced. The following responses illustrate how the elites pro,ptly assisted:

> I hereby undertake to pay the salary of one of the four teachers needed for the next five months, ie from February to June this year (2012). Pa, see to it that the teacher is hired to start from the 1st of February. I will cause the sum of 40,000frs to be deposited into the PS 20 account every month for the teacher's salary. Just let me know when is the best time (sic) for the money to be in the account.[8]

[7] Sent to PNF, 1 January 2012 (from Buea).

[8] Sent to PNF, 17 January 2012 (from Cape Town).

> I am particularly touched by the Computer crisis as I visited the site of that damage two days after. I hereby promise 05 PIV with LCD monitors on behalf of my enterprise Xxxx Xxxx.[9]
>
> … I would be donating 10 of these computers (CPUs only) to help in this crisis. My assumption here is that others can join me in getting them ready for use, by providing the other accessories (mouse, keyboards, speakers and monitors) needed.[10]
>
> I will send a widows might (sic) of 25000frs to add to the fund for part time staff.[11]

The exchanges above show how associations are institutions in their own right and are capable of planning their own courses of action. The danger of such a developmental plan is that it obfuscates the government's participation in development in the area and lends little or no credibility to the government. However, such infrastructural projects aim to celebrate migrants and the migration-development nexus as well as providing basic social amenities, enhancing human capital, the corollary to which would be more people contributing towards village development in the future. In the same vein, migrants are preparing their home village for their eventual retirement or voluntary return.

Fund raising is an integral part of both associations and is usually carried out for a village or quarter project. Alternatively, members may decide on a voluntary donation (*cour de Coeur*). MACUDA plans to launch a fund-raising gala night to generate funds to buy medical equipment for the upgraded district hospital in Mankon village and members have been asked to contribute in instalments towards the project. They have also contributed funds for the completion of the church hall in Mankon.

Photo 7.5 The church in Mamben, Pinyin that the MAKON sub-association in Cape Town raised funds to complete

9 Sent to PNF, 19 January 2012 (from Bamenda).

10 Sent to PNF, 22 January 2012 (from the US).

11 Sent to PNF, 10 February 2012 (from Ngoundere).

One of the sub-associations of PIFAM is Mamben/Kongfung Association (MAKON), which organized a twin event with an auction and a gala evening to raise funds for the church in Mamben (see also Trager 1993). Fund-raising activities are often arranged to coincide with times when a member is travelling home so the funds are handed over by a delegated member, who is usually a member of the elite. This confirms the worries of the younger migrants who claim that they are hardly recognized back home even though they contribute equally towards projects.

Researching the above-mentioned groups required attending their meetings, which I did regularly during my fieldwork in Cape Town, in conjunction with interviews with selected informants and (most of) the executives in both groups. I equally participated in various social and cultural events, attended wakes, birthdays and born houses, sporting events and parties, thus marking joyous and more solemn occasions. Of particular significance was the fact that on all these occasions, I met casual informants who always had interesting comments to share and made provocative remarks that whetted my curiosity. Unable to subscribe to the PNF virtual communication network, I relied on other informants to forward circulated emails, but I was able to subscribe to the Mankon Forum. Telephone conversations were inextricably linked to this section as I made frequent calls to negotiate interview sessions, find out more information and/or just to link up to maintain relationships. I also received forwarded SMSs sent to circulate information amongst PIFAM members, and I received all the SMSs from MACUDA when I was in Cape Town as my number was on MACUDA's SMS mailing list. Interviews were conducted with home-based leaders, village elders as well as HRH the paramount ruler of Pinyin.

'Na njangi be my all':[12] *Rotating Savings and Credit Association (ROSCA)*

ROSCAs or *njangis* are an integral part of migrants' daily lives and back up their economic activities. Ardener (1995: 1) defines a ROSCA as 'an association formed upon a core of participants who make regular contributions to a fund which is given, in whole or in part, to each contributor in rotation'. Money raised to contribute to a *njangi* comes from hustling in the informal economy and money raised from a *njangi* goes to boost their business. For the purpose of this study, the term *njangi* (*nchwah* in both dialects) is adopted instead of ROSCA because this is what such a micro-financing credit scheme is called locally in the home country and amongst the migrant population and their respective home villages. Whilst both communities have *njangis*, the various quarter sub-meetings affiliated to PIFAM also run *njangis*, with each group having its own by-laws. The significance of this variation, however, is that it is the articulation and multi-

[12] Max remarked on the importance of the *njangi* for him, Cape Town; 21/07/2011.

plicities between the categories that constitute the social life amongst these migrant communities, as everyone looks forward to the days *njangis* are held when they can meet others, eat Cameroonian food and hear the latest gossip in town and news from home.

The literature on ROSCAs is extensive, with some contending that its obituary will be written when financial/credit institutions become accessible to most people although others hold that they will thrive when credit institutions become unworthy. Cameroon's *njangis* have often triumphed, all the more so because of the 'insecurity of the banking system, but (they) have equally disrupted the banking system' (Niger-Thomas 1995: 96; Ardener 1995) because most people feel safer having their savings in *njangis* or credit unions. And *njangis* have already existed for some time in the villages where those who do not have access to banks get together on weekly market days, '*kontri* Sundays' (observed as rest days in villages in the Grassfields) and at the end of the month. They use them as a way of saving money towards projects and also for social gatherings. However, in relation to the past, the form of today's *njangis* practised by migrants, especially Pinyin, has evolved from physically contributing money in cash at meeting sessions to electronic money transfers using the Internet or cell-phone banking facilities. While it is true that loans can be had from banks, the speed with which *njangis* can raise a loan for a member (sometimes even large amounts) and at minimal interest cannot match that of banks. *Njangis* are fast becoming authentic financial houses for migrants. In this respect, the suggestion that *njangis* will die out when financial institutions are seen as being safe, such as those in South Africa, cannot be supported by the Pinyin and Mankon migrants in Cape Town. To most migrants, joining a *njangi* is the norm and not being a member is seen as unusual, although there are some members who are not part of one. Some members with white-collar jobs do not belong but are not frowned upon whereas a pitcherman who is not a registered member of a *njangi* loses respect amongst his peers. This alone does not account for the high rates of involvement in *njangis* but how can we account for the high rate of involvement in various *njangi* schemes by Pinyin and Mankon migrants?

Njangi as a way of life

As mentioned above, the *njangi* is a long-standing tradition in Cameroon (Ardener 1964, 1995; Niger-Thomas 1995; Rowlands 1995) and for most migrants they are a way of continuing with internalized habits. Some have been members of *njangi* groups at home prior to going to Cape Town and others have either continued to be contributing members of their group whilst in Cape Town or have withdrawn and now participate in one in the host country. But over and above this, most migrants are asylum seekers with temporary short-stay (6-12 months) residence permits whose renewability is not guaranteed. For these groups, *njangis* remain the safest way of keeping their money.

Both PIFAM and MACUDA run well-established *njangi* schemes of varying amounts. In addition to these groups, various cohorts of migrants have formed other *njangi* groups that are held either weekly or bi-weekly. For instance, the pitchermen at Epping have a bi-weekly *njangi* and the beneficiary receives the money after a secret ballot, while PIFAM's various sub-associations have their own *njangi*. Similarly, those with stalls in Parow have joined together to form a *njangi* group. As PIFAM is quite a large group, a separate day for *njangi* other than meeting days has been set (every Monday), while the MACUDA *njangi* takes place on meeting days just before the start of the meeting. Most people in both groups contribute more than one share. While MACUDA still contributes actual cash during *njangi* sessions, PIFAM has adopted the new technology to facilitate the process; although this has come as a result of having lost a huge amount of money.[13] All contributions are done via bank transfers into the beneficiary's account. Following each deposit, the recipient receives an SMS from the bank to notify him/her of the transaction and the amount deposited into his/her account. Two SMSs from the beneficiary of the *njangi* shed light on this point: 'FNB R 4000 paid to smart Account xxxxxx @ on line Banking. Ref. Ben + Doris' and 'FNB R 2000 paid to smart Account xxxxxx @ Gugulethu ATM. Ref. Andre 1 and 2'.[14] Except in cases where the beneficiary wants the money in cash, all transactions are done via the bank. Most of the members use their mobile phones for transactions or the nearest ATM at their place of business.

Although most have bank accounts in Cape Town, these are usually used to transfer money to Cameroon so they can be sure to be able to access it in the event of repatriation or it can be handed over to the family in the event of death. Almost all PIFAM members have credit-union accounts with the (Pinyin) Metayen Credit Union (MITACUL) back home so when they receive money from their *njangi*, the money is withdrawn from their bank account and placed in MITACUL by the trustee who has their Visa card. For instance, when Joyce received R 25,000 from the *njangi*, the money was withdrawn by her sister in Cameroon and put into her MITACUL credit-union account. The rapid Internet banking facilities have made it possible for migrants to transfer money to their credit-union accounts with relative ease and they can also monitor transactions from the SMS alerts sent by the bank each time money is withdrawn back home or in Cape Town.

---

13 In the course of one of their sessions, thieves made off with all the day's contributions and the members' cell phones.

14 SMS forwarded to this researcher, 19/05/2011.

Besides being self-employed in the informal sector where there is a high degree of unpredictability, most migrants have consciously chosen to belong to one or multiple *njangi* groups as a way of diversifying their savings and earnings. Conversely, having earned the name bush-faller by virtue of being in Cape Town, they are compelled to save to meet the demands of kin back home, given that those with Visa cards at home ask the trustee to withdraw a specific amount to respond to the needs of kin.

It is not only by being a member of the *njangi* that one gains respect from peers but the amount of money contributed is also significant. Those who contribute a great deal demonstrate that they are disciplined, hardworking and frugal. As such, a spirit of competition has been instilled amongst members (especially those in the informal sector), with each striving to be among the category of people who contribute the most. The salient fact is that these contributions underline how they are not reckless with their income and this serves as a way of establishing their creditworthiness within the group. *Njangis*, thus, provide an alternative to upward social and economic mobility.

Mutual benefits

The social and economic benefits of *njangis* to the migrants and their families in the home and host country cannot be understated. They serve as a measure of economic discipline to members (Ardener 1995: 9) as they compel them to save and avoid any form of lavish lifestyle or conspicuous consumption, which would be frowned upon. Max,[15] for instance, epitomises those who are hardworking and frugal. He is from Mankon and a pitcherman at Epping who left his wife and four children in Cameroon and went to Cape Town to make some money within two years and go back home to invest it. He did just that and he is now back in Cameroon. He belonged to various *njangis* and had more than one share in each. At the bi-weekly meetings in Epping, he contributed to two and a half shares (R 250) as each share costs R 100, and he belonged to another at Parow to which he contributed R 500 each week. In MACUDA, he contributed to a joint share with his cousin. Money received from these *njangis* was sent home via his cousin to his wife to open a credit union account and all the subsequent money he makes is channelled home through his cousin's bank account. The cardholder withdraws the money and gives it to Max's wife who then saves it with the credit union.

In this respect, the benefit of a *njangi* is seen in the self-abnegation of worldly pleasures in order to make a generous contribution to the *njangi* and to belong to different groups. *Njangis* thus define morally suitable goals and objectives concerning spending one's money, as members adhere to the social ethos of sending money home or saving part and investing the rest in their business.

---

15 Interview with Max, Cape Town: 16/06 & 21/07/2011.

In the event of death, the family of the deceased benefits from his shares in the *njangi*. All those he contributed to are requested to make a contribution and the money is sent home with whoever is accompanying the deceased, or it is wired. Such was the case when Ron passed away: all his contributions were collected and sent home and some of his asserts were sold. Prior to his posthumous migration, some of his money was sent home to the family so that they could begin making preparations for the funeral. When the corpse goes unaccompanied, the money is transferred to Cameroon to a relation.

Belonging to a *njangi* gives members access to credit that would not have been possible via banks given their asylum status. In the PIFAM *njangi*, each member is entitled to a loan of twice the amount of money s/he contributes and it could exceed that amount if the member is still to benefit from the *njangi* as the loan will be deducted when it is their turn to benefit. Njangis are a means of wealth accumulation that give their members a better chance of social mobility.

Status and social hierarchies in elite formation

As shown earlier, the notion of elite as it is used here defies the traditional norms of elite as being powerful against the powerless, and it is not about raw power but rather about respect, responsibility, selflessness, knowledge and wisdom. Through such qualities, members gain status and social hierarchy. Elites, as a result of social mobility, use their socioeconomic and cultural capital to create hierarchies deliberately or benefit from those ascribed by junior members to older members. It is such hierarchies that can cause friction.

How can we explain the formation of dominant structures/hierarchies in these groups? As noted by Chase (1980: 906), hierarchies 'emerge from the interaction among group members rather than being generated by differences among those individuals'. This presupposes that hierarchy formation is a developmental process in which outcomes of previous interactions influence the course of successive ones, and he concludes that: 'it indicates how patterns of interaction fit together to form the kinds of dominance hierarchies commonly observed'. In spite of this assertion, we cannot conclude definitively that hierarchies do not emerge from differences because these levels of economic and human capital amongst members breed hierarchies as well, given members' rise as elites are negotiated through their access to social, economic and cultural capital. Within both associations, there are those who have successfully acquired social and cultural capital that has enabled them to move up the social ladder. Others run successful businesses and this has catapulted them onto the higher echelons of the social ladder. Although participation in these associations is voluntary, they are not equal. As such, status and hierarchies tend to be advertently or inadvertently created. However, class here is not strictly subjected to the Marxist notion, in reference to economy. Instead, I seek to align myself here with the Weberian thinking of class in terms of status and prestige, although economic capital plays a vital role in projecting class. But I also contend that status is a cultural marker, which can be understood in the light of customary value systems in relation to services rendered to the community. In other words, status for the wealthy comes from their capacity to deal with people's needs – the joy of sharing and giving – that is made possible by their claims on goods that in turn derive from money. This highlights the point by MacGaffey (2005: 328; Chapter 6) that 'conversionary transactions have been widely used to transform ephemeral income into more enduring relational assets'. The idea of status in this case is not so much in terms of powerful dominance over others but is more about assuming responsibility for one's society and being willing to make contributions in an egalitarian manner. This then begs the question of what it is that makes one person more respected than others. What is the motivating factor behind making claims to status within the associational and village context? A response to this question would require us to look at both individual motivation and the expectations of the community. With regard to individual motivations, some migrants have laid more claims to the home village than others and, as such, dedicate time and resources towards its wellbeing. But community expectations too demand a sense of obligation and commitment for those who have achieved wealth as wealth brings with it status and prestige when used for the common good of community but could bring shame and scorn if one felt no moral obligation towards the community, especially if the person is considered to be wealthy. Such attention for the home village is in contrast with earlier studies that portrayed the jostling for political hierarchy and were marked by witchcraft, ethnic ambitions, divisions and difference. (Nyamnjoh & Geschiere 1998; Nyamnjoh 1999). Current shifts are being motivated by the prestige that members gain from assisting in village development.

Status and social hierarchy in both associations are thus differentiated along generational, gender and socioeconomic parameters. Often, the younger members feel that those at the top represent their interests and that, back home, they (the younger members) are not acknowledged when contributions to projects are made, as it tends to be the elite that represent the group.[16] This can breed some dissension among group members and, in turn, translates into a lack of interest in the engagement of certain sustainable development projects. Elites are believed to use the association as cultural capital in their quest for fame in the village by proposing projects that require substantial financial contributions (Woods 1994: 467). Younger members are therefore unhappy about the fact that flat rates are levied on contributions, irrespective of members' status and time of arrival. They feel that consideration should be given to new members who are still eking out a

[16] Interview with Desmond & Ben, Cape Town; 08/03/2011.

living. However, the association argues otherwise and claim that members who have met their financial obligations should be treated equally with regard to the association's constitution, and are thus eligible for identical treatment and have the same obligations. Such reasoning is in consonance with Trager (1995, 1998) that successful migrants are considered to have a greater level of obligation to their home-village association. This was the case when PIFAM decided to construct a morgue at the divisional hospital in Pinyin, a project that was championed by the elite. To this effect, a member visiting home was sent to assess the feasibility of the project from the elders and PDO representatives and he received financial remuneration to cover his transport costs. This was much to the disappointment of the group's younger members who see him as an elite and financially secure. They thus question his patriotism and loyalty to the group and the village, regarding the amount of money he was given for this mission.

The proposed morgue was turned down in the village and there were questions about why such huge amounts should be spent to take care of the dead when the living need more attention. An ambulance was then proposed but this was also rejected by the younger members of the association as they saw themselves as financially unstable. If they had money, their priority was to look after their families. The inclination for such huge projects, in the words of some informants, depicts how those who have risen to the ranks of the elite are more interested in carrying out large projects back home for self-aggrandizement than out of consideration for those who have just come or are still struggling financially. Although contributions for projects are often compulsory,[17] junior members feel that a flat rate for everyone is unfair and they expect the elite to contribute more (cf. Trager 1998). This is because contributions sent home are frequently attributed to the elite and so they subsequently receive a warm welcome whenever they go home. These younger migrants may have a point here because during my stay in Pinyin, I was constantly reminded of the development projects by those in South Africa and the names of the same persons were hailed for their selflessness. By the same token, we should not lose sight of the fact that these elites constantly visit home, and while there they participate in development projects. Joe's case is an example of this. Once when he arrived in Pinyin (from South Africa), the villagers were having a fund-raising occasion for the electrification project and he instantly donated FCFA 50,000 (approx. € 80), much to the villagers' delight. The same elite offered to pay the salary of a teacher, who would be hired by the PDO, for five months. Gestures like these from some of the PIFAM elite beg the question about why such scathing remarks are made about them by some

[17] The money is deducted from the members' sinking fund and they are expected to repay the amount within a given timeframe. If they do not, when they are faced with bereavement, the group will not come to the person's assistance because his/her finances are not up to date.

members. It could well be understood that the younger members or the newly arrived are still struggling to attain a level of financial stability and thus any call for a significant financial contribution from them is viewed negatively and they blame the elite for being too zealous and for having so little regard for their financial concerns. The sheer increase in numbers in the group, as noted by Jake (above), means that there will often be rifts between the elite and those who have just arrived. Also, as argued by Chase (1980: 922-923), hierarchy formation in larger groups emanates from the web of interactions in which interaction among individuals can be used to explain how macro-level structure and dominance hierarchy is produced. In this respect, we cannot consider the difference of scale among migrants as the cause of hierarchy formation. Nevertheless, it is the spontaneous gestures that leave indelible marks on the minds of those in the village and the elite are often called upon. To many, their contributions are seen as expressions of repayment to the community association for the support they received earlier in their lives. By and large, successful migrants have been looked up to as the elite and their generosity is taken to be the norm. And if they do not conform to cultural expectations of generosity, they will lose the respect of the community (Woods 1994: 471; Barkan *et al.* 1991: 479; Trager 1998: 370-375). Some of the younger migrants have stayed away from the village since they left home and resentment of the elite depicts the divisions among the younger migrants and the elite in PIFAM. This act of resentment is mostly manifested in the form of contestation against development projects and at times derogatory remarks are made of others.

Despite this subtle contention, the elite plays a significant role in fostering debate in PIFAM meetings, often taking the lead in heated debates by calming tempers when faced with a crisis, such as the death of a member in Cape Town or setting an example for the rest. When it comes to school projects, the decision to assist is unanimous and PIFAM (Cape Town and Johannesburg) was ranked as the highest contributor among all the diaspora communities to the village electrification project. Hierarchies are played out in various social fields and this should be understood as relational and situational.

Social life of PIFAM and MACUDA

Being away from home means that migrants want to be in constant communication with their families back home and to be aware of the social and cultural happenings in the villages. Their cultural and social orientation is not only informed by happenings back home but is also inspired by the re-imagination of the traditional habituation (Hickey 2004). Although home videos of funerals are watched in Cape Town to see whether the funeral was well organized, it is also in the hope of seeing family members and to learn traditional songs and dance patterns that have emerged back home so they can be re-enacted in Cape Town. It is common to find Pinyin migrants congregating at the home of a member on receiving the news that a video of a funeral from home has arrived. These tend to be watched on Sundays or in the evenings. The videos are often sent with someone travelling back to South Africa or very occasionally via DHL. While viewing them, a running commentary is given on the songs, dance styles and persons that the migrants have not seen or heard about in a long time.

The end of the year is marked by festivities and both groups organize a party to celebrate the results of their hard work over the course of the year. In each association, a committee is assigned to plan the party and report to the GA. PIFAM assumes full financial responsibility for its party, while MACUDA members have to make a contribution to theirs. The committee in charge of organizing PIFAM's party in 2011 opted for a catering service to the dismay of some of the members who would have loved to eat local Pinyin *achu* and other homemade food. As for MACUDA, they keep theirs local with *achu* and the sauce that Cameroonians always serve with it (a yellow and black soup).

On some other occasions, parties are complemented with picnics. Groups of people (sub-associations within PIFAM) or the entire group (MACUDA) organize a picnic at one of the city's tourist attractions (see also Thompson 2009). The date, venue and logistics are all transmitted by mobile phone, usually an SMS.

Photo 7.6 Eating achu during one of the festivities

Photo 7.7 A cross section of members at PIFAM party

As far as PIFAM is concerned, a party should mark the end of a football tournament between the various sub-associations following the awarding of the trophy and the prize-giving ceremony. Referees and linesmen come from within the Cameroonian community and the teams involved all take the tournament very seriously. Winning the cup will produce a huge sense of pride for the quarter that wins. Conversely, the event abides by strict rules and regulations, which highlights how nothing in this community is taken for granted. The tournament and the teams are essentially organized and coordinated via mobile telephony, as can be seen from these SMSs.

SMS 1

'Hi if u around Bellville pls check the jersies (sic) and the price of 15 pieces; top only and top with shorts. Pls let me know.'

SMS2

'Hope u have well rested from the 120mins match and now ready for the final on Sunday. We know the cup is ours, pls training is compulsory 2moro.'

SMS3

'Makong is threatening us with 2 campings (sic). Compulsory meeting 2night at Xxxx's place at 20h: 15 for game planning.pls do everything to come.'[18]

[18] SMS forwarded to me by Chris, Cape Town: 14/03/2011.

Photo 7.8 Members of the winning team present the trophy to the sub-association elite

PIFAM's ability to organize a league game among members (with five competing teams from different quarters) highlights the size of the group. These events do not only instil a sense of community and belonging but also uphold a celebration of the community that, as Thompson (2009: 374) puts it 'becomes embodied representation of the community within Cape Town'.

The winning team and other outstanding players receive trophies and medals on the night of the gala. When the 2011 champions, KWIANDA, received their trophy, they significantly handed it to their patron (elite) telling him: '*you be send we out for go hunt, na the beef this wey your pekin them don bring am back*'.[19]

Photo 7.9 Researcher presents a prize to one of the teams

[19] Field notes, Cape Town: 19/03/2011.

The importance of this statement is twofold: it is connotative and denotative. The strong image of a hunter was used to refer to their win and the trophy was the prized game. So they saw themselves as 'bushfallers' (Nyamnjoh 2011: 703). In this respect, the idea of bush, hunting and coming back with game is the language understood by migrants and their families, and this stresses the meaning of success and what it takes to be a successful bushfaller. By extension, it refers to the way migrants and migration are conceived back home, as hunters with the host country being seen as the 'bush'. Putting this into perspective, winning the trophy and/or whatever migrants set out to do have to be done diligently to produce results. Going away is thus for a purpose and failure is not condoned. Implicitly, this reflects the very essence of their being in Cape Town: they are out to succeed at whatever they set out to do. Goals must be set and met, as can be seen in their text messages. Over and above this, going all out to win the trophy is alluded to by going hunting and one has to come back with the prized trophy (antelope, buffalo or big game) to show prowess and demonstrate that one is a seasoned hunter. Being 'bushfallers', as they are considered back home and even amongst themselves in Cape Town, means they have to live up to the name and status they have achieved as good hunters who do not settle for less (or smaller animals). Presenting the trophy to the patron was a sign of respect and shows the high esteem and status he has earned in the group and in the quarter in particular. It also shows that, as hunters far away, they still abide by the same moral ethos and, as good children who have gone out to hunt, they have brought back booty that they present to the father with pride. Respect in both Pinyin and Mankon is due to elders and if it is lacking, the social values of the community are said to be crumbling, as was alluded to in the case of PIFAM Johannesburg by most informants. Although PIFAM organizes a football competition amongst its fold but they also respond to invitations from other associations for friendly matches and participate in inter-ethnic competitions organized by other associations. In addition to funerals and parties, football is the other main rallying activity among migrants.

Inter-cultural communication and associational networking

As many authors have affirmed (Evans 2010; Mercer & Page 2010; Mercer *et al.* 2008; Thompson 2009), the need for migrants to be ever more connected is suggested by the fact that, although they may be away from home, they still consider themselves to be part of the community of Cameroonians and/or Grassfielders. This is why newly formed tribal associations see the importance of introducing their association to other existing ones. Although the focus here is on the Pinyin and Mankon, these communities cannot be studied in isolation for they interact with other communities (Evans 2010; Mercer & Page 2010). They reach out to others or respond to invitations and requests for financial assistance from other groups. Such is the case of the Meta Cultural and Development Association (MECUDA) that sent a letter appealing for financial and material assistance to help the group realize a charity project in Cape Town, to which PIFAM and MACUDA responded positively.

The mobile and flexible nature of these associations and their members is made possible by ICTs that link members and play a role in building and maintaining a sense of community and connection with others. And, in a way, they have reorganized sites of social interaction and formed new collective configurations, and this has invariably presented new opportunities for collective affiliation and mobilization (Thompson 2009: 360). Is it possible to conclude therefore that with regard to associations, ICTs seam members together without any loose stitches? Despite the importance of ICTs to mobile and virtual communities, the absence of social cues (knowing who is talking) and the digital divide still seem to be an impediment that prevents everyone from fully participating in discussions. Nonetheless, in an effort to be connected, the Pinyin and Mankon have made their own telephone directory to ease communication and keep everyone under the radar. SMSs and phone calls have been combined to good effect to achieve effective communication. For instance, when a member is bereaved, the latter calls a member of the executive, preferably the president or the secretary general, to inform them as well as give information about the physical address of the member so that others can convene for a condolence visit. This message is immediately transmitted by SMS and phone calls and members are asked to pass it on to others.

As far as MACUDA is concerned, the group has a phone that was donated by a member and its number is given to all registered members. The phone is with the chief whip who is charged with taking and forwarding all calls and SMSs to the group. In case of absence from meetings, members are expected to call this number prior to the start of the meeting or to send an SMS to excuse themselves, otherwise they will be fined for 'wilful absenteeism'. Messages can also be sent personally via members. Connectivity is not, however, limited only to members of the association, and both associations strive to maintain a cordial relationship with other associations and with sister associations in Johannesburg and beyond.

Exchange visits between associations

The notion of interconnectedness ignores the perception of bounded and passive identity because flexible identity is part and parcel of migrants' associational life, with the continuous need to draw social and cultural capital from the association for various purposes. Piore (1995 cited in Cerulo 1997: 394) expresses a different opinion, contending that collective movements are rather 'communities of meaning' because they are narrowly focused, formed relative to distinction and are thus unable to have cross-boundary exchanges. While this may well explain the case in American ideology, extending it to a wider community would obliterate the sense of community and bonding that exist within the associations and the inter-associational visits. It is not often that associations exchange visits but, when they do, it demonstrates the cordial relationship that the groups enjoy. These visits can also be informative, for example to announce the creation of a new tribal association. The visit of a three-man delegation from the Cameroon North West Association of Cape Town (CANOWACAT) to PIFAM illustrates this point. Their arrival is acknowledged and later, during the course of the meeting, they are given time to address the house. Accordingly, CANOWACAT's president informed the groups about the envisaged changes to the constitution that would make the group more inclusive. In this respect, he visits all the associations from the Bamenda Grassfields to seek their approval. The most important revision that appeals to members concerns posthumous migration and this revision will make it possible for the deceased to be sent home, with levies being made per association and no longer individually. This point was particularly interesting as it is challenging to transport a corpse back to Cameroon, especially if the deceased did not belong to an association or was only a passive member. The idea of taking the deceased to the home village is a reflection of 'bridging capital' and 'bonding capital', which are the primordial definition of the home association. Such attitudes concerning bridging and bonding capital highlight the Cameroonians' solidarity and their sense of home and belonging. The notion of home is strongly invoked and articulated when faced with the death of a member from the Grassfields as can be seen in the following saying: '*we di bury Bamenda man only for place wey they bury yi nitong*',[20] which means one of theirs/ours cannot be buried away from home. This sentiment is echoed by the Pinyin elite back home who stress the importance of home where the umbilical cord is buried. 'Pinyin is my second home after paradise. I love Pinyin, where my nivel (sic) was buried.'[21]

The importance of posthumous migration was reiterated by MACUDA Johannesburg when they visited their Cape Town counterparts. The Johannesburg branch sent a delegation to Cape Town from 16-18 July 2011, a visit that culminated in a cultural jamboree. Both groups agreed to jointly repatriate the body of one of theirs, whether a Cape Town or Johannesburg member. Another achievement was the synchronization of the sinking fund in the event that members want to relocate to either place. Both groups pledged to cooperate on development pro-

---

[20] This is a common expression used by informants when faced with the death of a Grassfielder who has to be buried close to where his/her umbilical cord is buried, namely in the home village.

[21] Sent to PNF, 15 February 2012.

jects back home and the Cape Town branch promised to make a return visit as well.

Such levels of cooperation amongst migrants are not limited to those in the host countries. Operating within multiple spatialities gives 'mobiles' the opportunity to be virtually present but physically absent at multiple loci thanks to the appropriation of ICTs. In this vein, they are able to form transnational bonds with virtual home associations that are made possible by Internet-mediated spaces in most cases, and lately social networking sites, as can be seen below.

Transnational/trans-virtual associational networks

In the wake of increased transnational migration, home-village associations have become important nodal points within the wider diaspora network (Vertovic 2004), with PIFAM and MACUDA being no exception. Mobile communities are increasingly virtually connected to members of the diaspora and within the country through their respective virtual networks of Pinyin News Forum (PNF) and Mankon Forum (MF). In this respect, they are able to participate in multiple social spaces and within these spaces, migrants express a transnational/transcultural identity and bonding that revives and strengthens old forms of ethnicity and reflects the process of reactive formation (Popkin 1999). As Portes (2003; see also Vertovec 2004) maintains, it is only because of better networks that immigrants have become actively engaged in transnational debates, and their absolute numbers play a more important role than their social ties. Quite often, these are weak ties (Granovetter 1973) but they are glued together by a common goal, namely development in the home village and the safeguarding of their traditional village institutions. This is seen in the PNF's objectives that state that the 'Pinyin News Forum is a global meeting place for Pinyin people; its mission is to educate and inform Pinyin people, and as a team, Pinyin people can make a difference in the homeland'.[22]

All of this has been rendered possible by the new information and communication technologies, especially the Internet (Mercer & Page 2010; Mercer *et al.* 2008). Being connected suggests an awareness of events that happen simultaneously in different places. By the same token, such connectivity and awareness mean members get to be actively and emotionally engaged in events (Prelipceanu 2008). It could perhaps be inferred from this that, contrary to Putnam (2000), community is not so much in decline but rather that ICTs have created new alternative spaces for people to sustain meaningful communities at a distance through membership of associations and organizations such as the ones mentioned above. These public-sphere activities give migrants, miles away from home, the chance

[22] Moderator PNF, 20/10/2010.

to engage in critical discussion and self-expression through which participants (re)define their sense of identity, community and agency, and they can also create their own spaces, which Karim (2003 cited in Parham 2005: 373) refers to as the 'supraterritoriality … of diaspora is created and sustained by transforming a milieu: it is not a physical space but an existential location'. The redefinition of identity also charts multiple-identity affiliations that could implicitly or explicitly qualitatively change the nature of migrants' experiences (Cerulo 1997).

PNF and MF have thus created an online public space and this allows for an on-going dialogue largely unbounded by territory but rather by a common identity and shared values and ideas of a common place, i.e. home. This creates 'new social movements', collective initiatives that are self-reflexive and are sharply focused on the expressive actions of collective members (Melucci 1989: 60). He goes on to explain that such collective initiatives produce 'collective agency' because collectives consciously coordinate action. As shown at the beginning of this chapter, the coordination and assistance needed for the high school back home is done through the collective of PDO and channelled through PNF to attract funds. It is only through such Internet-mediated spaces that these mobile communities can make meaningful contributions to village development. Unlike the face-to-face meetings of PIFAM and MACUDA where actions are bounded by structures that enforce discipline, the same cannot be said of online public space.

While participation in PNF comes from all areas, Mankon Forum's is yet to witness as wide a participation as its Pinyin counterpart. The admission of new members is recommended and they need to be vouched for by existing members in both forums. Pinyin is open only to indigenes so my attempts at subscribing to it proved futile. However, I was duly forwarded emails from PNF by some informants/individuals in South Africa and beyond. Coming from Mankon, I am eligible to subscribe to the Mankon Forum. In each of these forums there are dedicated people in the home country and the various nodal points that keep the network informed about happenings at home and in their respective cities as well as about development projects that need to be carried out, although members in the diaspora may not be well versed in the realities on the ground or may want to pursue a different development agenda. By and large, the activities of the forum still remain linked to home-village development and the fostering of a collective sense of community and ownership, as is reiterated by this message posted on PNF: 'Citizen involvement/participation is a desired and necessary part for that community's development process. It is therefore a process that can meaningfully tie programmes and institutions to people and people to others.'[23]

[23] Sent to PNF, 22 January 2012.

In the sections that follow, a selection of emails and my observations are reviewed as a way into the level of virtual communication, transnational connectedness and development strategizing in order to adequately understand how colcollective identities direct participants to define the parameters and appropriate arenas of collective action. These mobile groupings are networking amongst themselves irrespective of distance and are constantly fed news from home on a daily basis.

Although members of MACUDA Cape Town do not post emails on MF, they are inspired by the development agenda within the forum and set to emulate it. One is the provision of scholarships, such as the MF Scholarship Initiative[24] for underprivileged children back in the home village. Scholarship funds are sent from the US but coordinated via MF and someone at home who identifies those most in need. Connection to this website provides various apps to view all contributions and donations and to make suggestions. News of each donation is reported on MF.

> Greetings,
>
> Once again, thank you for our generous donors. The start of 2012 has been nothing short of exciting for MFSI and the scholarship program. With 2012 version of our website protected pay link verified, MFSI continues to enjoy steady support from members, well wishers and multiple donors.
>
> We are very excited but before I get carried away, please let us acknowledge. ...
>
> 1. Dr. XXX today made a 150.00 dollars donation into the coffers of MFSI using our secured website link.
>
> 2. Dr. XXX at the University of Michigan has made a 200.00 dollars check payable to MFSI that was received in our office today. MFSI is making steady financial strides. Thank you to our wonderful donors.[25]

MACUDA Cape Town assisted financially towards the completion of the church hall of the Ntambeng Catholic mission in Mankon and started a project to assist the newly built divisional hospital with medical equipment. However, this project has stalled because of some non-Mankon members from Ngemba.

Development activities amongst the Pinyin are coordinated by the PDO via PNF, and participants respond very generously when they are informed of an urgent need by the translocal back home, as can be seen from the previous email regarding the supply of damaged infrastructure and the email below:

> Dear forumers,
>
> The buchi community wishes to thank all friends and well-wishers who turned out in their numbers to give them a push yesterday during the launching of GSS buchi project ... BUDECA[26] yde (Yaoundé) intends to construct and equip two classrooms worth 5,000 000.

[24] www.mankonscholarsh ips.org
[25] Posted on MF, 06/01/2012.
[26] Buchi (a quarter within Pinyin) Development and Cultural Association.

> During the launching yesterday, 800,000frs cash was realized … 1.5million was pledged. But like oliver twist (sic), we continue to ask for more.[27]

Nonetheless, we should be cautious about drawing hasty conclusions and overly praising these forums, and reassess them to understand whether they are as flawlessly seamed as they suggest. In this respect, some emails have caused quite a stir and have threatened groups' unity and focus. The circulation of an email with the subject, 'YOU LAZY (Intellectual) African Scum'! dominated the Mankon Forum for over a month (January/February 2012). In response to the email, Y wrote:

> As they say, If you stand for nothing, you'll fall for everything. Btw, (sic) as a Mankon man, writing to the Mankonforum, what possesses you to make a statement such as "I encourage you to read this piece right to the end before making your usual glib, noxious and self-mythologizing shallow decadent and culturally unpalatable commentaries?" Did I miss some demons on the Mankonforum or you just feel like using so many words to convey nothing?
>
> Anyway, I read (and reread) the article and I can't stand when people like Field Ruwe's Walters use racist undertones (and abusive language)against a race of people and claim they are not racist, irrespective of the good they want to make.[28]

Again Y wrote: 'I ask a basic question and you and X is too lazy to think except to infer censorship!' In response, the initial sender, X, wrote:

> I am totally capable of unleashing a cannonade of shock and awe artillery to defend myself when provoked, but fortunately, Mankonforum is neither the right venue nor the space for equivocation and palapala (*wrestling*)[29]. I sincerely think we are having a miscommunication problem here but let me state for the record what you classify as "Being TOO LAZY TO THINK". This your Mankon-Forum is a "Njoh" (*free*) medium (for now) offered by YA-HOO to Konto peepoo (*crafty people*) like you and myself to "Communicate". It is neither a classroom, amphitheatre nor an editorial enclave. I have no obligation to write long mails, short mails, intelligent mails, Njakri mails (jokish mails) or what have you...infact I owe no one any explanation whether I write or don,t (sic) write BUT I have an obligation not to be insultive or trample on other,s (sic) freedom of speech. This simply means I can write on any topic that pricks my fancy without fearing that the Ys of this world will criticize either my wowo (*bad*) English, my uneducated lazy thinking faculty with their temporaneous cynical injections[30].

The reason for dwelling on this is to show that although studies carried out on transnational and trans-virtual forums are often portrayed as representing a united front with the sole focus being on village development, they sometimes become public spaces for verbal and literary abuse with little regard for readers or the negative publicity it gives the forum. Such feuds often lead to the contestation of identity and belonging. This can reignite age-old divisions between autochthons and allochthons that are inimical to the unity of the association and concomitantly derail it from its development and welfare agenda. The messages are inter-

[27] Posted on the forum, 21/11/2011.
[28] Sent to MF, 22 January 2012.
[29] Italics are by this author.
[30] Sent to MF, 24 January 2012.

spersed with Pidgin English and the local dialect, limiting its consumption to the target group. It also underscores the fact that such Internet-mediated spaces provide few social cues as in the above exchange where a male member is addressed as female, as well as suggesting neglect of the moral ethos of respect. Such tantrums shift from the goals of the forum and destabilize the 'moral conviviality' (Mercer & Page 2010: 115-116), leading to some members asking the moderator to unsubscribe them. This also shows how participants in online discussions are sensitive to the use of offensive language. Despite the norms guiding discussions, as seen in the mission statements of both associations, a lack of social clue and distance makes it difficult to reinforce rules of communication.

Similarly, within PNF, some members discredit others concerning postings on the forum and tend to be arrogant as can be seen from this email; 'Your English and your reporting are terrible and unworthy of your public and you'.[31] In reply, another member cautions forgiveness:

> I want to encourage you X, to forgive and forget the reaction of our father, Y to your mail. Though he has an option to write to you, but he sincerely regretted to me when he met me a few minutes after he sent to the comment to the PNF (sic). I realised that he acted on impulse. Please, to err is human and to forgive is divine. Let all of us be more positive and forward looking on the PNF.[32]

There is also interest and research into the ICT debate and how it could be harnessed to benefit the community, with particular attention to cultural dissemination and education.

Contributions are made to PNF from all over the world by Pinyin indigenes who are connected to the Internet, while MF is dominated by those in the US and translocals. But news in both forums is diverse, from concerns about chieftaincy, land disputes, religion, cultural and heritage education, political discussions, death announcements, news back home and, of course, developmental projects.

Technology thus makes it possible for the diaspora migrants to create discursive spaces through which local and silenced identity narratives can be articulated (Mitra 2005). Such in-group discourse, as we have seen, provides the glue that holds the members together in virtual spaces, irrespective of location and physical contact and links them to the home village. Despite these linkages, such discursive spaces have also been spaces for animosity and discord. Although PIFAM Cape Town and Johannesburg enjoy cordial relations, they seem to have some doubts about naming. This stems from the fact that they each assert they were the first to be formed. Cape Town thus wants the Johannesburg branch to be called PIFAM Johannesburg and not South Africa as it suggests that Cape Town

---

[31] Sent to PNF, 9 August 2011.

[32] Sent to PNF, 10 August 2012.

is of less importance, which does not go down well with the Cape Town branch. To this effect, Jake said that:

I think it's more around the name because the people in Johannesburg they call themselves PIFAM SA … And we call ourselves PIFAM Cape Town. And our expectation here was that we would call ourselves PIFAM Cape Town, South Africa and they call themselves PIFAM Johannesburg, South Africa; so that ... it is clearly indicated that we are not national but just regional. But when they call themselves PIFAM SA they are sort of absorbing us.[33]

Despite such differences and discord among members and groups at times, associations and forums are guided by the need to develop the home village and be parents to members, as the title suggests.

Conclusion

Associations, as has been shown in this chapter, play a major role in the societal fabric and seek to establish a social cohesion amongst members as well as acting as a bridge between the host community and the home-village community. The social and cultural character of the association mimics that of the main associations back home and the activities that form part of their social life, namely sports, cultural activities, *njangis* and visits between groups. Such moments of cultural showcasing enable migrants to create a home-away-from-home, which reinvigorates their Pinyin and Mankon identity as well as their bond with the home village. In this regard, associations should be understood within the framework of agency in the production of structures because 'structure is both the medium and outcome of action' (Karp 1986: 135). Such structures define belonging, identity and exclusion and are a way in which actions can be enabled. For both communities, identification is first and foremost the projection of their Pinyin-ness and Mankon-ness over their Cameroonian-ness by emphasizing the language and re-enacting cultural activities. The projection of their Cameroonian identity is only noticed when they rally behind the national football team (Njubi 2002). As noted by Mohan & Hickey (2004 in Evans 2010: 418), the formation of the state in Africa was primarily based on the collation of ethnic identities where they are linked to a common place, i.e. 'home', through a shared meaning and culture. It could also be explained by the fact that migrants are more driven by the need to develop their village (Pinyin) to prepare a place for their eventual homecoming. As such, migrants and translocals, through these connections, play a role in the lives of those back at home because, as Parham (2005: 349) usefully notes, the creation of such mediated public spheres enables participants to 'move

[33] Interview with Jake, Cape Town; 11/08/2010.

from shared imagination to collective action', thanks to the 'Internet-mediated network public'.

This is in contrast to the situation in the Netherlands where migrants' Cameroonian identity is projected. The preference for nation rather than ethnicity could be tied to their limited numbers and is probably due to the long and odd hours they work to make ends meet and satisfy the incessant demands back home, which leaves them little time to create home associations. It could also be as a result of failed attempts to sustain an association. Nevertheless, urban-rural connections have challenged the modernization theory that forecasts a disconnection between urban migrants and ruralites. Migrants cope with home nostalgia by maintaining multiple-stranded virtual relationships with various associations and home associations, which is consistent with the Manyu case examined by Evans (2010).

Belonging is privileged, and although migrants tend to operate within multiple and virtual spatiality, their 'homeland' remains Pinyin and Mankon. In this respect, they have intensified their links with home by participating in rituals and development projects and requests, and also through the replication and redefinition of cultural activities in their host countries. Such multiple-stranded relationships and connections have been made possible by the omnipresence of communication technology and this has given rise to the convergence of virtual communities with a common focus: home.

While belonging is stressed and associations' Internet-mediated forums appear to be an umbrella that unites members, social cohesion is not a given as there are simmering tensions that threaten the unity and moral conviviality that bind members. Debates on autochthones and allocthtons, and identity and belonging that are part of the discourse in most associations back in Cameroon re-echo within the migrant community.

Although these associations aim to foster unity and promote their own society's cultural values, development at home is primordial. However when deciding on a project, the underlying problems of social hierarchy may re-emerge and associations can develop hierarchies, either advertently and inadvertently.

The advent of ICTs has increased migrants' virtual mobility and transnational/trans-virtual social ties with migrant associations that share a common 'place', meaning and identity. Through their respective forums, they have harnessed the Internet and, more recently, the new social media that have made a nonsense out of borders and rendered them porous, thus bringing the home village closer to migrants. Thanks to this, mobile communities are able to contribute to the development of the home village.

8

A mobile community as a fortress: Reinforcing the notion of belonging through 'life crisis'

Introduction

Chapters 6 and 7 discussed migrants' economic activities, their associational lives and how these are constructed in spatiality of place. They also considered how migrants seek to maintain their Pinyin and Mankon identities through the ties they maintain within the migrant spaces in the host country and ultimately back home through their re-enactment of cultural and ritual activities to reify their belonging to the home village.

To broaden our understanding of notions of home and belonging, this chapter offers an in-depth analysis of the ritualistic activities that migrants entertain in the host country to reinforce their notions of 'Pinyin-ness' and 'Mankon-ness'. The aim is to demonstrate how rituals are connecting devices, the glue of the mobile communities across borders, as well as how they connect the migrants to the home village. Contemporary African societies are increasingly mobile but this mobility does not prevent migrants from continuously seeking to partake in ritualistic activities related to birth, marriage and death in their host and home countries as the Pinyin and Mankon communities in South Africa and the Netherland do. The variety of rituals that van Gennep (1960) termed as 'life crisis' and their performance as 'rites of passage' celebrate transitions from one phase in an individual's or community's life to the next. Performing these rituals enhances their connectivity and implicitly expresses their sense of belonging and, by the same token, these rituals are a form of verbal expression and a means of attaining a greater sense of identity (Davies 1997). It is this difference in perception that disconnects them from other ethnic enclaves, especially from the host community. In other words, rituals connect people of similar orientation and disconnect them from others. Amongst the birth, marriage and death rituals that will be looked at in this chapter, the one that disconnects most is death, as the ritual varies within societies and 'outsiders' may find the performances meaningless because strands of meaning change from one culture to another. The funeral rites thus encapsulate and symbolize the society.

This chapter focuses on how mobile communities, although away from home, continue to uphold and participate in ritual practices and ceremonies in the host and home countries simultaneously. It shows how rituals are dynamic and reconfigured as a result of mobility and new communication technologies and looks at the ways in which these new technologies allow rituals to be transformed on the basis of migrants' desire to be involved.

The arrival of ICTs has made participation in these rituals and ceremonies virtual and gives them a fresh impetus because of their existing locus. In this respect, we consider what rituals mean to an ever-mobile community and why migrants seek to indulge in ritualistic or ceremonial performances or transnational/transcultural rituals. How do these rituals impact on social relations and to what extent do they form coherent patterns of internalized culture? What do the actors aim to achieve by participating in these ceremonies? And to what extent do life crises follow a society's cultural norms? How and to what extent do mobility and ICTs inform rituals or celebrations?

Understanding rituals and ceremonies

One way of answering these questions is by an ethnographic reading of the cultural practices that constitute rituals steeped in culture as a transnational habitus, as well as through an in-depth description of the various rituals. This chapter is largely informed by extended observation and participation as well as informal conversations. It equally draws from Faist's notion of 'translated people', where he perceives mobile communities as being 'continually engaged in translating language, cultures, norms, and social and symbolic ties' (Faist 1998: 239; see also van Beek & Blakely 1994: 3). It follows that rituals translate enduring messages, values and sentiments and it is through these translations that transcultural mobile communities 'forge their sense of identity and their community, not out of a loss or mere replication, but as something that is at once new and familiar. As a result of their mobility, there is the possibility of invariance' (Rappaport 1997: 36-37; Mbuy 2012). This is informed by the recognized and accepted pattern, or a '*bricolage* constructed of cultural elements from both the homeland and the receiving nation' (Kivisto 2001: 568). This is the case with birth and marriage rituals/ceremonies but is in contrast to Pleck's (2004) assertion that rituals are likely to vanish among migrants who do not live in an ethnic enclave or neighbourhood. Even though Pinyins and Mankon in the Netherlands do not live in such ethnic enclaves, they participate in rituals within an enlarged Cameroonian one, while participation in Cape Town is often in ethnic groupings. These rituals are important as they are ordered, with death and burial being the most important and ubiquitous of all the rituals. This is underscored by the fact that rituals that go with death are elaborate and involve the designation of a successor who continues to act as the social glue. Such rituals, for example the enthronement of a successor, declare the future responsibilities of the newly crowned and invest him/her with the authority to assume the intended role. Nevertheless, these various rituals hang together closely and involve personal and social crossings whose meanings are as important to mobile communities today as they were in the past. It is by performing these rituals and ceremonies that actors assume new roles in society. They portray a basic structure of separation, transition and/or incorporation but, over and above this, they define the social reality and offer a window on social order and people's understanding of this order as well as their position within it (Turner 1967).

Reference is made in this chapter to rituals and ceremonies but it is important to make a distinction between these two concepts and the context in which they take place. The distinction is based on that made by Wilson (1957: 9), which will serve as a continuous thread throughout the argument here. Using ritual to mean a religious action, for example an action directed at the secure blessing of some mystical powers, she maintained that, 'the action may be a negative, that is, avoidance or taboo, as well as a positive one. Symbols and concepts are employed in rituals, but are subordinated to practical ends.' Conversely, 'ceremonial is an elaborate conventional form of expression of feeling and is not confined to religious occasions and a ceremony is not enforced by mystical sanctions, only by conventional ones'. As Pickering (1974: 75) concurs, a ceremony is what in the 'actor's eyes is an important occasion'. It thus celebrates the born house or wakes to coincide with those back home as the most plausible thing to do. Shedding more light on this, Wilson (1954: 240) maintains that 'a ritual is often embedded in ceremonial which is not held to be necessary to the efficacy of the ritual but which is felt to be appropriate'. She goes on to ascertain that 'both ritual and ceremonial have a function in rousing and canalizing emotions, but ritual, by relating its symbols to some suppose transcendental reality, affects people more deeply than a ceremony'. What we can take from this is that rituals reveal a society's deepest values and are a gateway to understanding it. Here ritual is equated with religious action and mystical notions, while ceremonial is a more inclusive concept and refers to any 'elaborate conventional form', with no mystical notion attached to it (Goody 1961: 159; Norbeck 1963: 1255). In this regard, ceremonial may be used to refer to those collective actions required by custom and that are performed on occasions of change in the social life. Thus a 'ceremonial consists of a specific sequence of ritual acts performed in public' (Goody 1961: 159). From this perspective, I posit that, other than the death ritual that entails the presence of a corpse (as in the case of the Pinyin migrant who passed away or born house[1] where there is a representative of the family from home), most of the rites carried out in Cape Town and the Netherlands by the Cameroonian communities are more ceremonials than rituals. But by the same token, they are also informed by the nature of the rituals that go with such ceremonial. Let us now turn to the various ritual practices and how they are performed.

Birth

Birth can be viewed as the opposite of death as the two are the twin poles on which life revolves. As pointed out by Sawyerr (1974:78), 'birth is like setting up a plantation with new trees which increases the ecology of the family … Death, on the other hand, disrupts the balance of the plantation as the spirit of the deceased leaves the corpse to join the community of the ancestors in a new world.' While this metaphor is in order, I note that Sawyerr's views are informed by western norms as he omits the births of *abiku* or *obanje* (children of the spirit world who die shortly after birth and come back again). Such births call for rituals to stop the children from returning to their families again. What this suggests is that births of 'normal' children are often a blessing to the family and conversely that of *abiku/obanje* are unwelcome, and rituals are thus performed to stop them from returning to the family. Nonetheless, rites associated with marriage and birth (born house) could either be rituals or ceremonies depending on the significance or the objective. Quite often, born house is ceremonial but becomes a ritual in the event that it involves twins or the birth comes after a number of miscarriages and/or stillbirths as rites are performed for the child and the mother. Amongst the mobile community, this could become a ritual when a member of the family is present from home to represent the ancestors and blesses the child. Unlike the rituals surrounding death, born house has become dynamic and taken on other forms of celebration including, for example, baby showers that are celebrated before the birth of a child. This is in sharp contrast to the traditional norms and customs back home as a born house is only celebrated after a baby is born. This celebration is linked to the co-existence of cultures that have invariably influenced each other. I was fortunate to have attended both the baby shower for Jane and the born house. The ceremony shows how mobile communities have adopted other forms of ceremonials (the baby shower) but still keep the traditional form of born house. Migration has resulted in a hybridization of events. This form of ritual is seen to be dynamic and fluid, while it connects migrants to other

1 Ceremony marking the birth of a child in the Grassfields.

societies, transgressing borders. Belonging can therefore be expressed differently. Not everyone can afford to organize a baby shower as it is characterized by wealth and affluence and, by extension, leads to the formation of hierarchies. While such celebrations call for the showering of baby gifts on the mother-to-be, it is organized more as a social event that involves wining and dining, hence the financial implications.

What marks it a ritual are the utterances and the 'breath of men' (*ndjiŵi*), the art of breathing onto a child that is known as *fogho ndjiŵi-zo* (blessing the child with one's breath and the utterances that are made in the process) amongst the Pinyin and Mankon. This is meant to protect the child and the breath is used to impart into the child the spirit of both the dead and the living, as a form of protection and of committing the child into the care of the ancestors because they are considered as 'living dead'. It could also have a punishing effect, i.e. bad mouthing.

The birth of a child within both communities invites the family to choose a suitable ancestral name for the new-born. Before birth, the family phones home to ask for suitable names for both a boy and a girl or the message is passed on just after the baby's birth. In this connection, a name signifies an event, marks an occasion, is a prophetic insinuation or shows the striking sentiments of the father or the head of the matriarchal or patriarchal lineage by naming the child after a deceased member of the family or the father. If it is a first child, the name is given by the wife's family. Jane's case is worth highlighting.

> She is a young girl who got married in August 2010 and travelled to Cape Town the same month with her husband. In January 2011 she enrolled at the university and in September 2011 she gave birth to a son. Prior to her delivery, her friends organized a baby shower for her in August 2011 that attracted a substantial crowd with gifts for the baby. As part of the ceremony, women who had already experienced childbirth were called upon to advise her on what to expect, and motherhood. In the course of the celebrations, she and her husband were smeared with baby powder but he was immediately sent out afterwards. This is an all-women affair and the men who are invited are received in a separate room and are not allowed to join the women. This is because all the discussion is centred around labour, childbirth and motherhood and balancing this with the duties of a wife. Despite having borrowed this form of ceremony, the women have succeeded in enculturizing it by inviting men, although separating them from the women. Generally it is a women's affair and men are not invited, such as the one I attended in The Hague.
>
> Being her first child and having delivered by caesarean, the parents decided that her mother should travel to Cape Town and assist her for a few months. It was during her stay that Jane and her husband took the opportunity to organize a born house for the baby. This occasion bears the hallmark of both a ritual and a ceremonial: a ritual in the sense that ancestral blessings were adorned on the baby and a ceremonial because it involved the entire Mankon and Pinyin communities since the husband belonged to both associations. It was in the course of the celebrations that the ritual was performed.[2]

2 Fieldnotes: baby shower and born house for Paul and Jane's baby: Cape Town; 20/08/2011 and 05/11/2011 respectively.

Men have no role in this ceremony and they come merely to socialize. During the ritual of committing the child to the ancestors, the ancestors are called upon to 'watch over a new addition into the kin, while the living are called to direct and help the child to know his/her lineage, and that anyone who looks at the child with "bad eyes" will be dealt with by the ancestors', and the visiting grandmother is called upon to bless the child with her breath (*fogho ndjẅi*), which represents those at home too.[3] The breath as such is not hers but is that of the family and the ancestors that she represents. This subsumes that the breath of men imparted on the child is capable of punishing those with evil intentions towards the child.

Photo 8.1 Jane's mother about to bless the child with her breath, flanked by Jane and her husband in traditional regalia (*ntogho*)

For the born house, Paul and Jane hired a professional photographer and a video camera and there were also people present with mobile phones and digital cameras to immortalize the event. A photo album that covers Jane's mother's stay and that of the born house was given to her to take home. The mother was moved to tears by the way the Mankon and Pinyin communities celebrated the born house, paying attention to the ritual songs (that for the parents of twins come with a plate of salt and palm oil for everyone to take a pinch), especially as Jane is a twin. This is to celebrate and acknowledge the twins in the family and those present. Such attention to detail highlights their identity and belonging. For her mother, these photos will be testament to those at home that although the

3 See Fieldnote 1.

child-ren have 'fallen bush', they are still versed and entrenched in their cultural practices.

Birth in both communities is a rite of passage that calls for celebration but one that also calls on the women to maintain the habits and practices that they have internalized over the years. One such practice is when a woman gives birth and kin and/or friends take turns to visit early in the morning or spend a few days with the young mother to bathe the baby and to teach the mother what to do during the early days of motherhood. As well as being a researcher, I also participated in such practices and helped bath Henry's baby, and also attended born house and baby-shower ceremonies. When there is a new baby in the community, women close to the mother and her *nkaŋg* (age group) visit regularly but friends too come due to the strong network ties that have developed over the years, as do village Pinyin folk in the case of Pinyin women.

Mankon women's networks extend to their Nigerian friends who come to assist in bathing the baby. Henry's wife talked about how her Nigerian friend came each morning to assist in this, even on the day when it was my turn to bathe the baby. It is thanks to the actualization of such internalized practices that young mothers away from home can learn how to look after their baby. Through such practices, members are able to express a shared identity distinct from that of the host community. By contrast, when Charles's South African wife gave birth while her mother-in law was visiting from Cameroon, the latter was surprised that none of her relations came to assist the new mother. Nor would the wife allow her mother-in-law to help her because they had different cultural identities and belonging.

The residual of the umbilical cord (*nitong*) that falls off when the naval heals is regarded as a form of connection between the new-born and the ancestral land. This residual is often sent back to the village by migrants for their parents to bury under a plantain/banana grove in the compound. In extreme cases, the placenta is frozen and taken back to Cameroon by the migrant him/herself. This means that no matter where the child goes in later life, s/he will return to the place where their *nitong* is buried either in person or posthumously (see also Mebenga 1991 cited in Geschiere 2005: 53; van Gennep 1960: 51). To belong to a lineage is to be protected, and part of this protection is to send back the remainder of one's child's umbilical cord to Cameroon when it falls off. In the case of twins, a peace plant is equally planted behind the house. When plantain are produced by the tree(s) that it is buried under, the bunch is distributed among the members of both families. Although I am not aware of whether the plantain were given away in the case of Pinyin and Mankon children in Cape Town, the plantain that bore fruit from the *nitong* of my sister's son, which was sent from the US, was distributed to our aunts, uncles and maternal and paternal in-laws. What is significant here is not the quantity of the plantain but the acknowledgement that, as a family, they are able to celebrate their progeny. The eating of this plantain provides an opportunity for the ritual incantation of virtual blessings on the child by those who have shared in the plantains. This plantain grove and the fruit it produces link the child back to the homeland and the harvest are shared among family members who celebrate the health of the child, and the child, even as an adult, will be protected forever.

Marriage

The marriage ritual is similar to that surrounding birth. They have become dynamic but affluence has negatively influenced the traditional institution of marriage by substituting the bare minimum requirements with displays of extravagant consumerism (van Dijk 2012). It has resulted in migrants' marriages being known as 'bush-faller weddings' with most aspiring to marry a bush-faller in order to travel abroad (Nyamnjoh 2009). This has led to hierarchies in society due to the over-extravagance of the occasion. Others have chosen to set new standards, placing themselves apart from the rest. For example, Joe's brother was the first PIFAM member to have gone home and organized a white wedding. The marriage procedure begins with the son expressing his desire to get marry, which starts the family back home embarking on a search for a bride. Would-be Pinyin couples often rely on photos to get to know each other and some get married without ever having met each other. The groom sends his photo to his parents to show to any prospective bride, while the bride does the same. The groom coordinates the process of paying the bride price by phone and sends the money required. Most marriage arrangements are however done by parents and the bride is sent to meet her future spouse. Such arrangement and the objects involved have been described by van Dijk (2012: 146-151) as 'generational technologies of connectivity'. This refers to the 'ritual connection between the village and the city, and between the life worlds of emerging middle classes and the rural socio-economic milieu'. Similarly to this, in the Pinyin and Mankon communities, photos and the mobile phone have become the objects that connect the home village and the host country. The new age of technologies has transformed marriage rituals with regard to mobile communities as most marriage arrangements are done via the mobile phone and using wireless communication. Photos are scanned and sent electronically or are sent with folk traveling to Cape Town or Cameroon. Doris and Zoe, for instance, joined their spouses without formerly meeting them other than via the photos they exchanged. In this respect, both parents played a significant role in investigating prospective grooms before allowing their daughters to marry them. As Zoe was living in the South West Region at the time, she depended on the good judgment of her parents to check out her future spouse and his family. In her case, Doris communicated directly with her spouse and they exchanged photos. Her parents also investigated the prospective groom and made several phone calls to relatives in South Africa and Europe to ascertain his eligibility. Once they were satisfied, both women could get married.

Mobility and communication, in a way, have made it easier for migrants to bypass elaborate marriage rituals, such as *knock door* at the bride's home to introduce the idea of marriage and actually coming with palm wine to seek her hand in marriage, *muluh-me nkwenteh tsundah* and *muluh-me mbiteh maŋgie* respectively. In all of these rites, the migrant's absence is felt: he is physically absent but virtually present as he finances the project and makes regular phones call. For instance, after Mike in Cape Town let his parents know his wish to get married, all the errands related to meeting the various in-laws were done by his father, while Mike sent money to him to meet the demands of his future in-laws. After each visit, the father or mother beeped him and he called back to be updated. He also gave his mother money to enable her to prepare the food that has to be taken to the homes of the various in-laws. Due to his absence, certain customs were translated into money value and he sent the required amount. This made the process faster than it usually is. Families back home have also compromised certain processes because they want their daughters to be able to travel abroad. What underscores this ceremony is the level of networking and communication that takes place. When the migrant expresses the wish to get married, the family gets to work to find a suitable bride for him and performs the various rites that go with bride price.

Despite various compromises, migrants in the host country try to adhere to other marriage rites. Ikenga-Metuh (1985) elaborately described how, when a new bride is sent off, she does not go to stay in her husband's house immediately but instead goes to live with his mother before finally moving in with her husband. This ritual practice is still maintained among the Pinyin community in Cape Town and new brides stay with an elder kin of the husband for some time before joining him in his house. When Doris arrived in Cape Town, for instance, she did not go immediately to live with her husband but to live with his older brother for a week. It was the same for Mike's wife, who was a new bride in February 2012. She lived with Mike's maternal cousin and wife for two weeks before moving into Mike's house. This period is meant to act as initiation into the community (an educational period) and to allow her to get to know the ways of the community. It also shows respect for older kin and strengthens the kinship bonds by emphasizing that she has become a member of her husband's lineage. For Mike, this had double significance: it encouraged him to repair his strained relationship with his cousin and allowed his gradual re-integration into the PIFAM community following his earlier expulsion from it. Much as this is widely practised by the Pinyin community, not all grooms submit to such practices and their spouse moves in with them as soon as they arrive in Cape Town.

As far as the Mankon community is concerned, they are quite different from the Pinyin in this respect as the bride moves in with her husband immediately and friends of the groom may drop by to welcome her. The group performs a general welcoming reception for all the new brides and dishes were, for example, given to each of the four women who recently joined their spouses in Cape Town. Such a gift is significant. As part of the farewell (from her family) and her incorporation (into her husband's family), a bride is given a set of dishes meant for the food she prepares for her husband (*papa ee pan*). While some aspects of the above rituals have been minimized, they are fluid and dynamic, characterized by material wealth and have been transformed in such a way that they give new meanings, such as showing off affluence to prove themselves as successful 'bushfaller'. This emphasizes the importance of what is happening to the actor and to the participants as well. Seen from the perspective of bridging and bonding social relations, by choosing a bride from home and adhering to the process of bride price, having her live with a relation on Cape Town before moving in with her husband and subjecting to a consumerism-style of marriage, marriage ritual becomes transcultural (see also van Dijk 2012).

Death

Unlike birth and marriage ceremonies, the ritual of death is still sanctioned by habits that migrants have internalized over the years. By adhering to such rituals, they showcase their belonging and identity. Subscribing to the death ritual without seeking to alter it means a great deal to the mobile community. It connects them with the home village and garners respect for them (by repatriating corpses or the way they organize family funerals at home). In addition, they receive blessings from the village community and family, and the mobile community is presented as a fortress that is able to withstand adversity as the spirit of collectivism is exercised. Through this act, migrants ward off any bad luck or bad mouthing that would have come as a result of not performing the ritual. While such beliefs are widely regarded by others as witchcraft (see Geschiere 1997; Moore & Sanders 2001; Mavhungu 2012), this study distances itself from these tenets and, following the migrants, sees the performance of death ritual as a means of warding off bad luck, avoiding bad mouthing from the village and peers and creating a form of social control. According to the mobile community, people experience bad luck because of their failure to go home and organize a 'cry die' for close relatives. For example, Jerry's bad luck was attributed by the community to his failure to go home to see where his wife was buried and organize a cry die in her honour. Conversely, doing so is considered a means of gaining blessings.

People talk well about others and the power of words goes a long way as nice words pave the way for success and bad words breed ill luck. And one gains respect from one's peers and other villagers. These explanations thus have become plausible ways of comprehending and explaining the inexplicable and of perpetuating this myth. Ron's death in September 2011 and those of three Pinyin migrants who died in a car crash highlight the importance of performing the death ritual for the mobile community.

Ron's† story

> After he finished high school in Bamenda, Ron engaged in petty trade for two years until his elder sister negotiated a visa for him to travel to South Africa. He arrived there in 2002 and was received in Johannesburg by his cousin who later asked him to relocate to Cape Town where a bigger family circle lives. Here he was received by another cousin and was initiated into pitching, like the rest of the Pinyin. He married a South African, had two children and later separated. He then had a second relationship with another South African woman and had another child. During his nine-year stay in South Africa, he did not go back home once although he was planning to do so in grand style in December 2011. His pitching career ended abruptly when he was imprisoned for two years after being accused of attempting to buy goods with counterfeit money given to him by a customer. On his release he had no money to re-establish himself in pitching. An opportunity came when some tourists asked him to supply them with drugs (crack), which he did by investing his last R 50. He thus earned their trust and was asked to supply more. He supplied them with powdered 'Grand Pa' (paracetamol) and earned R 10,000. Amazed at how much he had earned on one deal and how much income one could generate from selling drugs, he settled down as a drug pusher with the aid of some Nigerian friends as his main suppliers. He also ran a small grocery shop as a façade[4] (see Chapter 4). In early September he had a car accident and died on 22 September 2011. The body was repatriated and buried in Pinyin on his land there where he had started to build a house.

PIFAM informants acknowledged that they had provided the very best casket and suit for Ron as a mark of respect. The body was handed over to an undertaker and a committee was set up to perform various tasks regarding the repatriation: liaising with the Cameroon embassy, processing emigration documents, maintaining contact with the undertaker and family back home, buying a suit for the deceased and arranging the entertainment committee. The men were charged with the logistics surrounding the repatriation, while the women attended to matters relating to food and organizing the social side of the events. Each member of the association contributed R 250 while the rest of the Cameroonian community contributed R 100 each. The aim was to supplement the sum the larger Cameroon community in charge of repatriating the body had raised and this allowed them to opt for a casket instead of a mere coffin, and a Markham's suit instead of one from a Chinese shop. Part of the money went into printing T-shirts with the photo of the deceased on it.

---

4 Interview, Cape Town; 12/07/2011.

Photo 8.2 Ron in his shop attending to clients and on the phone during our interview

Photo 8.3 Ron's grave beside his uncompleted house in Pinyin

Photo 8.4 Ron's uncompleted house in Pinyin

Members would converge every evening in a rented hall for a wake until the day when the corpse was repatriated, with PIFAM members providing food and drink for all the mourners. On the eve of the removal of the corpse, a grand wake was organized with an *mbaghalum* dance. In the meantime, the family back home was kept abreast of all the developments in Cape Town by Ron's cousin and the team leader of the group processing the documents. Money from the deceased's *njangi* funds was sent to the family at home to start preparations to receive the corpse and they were updated as the procedure developed by mobile phone calls to the family that, according to Joe who was in charge, were quite frequent. Direct communication with Ron's mother was suspended in case of a mental breakdown and was established instead with his elder sister who had originally helped him to migrate to South Africa. His corpse was briefly taken to his house in Cape Town for everyone to see for one last time and for his spirit to

Photo 8.5 T-shirt printed by PIFAM in honour of Ron

rejoin the body. One person was assigned to take photos at this point and on the day it was removed for repatriation. All the photos taken during the funeral were loaded onto a USB stick and sent with his partner for the family.[5] While the body lay there, his cousin announced that his creditors and debtors should declare in the presence of the corpse what they owed the deceased or what he owed them in order to pay or be paid (see also Jindra 2005; Sawyerr 1974). Only one person stepped forward, a creditor, who was paid the amount due in the presence of the body. It is believed that the deceased can witness this transaction and any lies are considered tantamount to bad luck. Swearing over a corpse is a very serious ritual that is practised among the Pinyin and Mankon people. Similarly, when this researcher's father passed away, a mason informed us about a job he had done but had not been paid for. Not sure about his story, my mother took him to my father's grave where she placed the money on it for him to pick up, which he did. With Ron's body at his home, a pastor was found to pray for him as he set off on his last, and most important, journey home. The entire PIFAM accompanied the body to the airport for a final farewell and the corpse was accompanied to Cameroon by his new partner and son, and was received there by his sister. The eye for detail shown by the Pinyin community is similar to what takes place in the village, with a particular focus on the appearance of the body, clothes and a casket for the deceased and the related social activities. All these details are emphasized by PIFAM and they thus send a message back home to show that they are still adept at their cultural and ritual practices. But importantly, it also reiterates the

5 Interview with Mark, Cape Town: 27/01/2012.

significance of home, as was mentioned by the deceased himself during my interview with him:

> … you know we stay here but we will never live here forever. There will come a time that you will say I've been staying in South Africa for 20 years, 30 years, now it's time for retirement, I must go stay home.[6]

And when the time came for Ron to go home, he fulfilled the promise he made to me (this researcher) on 12 July 2011, as the return of his body marked not only the final stage in his long-awaited journey back home (see also Sabar 2010) but also the very essence of internment at home (in home soil) that brings the deceased full circle to a final return to Pinyin. This is where he belongs and it confers a change of identity (Davies 1997; cf. Miller 2008). Ron's funeral was organized as a Pinyin (Grassfield) event and was a demonstration of the cultural values associated with ethnic enclaves. Mobile communities also have a moral obligation to send off one of theirs within the cultural code of conduct since '*na disgrace for bury any Bamenda pekin outside grafi land*'[7] (Grassfields), as was remarked by one informant.

1 January 2013: Three more deaths

While Ron's death may have garnered a lot of respect for PIFAM from the whole migrant community and the family back home, the death of three Pinyin migrants on 1 January 2013 in another car crash generated further admiration for PIFAM among the Cameroonian community. In local parlance, '*they wash hand for their back*', which can be loosely translated to mean 'they doff their hats to the group'. Taking financial responsibility for repatriating the bodies confirmed PIFAM's status as the most powerful Cameroonian association in Cape Town. Following the three deaths, members contributed R 800 each to cover the costs of repatriation and to support two of the sisters of one of the deceased. Usually, every Cameroonian is expected to contribute R 100, but PIFAM recognized that the bodies had to be sent home in the shortest possible time and, seeing that the Cameroon Association was not fully engaged in the process, they decided to take charge. Within ten days the bodies had been sent home. PIFAM's executive and the elders formed the committee charged with overseeing the repatriations and the work was subdivided into various committees. One person was commissioned with updating the families back home and a T-shirt was printed with the photos of all the deceased and was sold to members and friends of PIFAM.

A big wake was organized prior to the departure of the bodies and members danced to *Mbaghalum* and Christian songs. In the course of the dance, they

6 Interview, Cape Town: 12/07/2011.
7 Statement made by Mark: 28/11/2011.

Photo 8.6 T-shirt carrying photos of the deceased

Photo 8.7 Members bidding farewell to the bodies of three young PIFAM men in front of the airport freight terminal.

shouted and shamed death for being so cruel to them and as the names of the deceased were called out, members shouted and danced with even more frenzy and vigour as if to trample on death itself. Others took to mimicking the mannerisms of the deceased. This practice could be described, following Rattray (1954 cited in Norbeck 1961: 1268), as the 'deliberate anti-thesis of funeral rites' because they are occasions for jokes and hilarity 'to counter the sense of loss felt by the close relatives'. However, this is not exclusive to those in the diaspora as it is a common occurrence in the home village too. I should note here that I took part in similar activities when my father died: I had to wear his clothes and dance. This was also the case at Ron's funeral. Being a drug pusher, his colleagues in the trade dressed as scammers (with sunglasses, suits and hats), smoked *tik* (drugs) and puffed the fumes over the corpse, while others drank whisky directly from the bottle and invited the deceased to join them. All of this was reminiscent of their joint activities.

Drawing on Davies (1997: 4), such funerary rites are 'strategies of man against nature – words against death', whose consequence is 'human self-awareness'. Equally, they are adjustments to death through externalized, publicly accepted expression of grief, a passionate fighting of death itself, and given treatment in terms of ritual beliefs (Wilson 1954) and an 'assertion of life expressed in dancing, and feasting' (Norbeck 1963).

The following day, 11 January 2013, the undertaker took the bodies for viewing and transportation to the airport. Prior to this, everything had been arranged with South African Airways (SAA) and Ethiopian Airways that would ferry the bodies from Cape Town to Johannesburg and Johannesburg to Cameroon respectively. The day began with a series of cancellations and strikes. SAA called to cancel their engagement and there was also a strike by ground staff. Some PIFAM elders and the undertaker then started to make arrangements to transport the body by land from Cape Town to Johannesburg to catch the connecting flight. Their woes were further compounded by a call from Ethiopian Airways that they could only carry one body at a time from Johannesburg. PIFAM was beset by a series of complications and soon realized that their deceased brothers were in no hurry to leave them for their journey home. At this juncture, they decided to appeal directly to the corpses and ask that they 'unblock' the obstacles that they had created. The eldest PIFAM member (as custom demands) decided to perform a 'ritual of appeal' to the bodies. Placing his hand on each of the caskets, he appealed to them to make it possible for the group to send them back home. He said:

> XX you are lying here, you loved us and we also loved you but you are now in a different form and, following our tradition, we must send you back to the village. Why are you preventing us from sending your bodies to the village? Open the road; cover all those potholes you've dug on the way that are disturbing us, close them so that we can pass smoothly. Your parents are back at home; some are not eating, some are threatening to commit suicide, I mean they are not doing anything. Why should you die and then cause others to suffer? Then with us here; since you people died, we are not doing anything, we are not going to work, nothing. Must we suffer because you've died? We love you, that's true, and that's why we are doing what we are doing now. It's because of that love, so if you actually know that you love us, open the way so that we can send your corpse back home.[8]

[8] Appeal made by an elder to the bodies to pave the way for them to be repatriated. Field notes, 11/01/2013; Cape Town and interview with the elder, Cape Town: 20/01/2013.

While still standing by the caskets, a call came. This is what the elder had to say with respect to the call. 'Immediately I made that tradition, it didn't take up to 5 minutes, a message came that they've booked a special flight from Cape Town to Johannesburg. We believe that as I spoke like that, they understood and they opened the way.'[9]

After the appeal, he called on anyone who was a creditor or debtor to any of the deceased to come forward and declare the amount or item in the presence of the deceased. Eight persons came forward, and some declared the amount that was still owed to them by the deceased but turned down offers to be reimbursed because they valued their friendship with the deceased more than the money. One person declared that she owed one of them R 15,000 and promised to make the money available. It would be sent to his family once payment could be effected. The importance and meaning of this ritual lies in the 'performance utterances' (Davies 1997) that have been carried out by the elder and those of the creditors and debtors. The utterances carry with them the power of word/communication: words that are spoken in the presence of a corpse are testament and may be punitive if shrouded in lies. Also, these utterances do not constitute their own performances, but follow punctiliously orders that have been established by others (Rappaport 1997: 32). The ritualized elaboration of communication with each deceased highlighted the effectiveness of this performance and distinguished it from ordinary technical acts (*Ibid.*: 50).

At the end of these declarations, phone calls came through that SAA has made a special dispensation to transport all three bodies to Johannesburg, and Ethiopian Airways confirmed that they would carry the first two that day and the third one the following day. Members heaved a sigh of relief and attributed this to the appeal they had made to the deceased. Such an appeal finds expression in the idea that the departed soul is potentially malevolent and socially uncontrolled (Parry & Bloch 1982) and has to be reunited with the body for effective transcendentalism and avoidance of the ghost haunting and/or eventually visiting. It was believed that the spirits of the deceased were still wandering around in Cape Town and were not willing to go home which was why the group was faced with so many challenges. But equally it is believed that the dead are present amongst the living and hear all that they are told: they thus heeded the pleas being made.

Although the rituals associated with death have been standardized, they are dynamic in form and performance. Such occasions give mobile communities the opportunity to express dormant internalized ritual practices and showcase their connectivity and cultural knowledge. They can also show those in the home country their savvy and their adeptness at rituals. Transcultural participation in death rituals and intensified communication help to reinforce the community's

---

[9] See Footnote 7.

identity and belonging to the home village. Besides, the tendency for the group to help one another is because 'rain does not fall only on one person's roof; it falls on all of our roofs. When it is my turn everyone gathers to wish me well, and when it is the next person, we all gather there.'[10] Having side-lined the Cameroonian Association of Cape Town other groups organized collective donations that they gave directly to PIFAM. MACUDA, for instance, supported them to the sum of R5000 and attended the wake with their *Mbaghalum* instruments where they led the first half of the dance session during the cry die.

These deaths helped to reintegrate the community based in Cape Town back into that in Pinyin. In the process, social links were reproduced and places symbolically tied across space, as they revealed intense issues regarding 'identity and perception of place' (Gardener 2002: 198-199). The importance of the place of burial is emphasized to the deceased and the latter are directly appealed to, to make it possible for them to be sent home. Home to the mobile community is where they are buried and where their *nitoŋg* is buried, as was mentioned earlier. The host country is considered a temporary home ('bush') to which migrants are connected for financial reasons only.

Cry die in Cape Town

A cry die is a way in which members of mobile communities abroad choose to mourn a departed relative among their peers in the host country, especially if going home to attend the funeral is not possible. It is organized prior to the migrant's departure for the home village to assist in the funeral rites. These are moments of crisis for the bereaved and the community is expected to help them overcome this. For those for whom traveling home is not possible, a cry die is organized on the same day as the corpse is taken from the morgue in the home village and is buried.

When Paul's[11] father passed away, he could not travel back for the funeral because he had just returned from Cameroon a few months earlier. Having sent money for the funeral, he decided to organize a cry die in Cape Town to coincide with the day of burial. This saw a huge gathering as he belongs to PIFAM and MACUDA, and both groups came with their *Mbaghalum* dance and livened up the occasion. The event was given the widest publicity thanks to the incredible speed of the new communication technologies (calls, SMSs and emails). The bereaved communicate with any member of the executive, but preferably with the secretary who has the task of spreading the news by SMS. Members are also requested to give it additional publicity by forwarding it to others. A date is chosen by the bereaved when the entire group will come for the cry die but, prior to this,

---

[10] Interview with John, The Hague: 24/10/2010.

[11] He hails from Mankon and lives in Cape Town.

members go round to offer their condolences. The men take drinks while the women offer food or assist the bereaved by cooking at her home.

From the traditional Bamenda Grassfields perspective, the activities expected of women have gone unchallenged for too long and this has enabled the men to exert subtle control and ascribe roles to women and promulgate social taboos. The women are usually entrusted with the planning, preparing, cooking and serving of meals. Reiterating this notion, Pleck (2004; see also Mand 2002) holds that the domestication of rituals thrusts women into the role of 'queen of the home' or what Wilson (1939) refers to as 'owners of the death' in the case of death rituals.

In PIFAM, the bereaved are given R 3000 when a member loses a parent and the money is handed over during the cry die, while a bereaved member of MACUDA gives the name of a relative back home who will receive the money and it is sent via Western Union or MoneyGram. The idea is for the money to help pay for the funeral itself and this ensures that it is sent home and used there. The association wants family at home to know that the bereaved belonged to a group and this is the statutory contribution that is due to a member at such a time. The following SMS from PIFAM sheds light on this: 'PIFAM regrets to announce the death of Xxxx Xxxx's father. Wake keeping starts today at Xxxx Xxxx's place beginning from 7pm. Contribution is R36 and attendance is obligatory. Thanks.'

Cry die in The Netherlands

As mentioned in the previous chapters, there are no associations within the Pinyin and Mankon communities in the Netherlands so they belong to the wider Cameroonian community and have friends to support them at such times. One person in the Netherlands has opted to create a mailing list and assumed responsibility for sending out information by email to all members in the event of joyous or sad news. The email below elucidates this point:

> Dear All
>
> It is with deep regrets that Ms Xxxxx Xxxx announces the death of her Father
>
> Pa Xxxxx Xxxx.
>
> … Let us join the family to pray for soul (sic) of their father and souls of all faithfuls who have departed this world. She can be visited at the following address:
>
> Brueghelstraat XXX
>
> XXXXRG Den Haag

The first email sent out informs members of the community about the death, and a few weeks later, a second mail is sent informing members about the cry die:

Dear All,

The Funeral Program of the late father of Ms Xxxxx Xxxx (Pa Xxxxx Xxxx)
Program
Date: Saturday 20th October 2012

Phase I
Church Service
Catholic Church
Hoofdkade
The Hague
Time: 3.30 pm Prompt
Tram 12/11 near street market

Phase II
Wakekeeping
Brueghelstraat XXX
XXXXRG Den Haag[12]

Rituals are not limited to traditional and cultural practices but are coordinated with religious rituals and with respect to cry dies in the Netherlands. Most often, those who can invite an African priest (preferably a Cameroonian studying in Europe) to officiate at the Memorial Mass do not hesitate to do so (Jindra 2005). The bereaved in this case decided to invite a Cameroonian priest from Belgium as the celebrant. In Amsterdam, Marie,[13] a Pentecostal Christian, invited her Ghanaian pastor to her home to pray with them when she organized a cry die for her father who passed away in Cameroon. To those in the Netherlands, the church community becomes an extension of their social reality and this underscores the fact that the absence of village associations has resulted in the formation of a close Christian community known as the African community, with Cameroonians making up the majority of the congregation (see also van Dijk 2002). Olwig (2005) noted that such funerals can be seen to be a shared ritual in a religious fellowship of people who want to give the deceased a Christian burial. This is eased by their belief that the deceased was 'given to them on loan by God'[14] and the religious rituals are meant to thank God for his/her life and 'commit the deceased back to his/her creator'. In the same vein, Nieswand (2005: 246) holds that this also creates a 'public space for everyday relevance, which is a scarce resource for marginalized migrant groups'.

12 Email received on 08/10/2012.

13 I attended the wake held in Amsterdam on 16/10/2010 for Marie's father who passed away in Cameroon just before she left for Cameroon.

14 Excerpt from the eulogy when the bereaved organized a Catholic Mass for their mother in The Hague: 06/10/2012.

Burials and succession rites

The rituals performed after death in Mankon and Pinyin are the most elaborate of all their rituals and the themes, which are recurrent and implicit in others, are expressed most clearly. For the mortuary rituals to the burial, the family is expected to come from far and wide, especially those who are out of the country as they are seen to command the economic capital that the rest of the family will leverage to make the occasion a success. And every member of an association is expected to show up and make the stipulated contribution. Failure to do so is met with hefty fines and/or expulsion.

Social capital is an overriding factor that determines the success of the ceremony as well as the importance of the actors concerned. In other words, how well people can transform this capital into economic and cultural capital depends largely on the actors' social capital. Given the inequalities that exist in societies, acquisition of social capital is unequal too, with some migrants being able to command more important social capital than others. In addition to articulating the community's values, belief systems and customs, it is possible for ceremonials and funeral to have varying degrees of importance, which leads to hierarchies (see also Gardener 2002: 199; Olwig 2009; Brand 2000). As in the case of the Grassfields in Bamenda, some deaths have been labelled 'bush-faller die' as most or all of the deceased's children are living abroad and their economic capital draws in a great deal of social capital too. The death of this researcher's father is a case in point. Funerals are not as socially liberating because of the hierarchical formations they bring to bear on mobile communities, but funerals serve as outlets to extend inequalities across borders and thus reproduce the social class structures brought from the home country.

Death amongst the migrant community, as has been shown, is treated with deep reverence and the corpse has to be sent back home because burial in a foreign land signals social and cultural failure[15] on the part of the migrant community and they will be stigmatized by other communities (Geschiere 2005; Geschiere & Nyanmjoh 2000, 2001). But it is additionally an outward expression to those back at home that they (the migrants) still know the community's norms and values surrounding death. Self-gratification also plays a role here. Reverence is additionally guided by the notion of fear: fear of the dead, fear of bad mouthing and fear of ill luck that might befall them as a result of not performing the necessary rituals or of not performing them well (see also Wilson 1954; Wilson 1939). As such, attention is paid to dressing the corpse properly and taking it back into his/her house to reunite it with his/her spirit before the funeral rites. This begs the question as to why death can be so frightening and why a person commands

---

[15] The Congolese (DRC) are socially stigmatized by the Cameroonian community for burying their dead in Cape Town.

more respect in death than when they are alive. In a way, it could be inferred that the agency of the dead is stronger than that of the living. However, and as is argued by Jindra (2005: 358), 'it is not foreordained who lives on and who does not as ancestor, given it is posthumous ritual acts that establish this'. It is dependent on the type of burial accorded to the deceased and this marks the transition to being one of the ancestors. This is the case in most parts of the Grassfields and it is widely believed that death rituals in Pinyin and Mankon pave the way for the deceased to join the ancestors and become one of them. In the days after the burial in Mankon when the rest of the mourners have left, the immediate family decides on the successor to the deceased, hires the services of a diviner who says exactly what happened and who/what killed the deceased and prescribes sacrifices to ward off any eminent death or bad luck. This marks closure after the death and all the ceremonies. When Paul lost his first wife and took her body home from Cape Town, his parents-in-law requested that he move on and remarry as soon as possible in order to help him attain closure, especially as they did not have any children. In Pinyin, one of the last rituals is marked by sweeping the compound, as was the case when Ron's body was repatriated from Cape Town at the end of the funerary rites.

Succession is an important ritual for both Pinyin and Mankon, and is usually the last ritual. This is done immediately before the family disperses, or a year later. With the increasing financial and material accumulation of migrants today, the tendency is now to make a migrant the successor even though they recognize their long periods of absence. This has ceased to be a problem given the interconnectivity between migrants' host countries and their home country. The successor appoints an assistant and is constantly in touch with the home village and visits if important ceremonies demand his physical presence. In the absence of such ceremonies, his presence is felt virtually. In contrast to the past when migrants did not actively participate in family affairs, being a successor now connects them with kin back home. For instance, it was five years after Joe's father passed away that he and all his siblings travelled home and he was crowned as his father's successor. Although Joe goes home once a year, being his father's successor means that he has to be more attentive to what happens in the compound. He is also contemplating going back home permanently.

Transnational funerals

Migrants participate in these rituals be they at home or in the host country and seek to give one of theirs a 'decent farewell', which usually involves ritual practices. Others take over the role of the deceased as his successor. In this respect and as pointed out by Chesson (2001: 1), death is a 'sensuous arena in which the dead are mourned, social memories are created and, reasserted, social bonds are renewed, forged or broken, and individuals make claims for individual identities and group memberships'. As such, the death of a migrant explicitly reveals a great deal about the notion of place, in this case where the body is taken for burial. It also provides a context for reflection on the meaning of identity and belonging among the migrant community (Jackson 2005: 3; Geschiere 2005; Olwig 2009). By returning the body home for burial, the notion of transnationalism among migrants is rearticulated. Such journeys represent the final stage in many migrants' mobile lives (Gardener 2002; Geschiere & Nyamnjoh 2000, 2001). Funerals in Cape Town involve the entire migrant community but particularly the ethnic enclave from which the deceased hailed and this ensures the correct funerary rites and the eventual repatriation of the body. In this respect, sending the body home enables the migrant to be reinserted with the ancestors through all the rituals that will be carried out in his/her honour both in the host country and in the home country. They also help ensure closure for the family back home.

This notwithstanding, not everyone participates or readily avails themself for these rituals. Some choose what to perform and not perform or how to perform it, while others may readily join in out of fear of witchcraft, ill luck or bad mouthing. For many, performing these rituals satisfies a hidden need. What this presupposes is that participation in rituals is dependent on psychological needs, but do we not, by saying this, take a reductionist stance?

Although the power dynamics may not be evident, ritual performance is marked by subtle power and hierarchical formation that lends itself to how much social and economic capital can be harnessed and transformed into cultural capital (prestige), with the end result being the creation of a hierarchical society. This assumes that rituals find expression in a society that is endowed with networks.

By choosing to create overlapping networks and actively participating in home associations in the host country as well as maintaining a virtual and/or corporeal presence back home, mobile communities indulge in rituals and ceremonials to reaffirm their double engagement, to reminisce about what they have left behind and/or to invent new ways of performing that suit their current abode and are of equal significance and meaning.

ICTs as evidence of ritual autopsy

Performing these rituals is a statement in itself because of the significant way in which people use the new technology to access events. This might be as a result of anxiety, mistrust and/or concern about the way the funeral is organized, as well as anxiety about the cause of death. All of this is channelled through the use of ICTs, videos and photos. Proof or evidence is reminiscent of the past with regard to the death ritual when an autopsy, as an important death ritual, was performed on the body to establish the cause of death and to determine if there were any missing internal body parts as a result of witchcraft (Warnier 2012). Autopsy, where practiced, focuses on the passage from (the) throat to the belly and the body will be slit along this line to examine substances located around the liver for signs of poisoning or witchcraft substance' (Warnier 2012; Nyamnjoh & Rowlands 2013).

Today, among mobile communities, such rituals are no longer about performing physical autopsies but using film and technology (DVD and photos) as a form of autopsy provides evidence that the funeral ritual was done well. Such an autopsy is particularly important to the absent migrant to assure him/her that the money sent home has been well spent. Doing a ritual well informs harmony in the family and wards off bad luck.

It was for these reasons that PIFAM engaged people to take photos to send home with Ron's body as testament to those at home that the funeral they organized in Cape Town to send Ron's body home (Pinyin) was done according to the norms. The tendency is for the home village to send videos to relations abroad as evidence of this too. The aim is to reduce any anxiety amongst relations who could not attend and as proof that a befitting funeral was given to the deceased. News of the arrival of a DVD is always welcomed by friends and family abroad and their home will be full of friends who come to watch it. A running commentary will be given with regard to the form, content and performance of the various rituals (the mortuary and traditional rites). Conversely, the video is implicitly meant as evidence of proper financial management of the resources sent for the funeral. I watched the funeral videos of John's mother and father with him in The Hague and that of Ben's father and Joyce's mother in Cape Town and it was interesting to note the satisfaction they derived from it, judging by their comments. These were mostly about how well organized it was, from the mortuary to the internment and the feeding of the mourners as well as the cry die. For instance, Joyce talked of how she spent R 17,000 (approx. € 1600) on the funeral but she was satisfied with the quality of the casket, the hearse that was hired and the food provided for those present, and she told me how she had allocated the money. The video was sent via DHL by her sister is to allay her fears of any impropriety. Prior to watching the video with Joyce, I had gone to offer my condolences at their home.[16] Most of time I was with her she was crying and recalling all the things that she had discussed with her mother just a few days beforehand. But the most upsetting she felt was the fact that her sisters had let their mother die without talking to her (bidding her farewell), and thus deprived her of the chance to give them her blessings. Again, the importance of breath (*ndjiẅi*) is mentioned, as Joyce considered the last breath of her dying mother to have been full of blessings and there is no doubt that her mother would have *fogho ndjiẅi*

---

[16] Condolence visit at Joyce's, Cape Town: 17/11/2011.

on Joyce. Communication with her mother was thus seen as a form of verbal autopsy whereby she would have been able to assess the gravity of her health. By contrast, John in The Hague was fortunate to have spoken with his mother a day before she passed away. Through this verbal autopsy, he assessed the gravity of her illness and was psychologically prepared for any eventuality. Unable to travel home, he had to rely on his cousin's smart phone for a virtual presence. On the day of internment, he asked this cousin to put the phone on speaker so that he could follow and 'participate' in the burial rites. He also recorded what was said and the song that was sung became his new ring tone until he went home to perform a cry die for his mother (see Chapter 4). Through the various forms of ICTs, John, Ben and Joyce were able to assist in the funeral rites of parents back home and, more importantly could give their stamp of approval to the entire ritual performance and the organization of the funeral.

In summary, ICTs are transforming mobile communities' participation in rituals by making them accessible to all and they can offer proof to migrants that the ritual was performed properly.

Such reminiscences and the portrayal of vivid memories by Joyce could be seen as a way of channelling her sentiments of social bonding but, conversely, this made it difficult for her to come to terms with her mother's death. For Joyce to find closure following her mother's death, she had to go home and perform her own funeral rituals on the grave and symbolically bury her mother, despite the huge sums of money she had already sent. As part of the ritual, Joyce completely shaved her hair off. This act is seen as a way of self-cleansing and removing any bad luck that a death in the family is believed to bring. Joyce also felt it was the best way for her to mourn her mother given that she could not be in her home village like the rest of the family. Although shaving one's head is an age-old ritual, most families still adhere to it although a significant number today abstain from doing so.

Rituals as communication and the embodiment of society

Whether they are rituals or ceremonials, increased participation in them is due to the huge success of ICTs and increased connectedness among community members and kin back home that has provided a base for frequent communication and possibilities of more frequent visits. In addition, ICTs offers migrants the possibility to be virtually present in the home village and/or draw on the social capital that they have accumulated to make the occasion grand. The corollary is that social capital is thus transformed into economic and cultural capital.

Being away from home, life-crisis events such as marriage, born house, deaths and cry die (wakes) by mobile communities are contingent on language as 'language is the foundation of human life' (Rappaport 1997: 7; see also Davies 1997:

7). This is because 'human systems are cultural-organic systems constituted by linguistic information' (Rappaport 1997: 10). In this respect I concur with Rappaport (1997: 29) that in 'human rituals the utterances are usually predominantly verbal, and are symbolically signifying, and the acts, in being formalized, are *ipso facto*, invested with meaning'. It also follows that utterances are exuberant elaboration of identity as the form, content and performance of the actors 'generate a sense of identity, while others confer identity' (Davies 1997: 4). This is the desired effect of performing rituals, i.e. by defining ones' identity in relation to others. As attested by a Pinyin informant, performing the death ritual 'is part of our culture, it is who we are, and differentiates us from the Congolese (DRC) for whom sending the body home holds no meaning to them, hence they bury their dead in Cape Town and disperse immediately after'.[17] In this connection, the appeal made by the elder to the three corpses mentioned above to pave the way for them to be sent back home in conjunction with placing his hand on each of the caskets was quite meaningful. The utterances in this case were complemented by the action of placing his hand on the caskets to create direct communication with each of the deceased. This confirms the punctiliousness and repetitive nature of rituals that serve to bring out the power and efficacy embedded in the utterances. Migrants want to be seen to be doing the right thing and doing it correctly. Therefore any deviation may be regarded as neglecting traditional religious aspects and the corollary may be bad mouthing or bad luck. The defining intention is the response of the receiver (the deceased) who is expected to act on the statement. The language or ritual utterances used here are not 'mere words' but 'possess special effect' (Bloch 1973 in Rappaport 1997: 49) and the performance is intended to elicit a particular response from the receiver, for example in this case that the deceased would unblock all the obstacles that were inhibiting their repatriation.

Inasmuch as rituals are contingent upon language, it is important to note that rituals are internalized habits that migrants have learnt over the years and are able to re-enact in the home and the host countries. In this respect, the notion of transnational habitus (Guarnizo 1973; Vuerola 2002) provides a clue as to why mobile communities continue to observe these rituals in their host countries. (See Chapter 9 for an in-depth discussion on this subject.) While this concept illuminates the bifocal circuits and the scale of simultaneity of being home away from home, mobile communities seek to integrate cosmopolitan values while defending and maintaining particularistic values within the context of being both 'here' and 'there'. Away from home, mobile communities do not see themselves as having severed links, they have simply extended the locus of home. The host country is temporal and the home village is considered their permanent abode. These rituals

[17] Conversation with Mark, Cape Town: 28/11/2012.

retain their meaning within specific social fields, defined by the language that distinguishes them from other rituals as the language used is diametrically opposing. The notion of field represents the external structure of relations between historically defined positions and its structure is one that mobile communities seek to replicate by performing internalized habits during various birth, marriage and death rituals to affirm their belonging to their ancestral land. For instance, the blessings of the 'breath of men' in the born house are different to those given to a prospective migrant about to leave home by his/her parents. While they may both pass for blessings, they take place within different social fields and contexts.

Employing these concepts has provided a roadmap to understanding the practices and adaptations made to the various ritual activities by mobile societies. It also provides some leeway for me to navigate between the various ritualistic activities.

The concept of capital is intrinsic and complements the notions of transnational habitus and social fields in understanding ritualized or ceremonial practices and how they are embedded in social networks. I will not dwell on this in detail here as it was discussed in detail in Chapter 3 but what it is essential to note is the notion of strong ties that underpin ritualized practices. As was seen in Chapters 3, 4 and 5, neither the individual nor society can stand independently of each other. A complex web of networks comes into play at different intervals in accordance with the needs and demands being faced.

For rituals or ceremonies to occur, capital accumulation and the active engagement of social capital are required because what gives the distinctiveness of any of the above rituals is the crowd that a particular activity is able to generate. In other words 'the size of relations that can be effectively mobilized and the amount of economic, cultural, and symbolic capital of the members of that network determine the volume of social capital of a specific actor' (Brand 2000: 26). This is because the family and kin are the institution that initiates the socialization process. It is important to bear in mind that capital accumulation is often used or offers particular resources and may be transformed into economic capital. This presupposes that, amongst these mobile communities, people are investing in relationships in order to achieve capital and these have to be continuously maintained and lubricated 'by means of material and symbolic exchange' (*Ibid.*: 26). In this respect, even after withdrawing from the Cameroonian Association and taking full charge of repatriating their members' corpses, PIFAM was able to receive financial support from other sister associations because of the support it had given them over the years. The concept of 'rain on a person's roof' captures this reciprocity. By extension, this may also mean that the celebration and performance of rituals could be associated with social ranking (hierarchy) because of the economic, social and cultural capitals accumulated. The amalgamation of these capitals project people to higher positions or rankings and they use rituals (with those of migrants nicknamed bush-faller wedding or bush-faller die) as a way of distinguishing themselves from everyone else. For instance, when Emile[18] went home to Pinyin to organize a memorial service (cry die) for his father, the ceremony bore the hallmarks of a bush-faller cry die. Prior to his departure, he had shipped over drinks, disposable cups and plates that are not easy to find in the village. Relatives and the entire neighbourhood (quarter) and the church group that his father was a member of were invited for *alah* (celebrations with abundant food and drinks). These celebrations helped to catapult his social standing in the village and he received blessings from ancestors and relatives. This gesture is still a reference point for most in the neighbourhood and people continue to speak well of him in the village. It could however be argued that overspending on the ceremony was a way for him to compensate for his absence at the funeral.

Pleck's post-sentimental approach to rituals can enlighten us here. This approach reinvigorates rituals and ceremonials and is both a 'way of talking about ritual and a style of practicing it' (Pleck 2004: 52). She argues that post-sentimentality uses 'sentimentality as foil and it requires the persistence, not the total disappearance, of sentimental occasions'. This approach provides a rationale for overindulging in ceremonial events because such an event is therapeutic for the organizer and may also be a creative solution to a personal crisis as in the case of cry die in the mobile communities in both Cape Town and the Netherlands. This ceremony, as mentioned earlier, is derived from an existing ritual or better still can be seen as an authentic traditional ritual that the individual finds personally meaningful. In these communities, it is an anomaly when such standard practices are not adhered to and the repercussion is bad mouthing by the community. Concurring with the therapeutic value of rituals, Imber-Black & Roberts (1992: xvi-xvii) posit that it is a way to 'maintain and alter important relationships, to facilitate complicated life-cycle changes, to heal losses, to express (their) deepest beliefs, and to celebrate life' (see also Metcalf & Huntington 1991).

Post-sentimentalism is borne out of nostalgia and the growth of ethnic identity as a result of marginalization and exclusion from mainstream society. Pinyin and Mankon migrants, like most migrant communities, are often reduced to the margins of society, which leads to ethnic enclaves and identities. In a bid to retain what they have left behind, they develop a sense of communitarianism and new ways of celebrating traditional events, sometimes by adopting western-style celebrations (such as baby showers) but also by giving them a local twist to fit their new locus and to create a sense of ritual. Importantly, these ceremonies become

---

[18] He hails from Pinyin and lives in The Hague.

cathartic and act as insulation against their societal exclusion, but they also reify their absent presence in society. Drawn together, these remembrance-communities find in these symbolic rituals and ceremonies what they do not find in daily life and, in so doing, the community reinforces a certain 'public identity (of Pinyin-ness and Mankon-ness) that is ultimately communicated to both local citizens and the external publics' (Procter 2004: 132).

Bloch & Parry (1982) also see rituals as a form of social control and argue that the prescribed occasions of ritual function call for appropriate emotions by the participants, and can be seen as a technique in the acquisition and control of life (see Chapter 9). Similarly, fear of bad mouthing, especially in the mobile community, becomes an unspoken control that hangs over defaulters. While they may have a valid argument, the question that emerges concerns those who show no emotions or are indifferent, or those who have opted not to be controlled and pay little or no attention to rituals. We could also look at it from a reciprocated deterministic point of view, given that rituals and emotions mutually shape each other. It is not a given that rituals elicit the same emotions in everyone and they can range from no emotion to overwhelming emotion. Nevertheless, rituals of death elicit emotions and the 'power of words embedded in them to trigger and direct emotions are great' (Davies 1997: 8).

Conclusion

Rituals play an important role in the lives of migrants in the host country. However, of the life-crisis rituals related to death, birth and marriage, those concerning death constitute those that best define mobile communities and connect migrants most closely to the home country.

Born house and marriages, unlike death, are rituals that have become dynamic and have adopted new identities in conjunction with existing ceremonies. They are fluid and have incorporated aspects of western practices, such as interlinking utterances and performance with traditional ceremonies. The dynamic nature of these ceremonies can be attributed to mobility, with mobile communities seeking new ways of making a living in the host society. While these ceremonies are also premised on the notion of belonging, they are informed by the idea of wealth and success, which invariably results in a hierarchization in society. Capital accumulation gives these ceremonies a new twist as migrants have the tendency to spend heavily to prove themselves and it is then wealth and success that inform belonging, earning them the nickname 'bush-faller wedding' or 'bush-faller die'. Although the ritual of death is still tied to traditional forms, it is increasingly being defined by wealth too. How much economic and social capital accumulation one has managed to amass translates into a successful funeral and a cry die earns one cultural capital. As mentioned earlier, not everyone in the community has a baby shower or born house, just as not all are able to arrange a big group of people to welcome their bride to Cape Town.

Although many migrants make a conscious decision to leave home, it is a difficult one to sustain because home always travels with people through the various rituals and ceremonials that they maintain and reconfigure to suit their current locus, but also because these rituals are a form of social control and are, paradoxically, liberating. Being excluded from mainstream society as they tend to be, these communities develop ethnic enclaves that act or shield them from society. Part of this defence mechanism is nostalgic reminiscence of the rituals they have left behind and consequently want to uphold or reinvent to fit their present society. An active engagement in rituals thus lies in the post-sentimental present informed by the sentimental past (Pleck 2000), especially given the fact that mobile communities find them cathartic.

The ability of mobile communities to engage in the form and performance of rituals has been significantly aided by the appropriation of ICTs that make it possible for them to coordinate rituals between home and the host country, and amongst themselves in the host country. These are also periods of intense communication between the community and other groups. ICTs have become a new form of autopsy for migrants to assess and ensure that the funeral and rituals have been done well. Autopsies today are in sharp contrast to those of the past. SMSs have become the fastest mode of communication if one needs to reach a target audience, such as these messages from PIFAM and MACUDA respectively. 'The final wake keeping for our brothers will be on Thursday 10 at the Belleville hall from 7pm.' 'MACUDA. Program for born house today starts at 4pm. Be on time please. 9 aveneu Kensigten. (sic) Thanks.'[19]

Despite the grim Weberian forecast that modernization, industrialization and the pursuit of material comfort will result in a decline in rituals and a loss of value, interest in rituals has increased because rituals are what set communities apart from each other and highlight their collectiveness. By the same token, rituals connect mobile communities to the home village and serve as a platform for them to participate in rituals both in the home and the host countries concurrently, a phenomenon that was not possible in the past when the host countries' policies required total integration. Mobilities have consequently been enhanced by new communication technologies and the obituary of assimilationist policies has been written. The corollary has been increased participation in rituals in both the home and host countries.

---

[19] Message sent on 05/11/2011.

9

'I di beep na for beep, them di call':[1] Straddling relationships between the home and host country amongst kin and kith

Introduction

This chapter focuses on migrants' moral obligations regarding remittances and their virtual presence in the family circle they leave behind. It considers the micro individual experiences of migrants' kinship relations that highlight personalized ties towards kin and kith and the community. It moves away from the PIFAM and MACUDA communities and pays particular attention to the individual forms of relating with the family, looking at how migrants confront and deal with the emotional challenges of being away from the family, and also notions of witchcraft. Coming from societies and fondoms that have long been embedded in a history of migration, the notion of strangers ceases to be strange. The question is thus how they entertain multi-stranded relationships and how, as individuals, they are able to sustain relationships back home and cope with being apart from their family and having to adapt to the host society and yet hold on to the idea of being different. What accounts for individuals' engagement in transcultural lives and what role does this duality play in trans-border relationships between mobile migrants and their home communities? Individualism as opposed to community is at the heart of this chapter.

This penultimate chapter in this volume details how individual Cameroonian migrants living in South Africa and the Netherlands are actively engaged in the livelihoods of those in the home country. I consider the myriad of transcultural practices migrants are engaged in towards their respective families back home, and also the way families at home respond and initiate similar transcultural pro-

---

[1] Interview, Pinyin: 10/10/2011. It is they who call me, I only beep them.

cesses. This is informed by the fact that looking after kin and ageing parents are inculcated in the moral upbringing of children. From this perspective, it is plausible to view migration not only as a transnational social process (Guarnizo 1997) but also as a transcultural process, which enables us to assess the various dynamic processes that are inextricably linked to it, such as the emotions that are brought to bear in the absence of relations and how migrants cope. As individuals, they seek to maintain multi-stranded relationships that are entertained in the construction of cultural identities and are able to express this identity in the relationships they sustain back home and in the coping mechanisms adopted following the separation of families. This means that migrants' pathways of 'simultaneous incorporation' are less to nation-state activities but closely relate to the cultural activities that they engage in (Levitt & Glick Schiller 2004). This presupposes that the 'concept of transnational field takes us beyond the direct experience of migration into domains of interaction where individuals who move themselves maintain socio-cultural relations across borders through the various forms of rituals and communication' (Levitt & Glick Schiller 2003: 7).

In this respect, I maintain that mobilities have reified migrants' quests to increasingly take part in transcultural activities to prove their loyalty to their families and, in the same light, mobilities have ceased to be a hindrance to families not to consider their mobile kin as successors. From another perspective, transcultural processes may emerge from visceral feelings or from the fear of bad mouthing as a result of not assisting the family, and this leads to imminent danger or bad luck.

Negotiating transcultural social fields

Migrants straddle different social fields and, as a result, their lives are embedded in multi-layered, multi-sited transnational social fields (Levitt & Glick Schiller 2004). Those with the correct legal documentation move back and forth between Cameroon and Cape Town for vacations and/or to supervise projects they are engaged in back home. This is not to say that those without documentation are incapacitated: they 'travel while sitting down' (Archambault 2012). This form of virtual migration enables them to navigate the transnational social field despite their absence. Two case studies will elucidate this perspective. As regards the first, Jake, who readers will remember from previous chapters, has permanent residence status in South Africa with his family. They visit home as a family once in a while but he has been twice before because of the project he is involved in, namely building student accommodation. In 2009 he went home with his older daughter and, in April 2011, he went with his wife and two children. Although they are always happy to be united with family, his experiences and the lack of development at home does not motivate him to go home until he retires. He embarked on this building project over three years ago and is also constructing a small shop to rent out for the time that he is away but plans to eventually put it to use when he retires and goes home. While at home, the family enjoyed visiting relatives but were constantly struck by the lack of development, the bad dirt roads and the increased police harassment and number of road checks. They were travelling with children whose experience of 'home' is Cape Town and who did not find their country (Cameroon) interesting, and constantly asked when they would return 'home'. It is important to note the nuanced meaning of home here. To the children, home is Cape Town and they do not identify with Cameroon. This is in contrast to their parents' disgruntled identification with it and, for them, attachment to home is through relations and the picturesque photos of greenery that Jake took during his visit to satisfy his nostalgic feelings, once in a while, when back in South Africa. In this respect, home is defined locally and globally because, as immigrants, people tend to define and develop an outsider's perspective of home, thus challenging the social and developmental shortcomings that characterize home. According to Jake:

> I will never live in Cameroon until when I go back perhaps on retirement, I think because all this while I've been to Cameroon just twice and during these two periods, my experiences generally were ... I didn't experience the place as one that I want to live in. It's not just the physical environment but also the mentality of the people you meet … you go into a bank to do a transaction, a person expects you to act in a certain way, you don't act that way and they have the right to say I am not serving you … You sit in a vehicle, just because someone says something to someone in civilian attire who is a Captain in the Cameroonian army, they start beating him and they identify themselves '*Tu parle a un capitaine de l'armée Camerоonais*'.[2] So I saw all these things and I felt like mad, and you cannot say anything … And the physical environment, you get out of Douala airport you suddenly feel like you're on another planet.[3]

Other migrants visiting home have similarly concurred with Jake and, like him, would only retire home when the time is ripe, hence increasing the propensity of their connection back home. Prior to retiring, they maintain a 'dual frame of reference' (Guarnizo 1997: 310), whereby they constantly compare the situation in their home country to that in the host country.

The second case resonates with gender asymmetry and illustrates varying degrees of sentiments by female migrants. Female migrants take with them the family roles of care and structures across borders. The case of Joyce highlights a trend that exists within this community. She has four children and joined her husband in Cape Town in 2008. Of these children, the older two ones live with her elder sister in Bamenda, while the younger two boys live with her parents-in-law in Pinyin. In the school holidays, the children spend time with other sisters

---

[2] You are talking to the Captain of the Cameroonian Army (own translation).

[3] Interview, Cape Town: 09/04/2011. It should be added that after this interview in August 2012, Jake went home with the family for a third visit. His impressions have not changed.

and those who live in town or go to the village as well. Through these connections, Joyce maintains her parenting obligations by sending money to her sisters and parents-in-law who look after her children. In addition, she is able to assign roles to various family members who look after her children. Although she does not like the fact that the younger two are still in the village where there is a lack of basic services (water, good schooling and proper sanitation), she is nevertheless consoled by the fact that her children are well brought up (well-mannered) and are not in South Africa where they would be exposed to the moral decadence of society there. Others like Joyce who have opted to send or keep their children at home share similar concerns: they want the children to be disciplined and not to be exposed to bad influences around them. As such, the decision to keep children back home overrides concerns regarding basic services and infrastructure that seem to be the guiding factor for men. The women are more concerned with family welfare and cohesion. They fear being alienated from their children by incompatibility between their own cultural identity and that which the children will acquire in Cape Town. They worry about any erosion of their authority and control over their children. The idea of leaving or sending children back is underpinned by the notion that 'a child is one person's only in the womb', the belief that the support system of kin will help raise the child in desirable ways (see Nyamnjoh 2002). And for the first time since her arrival in Cape Town four years ago, Joyce was able to go home in October 2012 to visit her children.

We should not lose sight of the fact that what these individuals perceive as intolerable is by no means new in the country or village. These were typical of their experiences before they left home and they constantly allude to how things have not changed since their departure. It would require more than a lack of basic services for people to leave. We should bear in mind that, there had been a virtual transnational comparison for most migrants prior to their departure and which intensified upon arrival in the host country. Their ability to navigate between the host and home countries has changed their tastes and perceptions of basic services. Conversely, their standard of living in Cape Town, where the provision of basic services is good, has changed their attitudes and make them more critical of their home, where basic services tend to be lacking. At first glance, this may seem to account for migrants' unwillingness to go back home but we must remember that the primordial reason for emigration was economic or spousal reunion and, much later, transnational comparison redefined their reasons for not wanting to compromise their standard of living.

Home to Joyce and Jake and their children does not have the same significance for them all. For Joyce, it is a place of nurturing and somewhere she can bring up her children to be well-mannered and is remote from the social ills that plague South Africa and, for Jake, it is his attachment with home as his place of birth and where the rest of the family reside. In other words, they both have deep reservations about Cameroon's nation-state but great attachment to their village and place of origin. While this invariably affects their conception and attitudes towards home, it does not affect their emotional attachment to the place and relations there.

Despite misgivings about their country of origin, family ties are usually tightly maintained and nurtured, with families reaching out to each other in fascinating ways. Jake's desire to see his family after seven years in South Africa led him to adopt unorthodox means to go home with his elder daughter. Without a passport and as refugees, they used a *laissez-passé* to get into northern Nigeria and then went to the north of Cameroon and travelled by train to Yaoundé from where they continued to their village. For his part, Joe used a friend's passport that he pasted his photo into as a travel document to go to Cameroon the first time. Having spent over seven years without seeing the family, it was imperative, according to them, to use whatever means they had at their disposal to go home. Jake was very happy to have seen his mother before she passed away two years later. It would have been terrible for her to have died without seeing her granddaughter. Going home with his child was a way of connecting her to her roots and for Jake to re-establish his connections and sense of belonging.

In the event that those in Cape Town are unable to visit Cameroon, mothers are flown to Cape Town to visit. In the previous chapter we met Jane's mother who came to visit and to help her look after her first child. The visit of Charles's mother and Jake's mother-in-law were also planned to coincide with when their wives would give birth. Others arrange for sick relatives to travel to South Africa for medical attention or assist family with university admission. Despite the tendency for women to be more concerned and link up with home more regularly than men, the new technologies are increasingly involving men in linking home through the frequent calls they make.

Remittances have always been one of the ways for migrants to try to reconnect with home, as was expressed by the much-acclaimed actor Sidney Poitier (2000: 62) as far back as the 1930s when his host country was connected to his home village only by letters and 'it was an unwritten law in the Bahamas that people who go to America to live write home (and) they put a little something in the envelope'.

Remittances

The focus on remittances has often been on monetary remittances and how these impact on the social welfare and transformation of the recipients (Vertovec 2009: 103-118; Guarnizo 2003; World Bank 2001). Apart from the financial side, they entail tangible and intangible benefits (medication, education and goods) that have the same social welfare function and much more. Remittances also transcend the social imposition on migrants as they are embedded in the way social and kinship relations are constructed in the Grassfields where everyone (both distant and immediate relations) looks to the migrant for assistance. It is a given that when a family member migrates, the family will receive remittances and assistance because of the way migration and countries of destination have been construed, namely as a forest where migrants go to hunt and then send home the game. For example, Mike received a call from his maternal uncle asking for money to pay his children's school fees as the uncle sees himself as being entitled to what Mike makes as a bushfaller.

Gifts as a form of remittance

Remittances are not necessarily monetary, especially from parents and relations at home. As a way of being morally supportive to their migrant kin overseas, families send gifts of various size and significance. Lola, for example, sent a necklace from Pinyin to her spouse (Martin) and mothers often prepared small food parcels for me to take to their children living abroad. When Jake visited home, his father sent video messages for the other four children in Cape Town. He felt that this was the most valuable gift he could give them at this point in their careers: he sent advice, warm words and encouragement, things that money could never buy. As he put it; '*na the only thing that wey I fit sendam for ma pekin them – fine talk* (advice)'.[4] He wanted to send a recorded message that the children would be able to keep even when he was 'gone' so that they would still be able to listen to him. Death will not separate him from his children. Studies on remittances have tended to focus on monetary remittances more than on intangible gifts, which also constitute an immense form of remittance. These kinds of remittances underscore family feelings of togetherness despite the distance that separates them, for example thanks to Jake's mobility and his adeptness with new technologies, especially the camcorder.

Migrants often take advantage of someone going home to send gifts and money, no matter how small these may be. As a result of my travels back and forth between Cape Town and Cameroon, I often carried money and gifts for families in the village, with instructions to change the money from South African Rand to Central African CFA at the airport before giving it to the family. I also took five Visa cards along to hand over to friends, fiancée and relations. Amongst the things I took, Max's gift to his children caught my imagination in particular. My diary entry for 03/09/2011 attests to this. I had just returned from Cape Town to Bamenda to continue my fieldwork.

---

[4] Conversation with Jake's father in Pinyin: 11/10/2011.

> I arrived in Bamenda … and went out to distribute the parcels I brought from relations in Cape Town. Most had called me already given they had been given my number by the senders of these parcels. Some of the things sent were just to signify that they were out in Cape Town. Max, for example sent two packets of biscuits, some pens and some pirated cartoon DVDs for his children and a ball for the football club where he was a member. There was nothing special about what he sent. But I guess the significance is not so much about the things sent but rather the meaning attached to them and the bond that keeps them together. The things cement the relationship and maintain his virtual presence. Also, the meaning the children attribute to the things, anything foreign and coming from family abroad and no matter how small it is, plays an important role in the mind of the receiver.

Most migrants with ailing parents or family on medication for a chronic illness have found ways of consulting a doctor on their behalf in Cape Town and they then purchase medicine and send it home whenever they can find someone willing to take it. Bianca (from Pinyin), a nursing student in Cape Town, has been able to get her mother's drugs this way. Her mother lives in the village, is diabetic and also suffers from hypertension. Whenever she knows somebody will be travelling back to the village, she contacts her mother about the state of her health and consults her GP on her mother's behalf. She then buys the drugs prescribed and sends them with the person travelling. Bianca always prefers to send drugs to money for the drugs because the South African drugs are much better than what her mother can buy at home. This is her story:

> … we communicate on a weekly basis and my mother is a patient. She's a diabetic patient and has high blood pressure too. What I do is that every two months, I can send her at least, FCFA 25,000 to FCFA 50, 000 to keep her constantly on medication. And sometimes when I see someone who is leaving for Cameroon, I go and consult a doctor ... My own doctor who I consult is a doctor in this new medical centre, Dr X. I consult him, then buy medicines and send (sic).[5]

Monetary remittances

As discussed in previous chapters, these 'mobiles' have established their own form of monetary transfers regarding remittances. While M-pesa is the obvious means of sending money today for most in Southern and East Africa, this method is still to appeal to Pinyin and Mankon migrants. Sending money via mobile phone is unpopular with them because mobile money is a fairly new way of transferring money and was not popular in Cameroon at the start of this research. By the time it gained currency, migrants had already turned to using Visa cards. And most parents in the village are not literate and can only manage to receive and make calls on a mobile phone. Quite a number have electronic Visa cards in the keeping of friends and relations who are frequently instructed on how much money to give to relatives. Those without cards at home use those of others to send money. The amount is deposited in the account with sufficient to cover any

---

5 Interview, Cape Town: 17/08/2010.

bank charges, and the recipient back home is requested to meet the cardholder in Cameroon.

Whether in cash or kind, remittances go a long way to alleviating the social and physical woes of recipients, as well as compensating for their children's absence. While some send money periodically or in times of crisis, others have felt it is their duty to place parents on a monthly subsistence allowance. Emile,[6] who is from Pinyin but lives in The Hague, gives his widowed mother an allowance of FCFA 10,000 a month, over and above anything he might send her if there is a serious family crisis. Emile's mother told me about the financial assistance she receives from him and his sister in Cape Town. Although she rarely calls them, she beeps so they can call her back. And if in dire need, she beeps and waits for them to return her call with a simple message '*I di call for tell them say dat thing* (money) *don finish, I just be me empty hand*'.[7] She goes on to describe how the money is sent to her from the Netherlands; 'Emile sends the money to his sister in Limbe (South West Region), who then sends it to the Mitayen Credit Union in Santa where the mother goes to fetch it, or the money is forwarded to the one in Pinyin'. She expresses her satisfaction with the level of financial assistance she receives from them (Emile's sister in Cape Town as well). And to mark her gratitude, every time they send her some money '*I di give special offertory for church for thank God for them and di beg yi* (God) *for them for continue di bless them*'.[8] Unlike some parents who can place excessive demands on their children (noted from interviews with the children in Cape Town), Emile's mother is cautious about being too demanding and would not want to encourage them to engage in illegal activities in order to send her money. According to her, '*if pekin go out you di ask money all the time na for send yi say make yi go enter coupé* (*nyongo*). *Better yi give the small one wey ee get.*'[9] Some of this money goes to sponsor funerals at home. Being cautious about her demands from her children is her way of morally protecting them from indulging in unlawful acts to satisfy her demands.

Partaking in virtual and transcultural funerals

Chapter 8 dwelt on how associations assist with funerals in the host country when a member passes away. In this section, I focus on how individual migrants relate to funerals in their respective families, the trauma of absence and the emotions that this brings to bear on them. Funerals and deaths back home account for

6 Interview, The Hague: 28/11/2010.
7 Conversation, Pinyin with Emile's mother; 10/10/2011 (see Footnote 1). Word in brackets is the researcher's.
8 Conversation, Pinyin with Emile's mother; 10/10/2011. Word in brackets is the researcher's.
9 See Footnote 8.

the greatest part of migrants' remittances and engagement in family circles (Mazzucato 2008; Mazzucato *et al.* 2006; Gugler 2002). And they also herald a time for increased telephone connectivity to make arrangements for a decent funeral and for rituals to be performed. Increasingly, they are becoming multi-sited events. Funerals and rituals form part of the code of honour that keeps migrants attached to the family leash, fearing bad luck if they do not participate financially in funerals at home. As was shown in Chapters 7 and 8, participation in such events has a multifocal role as it reifies identity and belonging to the community and the home village. It is an expression of wealth as it proves how much social and cultural capital one has garnered. In this light, the death of members with kin and children can either adopt the appellation '*bush-faller die*' (*ala-ah*) or '*yeye die*'. The former denotes that mourners were well catered for, and the latter means that there was barely enough food and drink to feed the guests. By extension, the quality of the funeral organized can render pride or bring shame to the family: pride that mourners are well looked after or shame if the home community expects a grand funeral and it turns out to be a '*yeye die*'. As such, funerals can sometimes be delayed for up to two months, with the body lying in the mortuary waiting for the arrival of all the deceased's children, to provide adequate time for decent funeral arrangements to be made. Bad mouthing and village gossip take centre stage, with people expressing their indignation. For a migrant, assent is important to their dignity as well as the entire family. John's story clarities to this point.

John's story

John hails from Mankon and emigrated to the Netherlands in 1999 to study. Upon completion of his studies, he could not find work and decided to continue studying. He graduated in 2003 with a second MBA, this time in Management and Marketing and since then has only had secured work in low-paid jobs that do not match his qualifications. In the meantime, he acquired Dutch citizenship in the hope that this would open more doors for him. But he soon realized that a Dutch passport would only give him the opportunity to live in the country, as he was constantly asked at job interviews why he did not go back to Cameroon after completing his studies. He got married in 2005 to a Cameroonian with Dutch nationality who he met in Holland and in 2008 she passed away. Despite her having Dutch nationality, he took the body home for burial. Prior to his departure, the deceased's cousin and other friends spent weeks making funeral arrangements in the Netherlands, processing repatriation documents and liaising with family back home. A cry die was also organized in The Hague, when sympathizers and the church offered donations to help towards the costs of sending the body home.

In 2010 John's mother passed away and he sent money home for the funeral, but still travelled home two weeks later because he had initially bought a non-refundable ticket to visit her while she was sick. At home he invited the choir that his mother was a member of to sing at a church service that he organized in her memory. After the service there was a meal (provided by John). Going home was the right thing for him to do and was a way of mourning his mother, despite the fact that he had already sent money home for the actual funeral. He maintained that: 'I really need to satisfy myself that I was at the burial because when I go

I will make something, I will throw a feast and it will satisfy me psychologically because people will come around; it will be as if am making my own service.' Prior to going home, he had reprinted and enlarged photos of her and hung them on his wall, and organized a cry die in The Hague among the Cameroonian community. On the day of the burial, he followed the ceremony virtually from the Netherlands, i.e. when his mother was being interred, he phoned his brother and asked that he put his phone on speaker so he could participate, while simultaneously recording the song that was being sung then and that he later used for a while as his ring tone in memory of her. Getting his brother to put the phone on speaker was the only way he could be part of the ceremony. This gave him some emotional consolation and the gesture was significant:

'I did that because I wanted to have some presence at the graveside, that's all ... I wanted to feel that atmosphere, to listen to the types of song they were singing at the grave, to listen to that melody, it's very important because I will never ever forget the songs. It gives me the idea of presence, as if I was also participating in the funeral. I felt that I was present there.[10]

Despite this virtual presence, he went home two weeks later, as mentioned above. It was during this visit that he was given a title in the Fon's palace, loosely translated as the 'Fon's Listener'. Before returning to the Netherlands, he made arrangements for his mother's grave to be built and for a tombstone to be erected.

In 2011 his father passed away and this time he travelled home for the funeral and spent a month there. The body was put in the mortuary while he made funeral arrangements with the family. During this period, communication between him and home intensified, as well as communication in the host country, with friends calling to offer their condolences and gathering at his home every evening. (I was in Cape Town then and had a series of Skype calls with him and when he travelled home, I sent my mother to represent me at the funeral.) But before travelling, he again organized a cry die in The Hague. Having never attended a funeral back home since he left in 1999, this was his first and he was amazed at the elaborate expenditure on food, drink and clothes (*aswhabi*) for family and friends. Nevertheless, he was very happy to be home and to participate in the funeral.

John's story aptly shows, on one the hand, migrants' transnational virtual and corporeal engagement in funerals back home and, on the other, how living in the Netherlands for twelve years has not changed his notion about participating in funerals. If anything, his long absence had heightened his craving to be part of the family when his father passed away, which was why the body was put in the morgue until he arrived. Migration has played an important role in changing the face of funerals, in sharp contrast to the low-key and modest ceremony I used to see in the village when I was a child. Within hours of the death of both his parents, the news was communicated to him by phone. While such swift information was partly to let him know about their deaths, it was also in expectation that he would send money or bring it with him in person. When his mother passed away, the family did not want him to attend the funeral because of the earlier expenses he had incurred. Acknowledging his financial contributions during her period of hospitalization was a way of allowing John to recuperate financially. What also marks this moment for migrants is the emotional trauma they have to endure if they do not travel home or if they have not seen the deceased for a long time before his/her death.

[10] Interview with John, The Hague: 29/09/2010.

Attending funerals and organizing simultaneous ceremonies in the host country helps individuals to deal with the situation psychologically and overcome the emotional trauma that comes with it. By organizing huge funerals, it is easier for them to come to terms with the death and it absolves them of their long absence and re-inserts them back into the family fold. But these are habits that they have been brought up with. However, going home gives them the opportunity to organize a 'bush-faller die'. Besides the emotions and the 'politics of status' (van der Geest 2000), fear is often at the heart of these elaborate funerals, fear of ill luck and that a poor funeral will bring shame to the family. As Bosman (1967[11] cited in van der Geest 2000: 104) noted with reference to funerals in the Akan society of Ghana, funerals are 'a point of honour to make a great show at their funeral customs, and they vie with each other in performing these expensive burials. Even the poorest will pawn and enslave themselves to obtain the means of burying a relation decently, according to the ideas of the country.' A Pinyin or Mankon would immediately identify with this process. This is testament to the similarity of funerals in both societies because death has come to symbolize munificence over the dead body in Pinyin and Mankon as well as in most African societies. And this munificent is escalated by the number of family members abroad who want to attend and request that the body be put in the mortuary pending their arrival. Their presence is usually remembered for their excessive spending at the funeral itself. On such occasions, it is common for families to be levied, with bushfallers paying the highest levy.

The case of Terence, who originates from Pinyin but lives in Cape Town, illustrates this point. His maternal grandfather passed away in the village in October 2011 and as the deceased was considered by all in the village to be a man of status (a 'baobab' in everyday parlance), the family was eager to maintain the status quo. A family meeting was held to plan the funeral and his mother was charged with communicating to him his levy of FCFA 300,000 (approx. € 500 or R 5000) and other obligations (*kontri fashion*) that he had to perform. This was a disappointment because he thought the amount was too high. He wondered aloud how much he would have to spend when his father died. However, the money was duly sent to allow his parents to entertain their own guests and when I arrived in the village a few days after the funeral, I heard stories about how grand it had all been. His father was very proud of him and claimed that such funeral remittances helped to give him and his peers a voice in the village because he could slaughter a pig for the occasion, pay for gunpowder and entertain his guests. Respect is not only gained by the migrants when they send remittances but the receivers of these remittances equally garner prestige and respect from their peers

[11] This is as relevant today as it was when it was written over four decades ago.

as it increases their spending capabilities and compensates for the migrant's absence.

Channelling emotions

> ... it's not just about the demands, it's about the attitude. The more you give, the happier the family is with you, the more responsive they are to your communication and phone calls. The more withdrawn you are financially, the more withdrawn they are, so sometimes you don't even feel like phoning when you're not giving. You know ... for me it's not just about phoning and getting these complains and demands, for me it's about phoning and speaking to people who really show emotions of joy and excitement that you have phoned, even over the phone. You can sense them. The more distance you have financially, the more some people will be distant even in their communication approaches so it puts me off and I just do not feel like phoning those people.
>
> It should be reciprocal especially when you have a child out of the country ... The only thing you can give them is when they phone and you speak to them in a manner that they would go away knowing that I've connected with my family. But when they phone and they are distant with you, give more monosyllabic answers to their questions and there's no communication, then it puts one off and it makes you miserable being that far away.[12]

Using this quote as our point of departure and to give insight into how human interactions are characterized by social relationships and may lead to 'memorable and meaningful aspects of their lives' (Leach & Tiedens 2004), we can assess the impact of migrants' mental states when they feel disconnected from family bonds. It also shows how, if remittances are not properly managed, the situation leads to anger and unhappiness, and conversely to happiness when everything is well managed, as was shown above. Through Jake I aim to show the inextricable relationship between remittances and emotions and how, together, they lead to disruptive or interactive social relationships because 'social interaction and relationships are so important, and emotions respond to them' (*Ibid.*: 7). As such, it is difficult to talk of transnational families without incorporating the emotions expressed and the way migrants talk about these feelings because they form a 'constitutive part of the transnational family and migration experiences – experiences of adjustments, nostalgia, shattered sense of belonging, loss, abrupt ending' – all of which form salient sources of emotions' (Skrbiš 2008: 236). The emphasis on transnational studies has been less about remittances, double engagement and funeral participation than emotions. Empirical findings from this study show just how much emotional content pervades relationships within family circuits. These emotions are couched in the longing for home. In their absence, migrants substitute home for the mobile communities in place. In this respect, I contend that migration is an on-going emotional journey for both the migrants and those left behind in the home country, and an emotional challenge that they seek to mitigate. While ICTs have made nonsense of physical boundaries, match-

---

[12] Interview with Jake, Cape Town: 09/04/2011.

ing this with respect to emotional boundaries still requires co-presence and face-to-face interaction. However much emotional investment is made in maintaining contact with those left behind, those left behind also attempt to reciprocate with their kin abroad, making them 'local transnationals' (*Ibid.*: 238). While studies have primarily focused on migrants' involvement back home, as though communication was one-way traffic, I show that although calls from home are not necessarily frequent, they do have an effect on migrants. This sheds light on the emotional anxiety that occurs as a result of emigration and family separation.

As Turner & Stets (2005: 1) have so well argued, emotions can be seen as the 'glue' that binds people together ... to large-scale social and cultural structures, and by the same token emotions are also what can drive people apart and push them to tear down the social structures'. They extend their argument by asserting that emotions are 'socially constructed' given the fact that what people feel is 'conditioned by socialization into culture and by participation in social structures' (*Ibid.*: 2), namely ideologies, beliefs and norms as they impinge on social structures. We could thus infer that transcultural habitus is equally underpinned by the 'emotional culture' (Ryan 2008: 300) that defines how people should feel on various occasions, a feeling they have internalized throughout childhood and their adult lives and that forms part of their cultural ethos of 'feeling rules'. It follows that emotions emerge from situations that are intimately social, with individuals learning appropriate feelings and how to use them in different types of relationships (Turner & Stets 2005). The overriding question here is to what extent ICTs have enhanced or mitigated the channelling of emotions amongst transnational migrants. What role do emotions play in the negotiation of transcultural family life?

Imminent return visits and family reunions are, therefore, events that ignite deep emotional anxiety and joy. These visits and reunions, while leaving an indelible mark on the migrant, take place in the continuum and provide a window into their accumulation or degeneration of emotional capital. While such emotional contents have been made vivid as a result of enhanced communication technologies so that migrants can easily express joy or vent their frustrations about relationships with family, it does not preclude the fact that these expressed emotions date back to the period of rural-rural mobility when migrants eagerly awaited market days to hear news of family members from traders who came from the village to sell products at the market. Pa Muluh from Pinyin, for instance, used to live in Njinikom where he would go to the neighbouring markets in Nkambe, Wum and Fundong to buy goats to resell in Njinikom. On market days in Njinikom, he looked forward to receiving a message from his father and to getting news of his village. The last such message was an urgent message from his father requesting him to come home.

Letters in the past were seen as methods to channel such emotions among transnational families, as was shown by Thomas & Znaniecki (1918/1974), whose seminal work provided an embryonic understanding of transnational family emotions amongst Polish migrants in the US. It is in this light that Skrbiš (2008: 232) opines that these letters contain information as well as being repositories of emotions that express 'sadness, longing, dread, pain, happiness, love, faithfulness, betrayal and emotionalised claims about truth-making all competing for attention'. Returning to the opening quote above by Jake, this attitude could be read as emotions of longing and seeking attention. Today, such emotions are expressed more in phone conversations and emails than in letters, which used to take a long time before they reached recipients. Migrants and families have the agentic role and can choose which particular communication tool to use to best convey a specific emotion. For instance, when Joe's mother passed away, the family at home chose not to inform him by phone even though he made frequent calls home. Instead, an uncle insisted on having his postal address and a letter was sent informing him about the death. According to Joe, it was the right decision as the emotional shock would have been too great if the news had been given in one of their phone calls. The carefully worded letter had a more soothing effect than any phone call could ever have had.

The opening quote is illustrative of the expected 'emotional reciprocity' (Wolfe 2002) that migrants constantly negotiate in terms of psychological balance, wellbeing and belonging. Additional information from the voice, tone, expression and silences that is derived from phone calls conveys lucid emotions from both parties. However, migrants like Jake are quick to notice when the voice and tone do not convey the usual emotional concern and this has a telling impact on the migrant. While phone conversations in the context of long-distance communication aims at calming fears of disconnection from the family fold or sustaining virtual intimacies, monosyllabic answers or long silences and lapses only help to raise the tension. What this equally affirms is that transnational family relationships are dynamic and fluid, shifting according to life-cycle events as well as responding to the flow of remittances and perceptions of emotional closeness. As such, families exhibit a variety of communication patterns over time and context.

Family reunions and planned visits

The imminent visit of a migrant triggers anxieties and emotions from when the journey is planned until the actual journey takes place, and any impending face-to-face meetings heighten emotional arousal and signify attachment to the home village. Gordon (1961, 1963 cited in Turner & Stets 2005: 27-28) conceives such face-to-face meetings as an 'encounter' that bespeaks 'an emergent "we" feeling of solidarity and flow of feeling and a ritual of ceremonial punctuation of openings, closings, entrances and exits'. He carries on to say that these encounters are embedded within an occasion that is part of a larger gathering. Such gatherings are meticulously arranged using any of the various forms of communication technologies available. As such, and prior to the gathering, there is endless communication back and forth to make advance arrangements. Longing for and requesting particular food to be prepared, mental notes of must-see members of the family and places to visit as well as what to take back are some other focal points. Similar to Gordon's 'encounter', Urry (2003: 156) affirms that such 'increasingly extensive networks, made further extensive through communication revolution, do need occasioned meetings for its survival', which he refers to as 'meetingness'. This 'meetingness' or 'encounter' is underpinned by emotions and the desire to reconnect with the family fold, and the affordable and myriad forms of travel that have made it possible for these reunions to take place at all.

The returning home of Rachael & Joe, as we will see below, was orchestrated by cultural obligations (Urry 2003) that were prescribed (*kontri fashion*) but also carried strong normative expectations of presence and attention from home. It was in this context that Goffman (1963: 22 cited in Baldassar 2008: 262) maintained that 'co-presence renders persons uniquely accessible, available, and subject to one another'. Rachael's case is a good example here. She originates from Mankon and migrated to Cape Town in 2007 for her studies. After she gained her Bachelors, she got married, had a baby and then informed the family back home. In 2010 her dad passed away and although she was unable to attend the funeral, she sent money. Communicating with her mother back home, they scheduled the memorial service for April 2012, and after that there was endless communication back and forth to meticulously plan the ceremony. She and her sister (also in Cape Town) sent money for the mother to begin renovation work on the house according to the standards that her father had mentioned in discussions before he died and also for the grave to be built. Money was sent to hire canopies and chairs and their mother was charged with asking the presiding pastor for the occasion, in advance, to diarize it. By the same token, Rachael was looking forward to 'just hugging her mother and crying with her'. 'I have missed her so much and can't even imagine what she is going through since my father died ... I also have a lot of things to tell my father when I go, I will have to talk to him over his grave'.[13] When talking about her father, she tries hard to maintain her composure as her voice cracks in the course of our interview. Going home was very important for both of them as they needed to 'face the place' (their father's grave) with an emotional and corporeal co-presence that until now had only been virtual. Rachael was then also still a student (pursuing an MA) and a businesswoman,

[13] Interviews: Cape Town with Rachael: 27/08/2011, and in Bamenda with her mother: 19/09/2011.

and had made a list of all the things she wanted to bring back for her shop as well as foodstuffs and a traditional dress (*ntogho*). The dress, which is a must-have in both communities, reinforces a woman's identity and sense of belonging to the home village as well as exhibiting her emotional attachment, judging by her insistence on buying the best cloth for hers. Rachael had been able to make all the necessary arrangements prior to her departure simply by pressing the call button on her phone and through email correspondence with one of her aunts who is good on the Internet.

Reunions, like return visits, are laden with emotions and can be seen to symbolize congregations of transnational family members who are of significant importance in the lives of transmigrants (Skrbiš 2008). While their importance should not be underestimated, what we should remember is that prior to these reunions, like return visits, a lot of behind-the-scenes phone calls and other forms of interpersonal communication are going on. In a Skype conversation from Leiden on 26/10/2012 with Joe in Cape Town, he told me he was delighted to inform me about the upcoming events in his family:

> Q: What is news at home?
> A: The good news is that the entire family will be at home this December; Esther, the baby and I will travel on 27/12/2012. Raymond will be flying in from Sweden, Alphonse is coming to get married and Chris will be going home for the first time since he arrived in South Africa (six years ago) to go and look for a wife with Raymond.[14]

Raymond is in Sweden and Joe and his two brothers are based in Cape Town so this would be the first time all of them were converging on home at the same time and the first time that Chris and Raymond had gone home since their father died in 2008. Joe was going home for the second time with his South African wife but also with a new addition to the family – their twenty-month-old daughter who is named after Joe's late mother.[15] She will be presented to the family for the first time. In addition, Joe will be crowned as his father's successor, as was discussed in the previous chapter. Joe had been communicating with Raymond in Sweden via Skype and they will arrive in Cameroon on the same day and travel on to Bamenda together. Back in the village, he is in communication with his family, his uncle and the quarter's head (*ndei*) who was going to preside over the ceremony. Straddling different social fields, Joe was able to make arrangements with respect to the family gathering thanks to cheap calls to the home country and free Skype calls on the Internet to Sweden, as well as face-to-face meetings with two siblings in Cape Town that were interspersed with phone calls. This experience will stay with the entire family at home forever; as it is a once-in-a-lifetime gathering. It would be reminiscent of their childhood days and provide

14 Skype conversation with Joe: 26/10/2012.

15 After naming the child after his mother, he felt that he had brought his mother back to life.

fulfilment of their intense emotional longing and nostalgia. Earlier conversations with Chris highlighted how he longed for the day when he would go home and eat 'proper *achu*' and drink palm wine. All the stories of their return visit and imminent family reunion bear the hallmarks of nostalgia and experiences of longing and absence. What this also reiterates is the importance of mobile communication in arranging such a family gathering. First, the timing has to be right for those at home and abroad and it is only by telephone communication that the reunion of the dispersed – in Sweden, Cape Town, Nkambe and Pinyin – was possible. Co-presence and emotional feelings were channelled through telephonic communication to make the gathering possible.

Photos and memorabilia as co-presence

> When I came to South Africa, I brought photos of my wife and children. These photos have been quite helpful especially on days when I miss them very much. I take them out and go through them, after which I call them while looking at the photos. It makes me feel as if they are next to me and after that I can sleep calmly.[16]

The opening quote in the title is reminiscent of when workers left the Grassfields to go to the coast as plantation labourers (Nkwi 2011) or when township workers left for the mines in South Africa (Feyder 2009) and all they had to remind them of their relations were their photos. Photos have long played a cathartic role in the lives of migrants and the families who stay behind. However, today, the story is one of evolution as cameras are common, have become digital and moved on from the black and white pictures of the past to today's colour ones. And, importantly, people do not have to wait for three months for colour pictures to be sent to Europe for printing. Photos taken impulsively to capture unexpected moments are increasingly popular thanks to the availability of mobile phones that incorporate cameras as well. Photos have become a daily reality in the lives of migrants and their families. Joe's mobile phone has allowed him to document his daughter's progress from her birth to today, as well as their visit to Cameroon. Part of my fieldwork entailed immortalizing events by photographing families and individuals that I interviewed, as well as things that struck me as memorable in the course of my research. In Cape Town, Pinyin and Mankon, these photos of relations were the most sought-after. Families in the village would ask whether I had photos of their relations and would want to print and enlarge them. In this respect, I asked individuals interested to give me brand a new USB stick so I could transfer their photos onto them (and protect my own computer). Others (for example Joyce) printed the photos directly from my camera's memory card. I

[16] Interview with Max, Cape Town: 31/07/2011.

Photo 9.1 Joyce's children in Pinyin

Photo 9.2 Joyce printing out the photos from the researcher's memory card

also took back lots of photos from family in Pinyin to show to kin. Some were recent photos, while others were old ones.

Bourdieu ([1965] 1990: 19-20) likened photography to a process of 'solemnizing and immortalizing the high points of family life, of reinforcing the integration of the family group by reasserting the sense that it has both of itself and of its unity'. This was seen as the main purpose and *raison d'être* of popular photography, but it is just one aspect of the story that 'expresses the celebratory sense of family events; the internalisation of the social function *of* this practice' (*Ibid.*). In the context of global mobility, I maintain that photos also play an emotional role in calming the emotional stress of absence and longing and also perpetuate family memories (see also Feyder 2009). While its role bears witness to events, it is

Photo 9.3 Max, just before boarding a flight from Cameroon to South Africa.

Photo 9.4 Most of Max's family, except for his mother, travelled to Douala to see him off at the airport

not as imminent as before. Unexpected photo shoots today tell the stories better given the proliferation of digital appliances and mobile cameras.

When I visited Max's family in Cameroon, his two children (aged three and five) were keen to show me photos of their father and were not happy when I asked if I could borrow some of them to scan before returning them. They cried and said: 'Don't take our daddy away, give back our daddy'.

The outpouring of emotion by the children made me return the photos but, understanding the importance of my research, their mother found a way of lending me the photos without their knowledge. Looking at the photos, I noticed Max's mother was not present among those who went to see Max off at the airport and I asked why. She said, 'going to the airport would have killed me, it broke my heart to see him leave the house on that day, and it was difficult to see my first son depart'. As she spoke, tears rolled down her cheeks and she is still bereft over his departure. On a second visit, the children told me: 'we will hide our daddy so that you do not take him away'. Incidentally, the opening quote is by their father Max, in Cape Town, and shows the extent to which emotions are embedded in photos and the soothing role they have on him and his children. Being miles apart, what keeps them together are the photos of each other that they both have in their possession, and the phone calls he makes to complement the photos. But his mother's emotions are still raw and all she wants is for him to come home. Through these memorabilia and his virtual co-presence, members seek to keep a 'sense of family' (Baldassar 2007: 392) across geographic boundaries and are updated with each other's developments. In other words, while these photos visually and virtually unite spouses and families within the abstract space and endless time zone of the oval frame (*Ibid.*), the outpouring of emotions witnessed (and captured) in this researcher's interaction 'describes an interpersonal dynamic in a way that no other description can, hence, constitutes the relationship' (Leach & Tiedens 2004: 8).

I arrived from the Netherlands with photos of Emile's son for his mother in Pinyin. Having called earlier to inform her about my arrival, she postponed her farming activities for that day and was at home waiting for me. When I gave her the photos, she looked at them, danced with them held tightly to her chest, and after a few turns kissed the photos and lifted them up to the sky and said 'Thank you Lord'. She then resumed her dance with the photos clasped to her, paused for a while, took another look at them and continued dancing (Photo 9.5).

This came to an abrupt end and she then sent for her eldest step-daughter to come and see the photos. When she arrived and they looked at the photos together, they both started dancing again. This was the first time they had seen pictures of their five-year-old (grand) son, which helps account for the outbursts of

Photo 9.5 Emile's mother looking at the photos of her grandson sent by Emile with this researcher (from Holland to Pinyin)

emotion. Photos, like ICTs, instil a sense that loved ones are present despite the geographic distance that separates them (Brinkerhoff 2009).

There are strong parallels with photos of the living and of deceased parents. All the photos are kept with utmost care and offer a sense of assurance to migrants that they are still connected with kin and that deceased relatives are still watching over them. As mentioned earlier, once migrants receive news of the death of a parent, they immediately enlarge his/her photo and put it on display, as was the case with John in The Hague and CJ, Paul, Peter and Joyce in Cape Town. When I visited John in The Hague, I noticed that the photos of his deceased parents had been moved from his living room to beside his bed. Asked why he chose to keep them so close, he said 'I want to enjoy them for a while, I get inspiration with them being so close to me'.[17] A continuing bond with them and no feelings of detachment from the deceased are common (cf. Klass 2001), and this has psychological benefits for the bereaved persons as in John's case.

These 'transnational objects' (Baldassar 2008: 257), such as photos, memorabilia, recorded messages and gifts, are significant to givers and recipients because of their tangibility and this takes the physical place of the longed-for person or location. They stand for the absence of being. For instance, Jake's father's video message to his children has helped them to connect to their father by constantly

[17] Conversation with John, The Hague: 25/11/2012.

watching and listening to his message. Placing their photos close to their chests (Emile's mother and sister) or like John beside his bed illustrates their commitment to their respective relationships. Although the photos and video become the 'co-virtual presence', they are the tangible things people can hang on to as a replacement for a 'physical co-presence' (*Ibid.*). But what we must remember is that the availability of these photos was as a result of me navigating between the Netherlands, South Africa and Cameroon because of improved airline travel. I was able to be at these loci at various times but there was also the possibility of sending the photos by Internet. The fact that John was able to enlarge old black-and-white photos of his parents to a restorative state signals great strides in digital technology that have helped him to immortalize his parents. Conversely, Jake going home with a portable camcorder that can record for a long time before recharging is indicative of their evolution from the large and cumbersome equipment of the past to today's sleek ones that double up as both camcorder and camera. As a result, migrants and their families are able to express simultaneous emotions of proxy, missing and yearning, thus becoming transnational and transcultural families. The phone conversations, photos and memorabilia allow families to act out and express their sense of closeness and intimacy.

Mediating long-distance relationships (LDRs) and co-presence

> Apart from these photos, I also have lots of photos of the children, my wife and mother on my phone. Even my screen saver (*photo displayed on the mobile phone*) is a photo of my children. On days when business is slow, I keep myself busy by scrolling through the photos.[18]

Max's coping mechanisms for dealing with the distance that separates him and the family are the phone calls he makes home and his kids' photos that are on his mobile phone. His phone is like a 'Swiss Army Knife' (Castells *et al.* 2007) that, besides calling, gives him emotional satisfaction and the feeling that his family is just a click away. Having been with Max in the field (see Chapter 4), his free moments are times to scroll through the photos on his phone to bring his family closer. He has given clear instruction to his siblings never to phone or beep him in the mornings and the only persons authorized to do so are his wife and his mother. For her part, his wife gives him a brief call in the morning just to 'say hello', especially if he has gone for two days without calling. She admits it is difficult for her and the children to be without him but at the same time they are happy as their lives have changed considerably for the better since his departure. Despite an improved livelihood, the strain on her is betrayed by long moments of silence, a cracking voice and watery eyes during the course of the interview. In this context, mobile phones do not replace face-to-face interaction but provide

[18] Conversation with Max, Cape Town: 27/07/2011.

new opportunities for constructing a co-presence despite the distance (Horst & Miller 2006), as well as mitigating emotional worries of absence and loss. The use of SMS for both Max and his wife has become part and parcel of their daily routine. She informs him about developments in the family, seeks his opinion before any major decision is taken in the family and this, in turn, gives Max the feeling of still being in charge in his family although he is faraway. But text messages too have created a virtual space for both of them to affirm their loyalty to each other. For Max and his family, the mobile phone has become the tool that maintains the LDRs and 'relational continuity' and is a way to define the relationship as continuous. The Internet is not yet a part of their communication repertoire, perhaps because neither of them is well-versed in how to use it.

This connectedness enables us to understand how the proliferation of mobile phones has mitigated distance and enhanced spousal relationships. This begs the question about how those who travelled in the late 1990s and left their families back home, like Joe and Jake, connected with them. For Jake, the Internet, though scarce and slow back then in Cameroon, was his God-sent medium of communication because of its relatively inexpensive nature, but above all because a friend (who had an Internet connection then) served as the link between him and his family. According to him,

> ... In my case, I also had this friend, Simon, who has this NGO in Bamenda; one Pinyin man ... most often I'll write emails to him, he'll print them and give them to my family then they will come and give him messages, then he will write them and send them to me.
>
> Q: OK he will print emails and send them to the village?
> A: Yes ... then when they have a message, they will give him mine, then he will type out an email and send it to me.
>
> Q: So somebody will leave the village and come to Bamenda to give him the letter?
> A: Yes ... that's how I communicated with people in the village at that stage and I still have some of those emails (laughs) that he sent to me ... even when I started communicating with my wife, that's how I communicated in Cameroon ... when she was in Cameroon then. Because she didn't have a phone initially and she didn't even have an email. So that's how we communicated initially.[19]

Although Wilding (2006) contends that such 'virtual intimacies' are characteristics of transnational families, it is equally important to highlight that virtual intimacies cannot go on indefinitely but need to be greased by a corporeal presence.

Prior to the proliferation of mobile telephony, the Internet was the main form of communication between families and kin abroad, and sending or receiving an email would cost FCFA 1500 (approx. € 2.50) at the only Internet cafe – Allied Bamenda – back then. But for someone like Jake, having an anchor person in Bamenda who could deliver emails to his family and wife meant that the distance

[19] Interview with Jake, Cape Town: 11/08/2010.

between them was less due to the ease of communication. And for urgent matters, he would phone his uncle in Yaoundé who had a fixed phone at home. Today, as a result of the reduced tariffs for telephone calls, migrants call more frequently and are able to 'lie on their beds and make calls home'.

On the whole, the introduction of ICTs in the lives of families makes migration and mobility more acceptable than ever before and gives 'distant individuals the means to manage and maintain their connections as well as to negotiate their roles through time' (Bacigalpe & Cámara 2012; see also Wilding 2006: 132). While ritualized communication helps to manage LDRs, it also fosters the transmission of belief systems and inculcates the notion of witchcraft.

Transcultural ↔ transnational belief systems and the notion of witchcraft

> *Since wey yi woman sick sote PIFAM contribute money senam back for Cameroon, the woman go sick sote die, ee no go for buryam. Till today ee never go for see whosai wey them bury ee woman, ee never go make ee own die, na yi bad luck ee don carryam so noh! Ee be among the first Pinyin people them for this kontri, but ee don fall (business) sote ee no fit pay yi rents, even for play njangi don pass yi. Na daso the bad luck di follow yi so.*[20]

The above except includes scathing remarks made by a female informant about another migrant (Jerry).[21] He is considered by most as a failure because he is plagued by ill luck. He is thought to have a curse of bad luck hanging over his head as a result of his inability to travel home for his wife's funeral and the fact that he has not even gone home since her death to see her grave or organize a cry die in her honour. Such perceptions are deeply embedded in migrants' cosmic belief system and negligence or the inability to yield to the dogma of tradition is believed to be sanctioned by ill luck. It is believed that whatever Jerry does will meet with failure. The ripple effect of not going home to bury his wife has stirred up feelings of resentment amongst the village folk, especially the women. And this resentment may leave him vulnerable to real or imagined curses of *kontri fashion* (Nyamnjoh 2005). The term witchcraft is used here in its widest sense and includes sorcery, evil spirits, bad luck and jealousy when referring to the Grassfields, but also in the way it is used by my informants.

Most informants believe that Jerry's shop and his attempts to get into pitching failed as a result of not him going home in eight years, and more specifically his failure to go home to bury or organize a cry die for his late wife. If he goes home to *make kontri fashion*, he will have some respite and '*yi road go di shine, anything wey ee want do-am go di waka fine*'.[22] These beliefs also mirror those held in Pinyin by parents who urge kin to come home and participate in cry dies. Both Jake's father and Emile's mother alluded to this bad luck but maintain that the

---

20 Interview with Zora, Cape Town: 24/08/2010. Insertion in brackets is by this author.

21 Incidentally this migrant has emigrated to Europe.

22 Interview with papa khan, Pinyin: 10/10/2011.

only way it can be warded off is by yielding to tradition and performing the ritual that is required because '*na kontri fashion; ee get to call dance, shot gun so that village go know say ee don come cry yi mami or papa*'.[23] Commenting on this, Pa Khan (Jake's father) says that failure to perform a cry die ritual in Pinyin and to see the grave '*na some bad luck, something go di happen say ee no be cry mami yi die*'.[24] He ends by alluding to himself and how he is still performing certain traditional rites following the death of his mother-in-law. These rituals are done less out of respect than as an obligation and out of fear of witchcraft or bad luck. In Jerry's case, the human body has become a vehicle of the spirit and its manifestation has occurred through ill luck and other forms of misfortune (Sabar 2010). How are traditional belief systems and the notion of witchcraft reinventing and refashioning themselves in the era of mobilities and new technologies?

Migration and mobility have not only fostered transnational communities but also illuminated the fluidity in transcultural beliefs and in witchcraft, thus expanding the discourse on belief systems. While the study of witchcraft has tended to be dichotomized and either focuses on the destructive nature that results in death or the aspect that stresses 'interdependence and conviviality' (Nyamnjoh 2005: 241), this section marries both trends of witchcraft: the destructive and the convivial. It shows how, although being away from home, the notion of witchcraft is still embedded in the migrant communities' repertoire because of the 'principle of simultaneous multiplicity' where the physical presence or the absence of the victim is inconsequential (Nyamnjoh 2001: 30). Migrants will do whatever it takes to avoid it chasing after them in the host country by negotiating 'conviviality between competing and conflicting expectations on their labour and its proceeds' (Nyamnjoh 2005: 245; 2001). Such notions of witchcraft and beliefs, as argued by Fisiy & Geschiere (1990, 1991), have injected postcolonial witchcraft discourses with a new dynamic reflecting the ability of witchcraft beliefs to adapt to the modern nation-state and new types of entrepreneurship. In this respect, failure in business is due to bad luck or witchcraft from home that pursues the migrant even in the host country or as a result of having jealous migrants in the host country (see also Parish 2000). This therefore assumes that witchcraft is not bounded by space and is capable of assuming new forms. Concurring with Nyamnjoh (2005), Ciekawy & Geschiere (1988: 3) posit that

> Understanding witchcraft is an essential element in any attempt to comprehend people's mundane realities and thoughts ... In everyday life in Africa and elsewhere it is a discourse about action and the urgent necessity to handle these dangerous but hidden forces.

Such notions of witchcraft are as mobile as the migrants and are fanned by ICTs thanks to the myriad of mediated techno-spaces available to both migrants

---

[23] Interview with Emile's mother, Pinyin: 10/10/2011.

[24] See Footnote 22.

and their families in the home country who have taken them to distant countries and virtual spaces. Migrants are perpetually reminded of the number of cry dies that await them at home and failure to perform this ritual may lead to bad luck. For instance, after seven years without going home, Jonas mentioned that 15 cry dies awaited him as he prepared a visit home, and he vowed to perform all the ceremonies before returning to Cape Town. He achieved this. But also, the migrant community remind themselves of their obligations regarding these ceremonies. The possibility of going to distant places means that those at home increasingly have the potential to call and transmit messages about witchcraft or about complying with certain traditional values. This suggests that those at home are no longer on the fringes of technology waiting for calls from their migrant kin but can reach the individual either by phoning or constantly beeping until the desired effect is attained, namely a call from the migrant. Despite the fact that Jake was present at his mother's funeral without his wife, she still had to do something (weed and sweep the grave) to show that she too had buried her mother-in-law. For his part, Emile, as mentioned before, sent money for the funeral and planned to go home a year later to organize a memorial service. Even though phone calls play a role in increased communication during these periods, one's co-presence is vital for certain rituals. Phone calls compliment one's co-presence but cannot replace it.

Bad luck and misfortune are not only a result of not performing certain ritual practices but are also attributed to migrants' inability to send remittances (see also Sabar 2010: 121-23; Nyamnjoh 2005, 2011). As a result of some migrants' economic inability to respond to family demands, they temporarily disconnect from the family (see Chapter 4). This may at times be met with defiance from family who repeatedly call demanding money and talking ill about those concerned. This kind of bad mouthing, it is believed, can be picked up by evil spirits and used against a person. Recalling the reasons why Ron sent money to his mother to buy groceries (salt, palm oil, Maggi cubes and soap) to distribute to his aunts and uncles' spouses can be seen as a way of receiving his blessing through their prayers and by thinking positively about him.

While witchcraft is increasingly associated with new forms of consumption and the accumulation of wealth (Geschiere & Nyamnjoh 1998), the new technologies have strengthened notions of witchcraft and belief systems. They act as a leash that keeps migrants in check and does not let them stray too far. Migrants are trapped by the notion of being caught in the web of *kontri fashion* in which evil, jealousy and fear thrive and can cause bodily harm or bad luck (Nyamnjoh 2005). But it is also a domineering form of power and control over migrants. Nonetheless and as is argued by Geschiere & Nyamnjoh (1998), despite the negativity and violent acts involved in witchcraft, it is seen to strengthen relations between migrants and family back home and promote a transnational relationship. It is in this sense of connectedness that stories about witchcraft and their mode of transmission are of interest here. For instance, even though Jones was unhappy with his family for being unappreciative of what he had sent home (they never call or send a text message to acknowledge receipt of anything he sends) and never once have called just to say hello, he cannot escape them because (shrugging his shoulders) '*if I no call them, small time I want see my things them no di waka because say I don forget family*'.[25] In light of the above discourses by migrants, we begin to understand their perception that witchcraft and bad luck have no borders and how it can travel across the ocean and long distances even though Grassfielders previously believed that witchcraft did not cross oceans. Witchcraft thus 'functions' (Moore & Sanders 2001: 7) and maintains the status quo. In this regard, the mobile concept is not limited to migrants but to their belief system as well, as it inherently forms part of their cultural repertoire.

It is easy for the concept of transnational habitus to capture the notion of remittances (gifts, social and physical development as well as village development because they are tangible) but too often we miss out on belief systems and how they are practised and impact on the wellbeing of mobile communities and their relations back home. Witchcraft, as has been shown by Geschiere & Nyamnjoh (1998), Fisiy & Geschiere (2001) and Moore & Sanders (2001), is elusive yet people are prosecuted for it. Crick (1970 in Moore & Sanders 2001: 4) suggested that witchcraft can be better understood from an historical perspective and from the vantage point of those living with and within it. It is 'morally tied to systems and must be understood in context given the fact that it cannot be transferred to another context', nor can it be extricated from a given context. In these social fields, the narratives on witchcraft are poignant and involve bad luck, jealousy, bewitching and evil. However, migrants have learnt to come up with creative coping strategies to ward off such elements by sending remittances or calling home when it matters most.

Perpetuating perceived notions of witchcraft

Some calls from home fan accusations of witchcraft through rumours and gossip. Before her mother's health deteriorated, Joyce was in constant communication with her and her sisters who informed her that their mother had been bewitched. Her mother also believed this and sent anointing oil to Joyce for protection. For Joyce, her mother's death was the result of sorcery by jealous co-wives. According to her, she was weary of her stepmothers/sisters. The dispute with her stepniece stemmed from earlier tensions in the family that had been transposed to

[25] Conversation with Jones, Cape Town; 26/12/2011.

Cape Town. While migration is inclusive of migrants' habits and tends to reproduce and even reinforce old family tensions, class and gender hierarchies are a hangover from tensions back in the home village and have not been dealt with but instead are transported with migrants to their new abode. Fed with information from her sisters, Joyce believed her mother's death[26] was due to envy among her co-wives and she gave vivid details of the information passed on to her concerning her mother's death.

> My mum was bewitched by her co-wives out of jealousy. They are jealous each time a car comes and pulls over in front of her house and brings her lots of gifts and foodstuffs. They had blown off her roof before through witchcraft. This prompted her brother to go to a native doctor to find out. He was told that the first wife is very jealous of her ... that she said she laboured on the coffee farm and when our dad sold the coffee, he used the money – FCFA 150,000 – to marry our mother. My sisters gave the money to our uncle to give to our father and he gave her the money as if it were from him. It was hoped that things would be quiet after this but she still continued bewitching her. The death of my mum has severed all links with my step-mums and even their children. My mum was aware of this and cautioned my sisters against any skirmishes after her death.[27]

Witchcraft has become a cancer that eats deep into the heart of kinship relations and destroys the social cohesion and conviviality of the family and society. I refer to this as transcultural beliefs and tensions.

These accusations of witchcraft and family tensions are fed and reinforced by telephone conversations and dreams. It should be noted that at times of crisis back home, the velocity of communication and wanting to know all that is happening intensifies. Whenever Joyce quarrels with her step-sister, this is communicated to her sisters at home and only helps to fuel divisions within the family. After her mother's death, Joyce dreamt that 'she was involved in a fight with someone and her mother intervened with a huge stick and chased away her adversary, and told Joyce that anyone who wants to attack her, had to confront her first'. Although migrants take solace from the fact that they are guarded by deceased relative, such endemic faith in witchcraft confirms beliefs, fear and the popular epistemology about witchcraft. After this dream, she phoned her sisters to warn them to be on their guard as well. ICTs, especially the mobile phone, play a role in perpetuating beliefs in accusations of witchcraft and in fermenting conflicts within families.

While the mobile phone serves as a conduit for allegations, it also shields migrant members from the brunt of emotional torture. In this connection, the family at home sees it as their duty to emotionally protect those who are abroad by sifting through information before deciding what to pass on (Baldassar 2008). But periods of silence or unfinished sentences in telephone conversations only serve to ignite suspicion. For instance, in the days leading up to her mother's death,

---

26 Her mother died in November 2011.

27 Conversation with Joyce, Cape Town: 17/11/2011.

Joyce made several calls to her but her sisters did not want her to talk to their mother because her condition was deteriorating, and talking to Joyce would make her realize the gravity of the situation. However this only escalated Joyce's emotional state and made her fear the worst. At such a time, migrants turn to the church to seek solace and to purge the emotional trauma and fright that they are going through.

Pentecostalism and recasting witchcraft

The flow of migration has equally seen an increase in Pentecostal churches, which have mushroomed to cater for migrants' spiritual needs and allay fears of unfulfilled desires (Sabar 2010; Meyer 1998; van Dijk 2001). Pentecostal churches recognize witchcraft practices and the struggle between good and evil, and provide believers with the solution to any ambivalence they may have about witchcraft. They take the fear of witchcraft amongst the faithful seriously and provide special prayer sessions to 'rebuke and shame the devil' (Sabar 2010). By imbibing these notions of witchcraft into their teachings, they make themselves relevant to modern-day Christianity and provide a new context in which they make sense and are relevant to migrants who are continually faced with the Herculean task of meeting never-ending demands for remittances and the consequences of being bewitched through bad mouthing or being caught up in the gimmicks of 'kontri *fashion go catch them'*. In this regard, migrants are adopting global religious institutions and asserting their dual membership of spiritual arenas. They are thus broadening and extending the globalization of religious life and practices (Levitt 2004: 2). Joyce, for instance, was a practising Protestant but joined the Jehovah Witness Fellowship in Cape Town. She is not a regular churchgoer but admits that they provide her with the strength she needs to overcome her daily struggles in life. For others, mobile communication plays a significant role, facilitating therapeutic sessions for migrants by offering chain prayers between migrants and pastors, amongst migrants and among family back home and a pastor, i.e. a triangular prayer session. Jane's story again illustrates this point. When her parents received news at home that she had gone into labour, her father was driving from Yaoundé to Bamenda and immediately phoned his pastors to convey the news. They agreed to immediately start a chain prayer on her behalf and he stopped driving for an hour of prayers with the others. By the time he arrived Bamenda, Jane had given birth. While the idea was to pray for a safe delivery, it was also to ward off any sorcerer that might take advantage of the situation and harm her or the baby, given the perceived notion that sorcerers are opportunistic, invisible and can see into the future (Nyamnjoh forthcoming).

Groups of prayer warriors have emerged in Cape Town to assist members with prayers to ward off witchcraft and deliverance. Accounts of Neh's story highlight this phenomenon:

> She joined her husband (Henry from Mankon) in Cape Town in 2010. While she was a practising Catholic at home, she worships with the Synagogue of all Nations' Church in Cape Town (which has its headquarters in Nigeria). According to her, it was a personal decision to 'give her life to Christ and to receive Him and know God ... I want to spend the rest of my life with Christ because I am far from home and the things I see happening worldwide'. In April 2011, she received a phone call that her brother was gravely ill and that the doctors were unable to offer any diagnosis. The family turned to a traditional healer who said that he had been poisoned because he was doing well in business. First she asked for a photo of the brother to be scanned and sent to her, and she printed it out and took it to church for a prayer warrior session. After this, the photo was posted on the church wall along with others for continuous prayer intercessions. Later, she called and offered to pray with him over the phone. But before she began, she invited him to believe that God could heal him, and if he had done any 'bad' to people, he should recant and change his way of life. After submitting himself to God, she prayed with him over the phone and implored him to continue praying while she did the same with the prayer warrior group. Two days after the phone prayers, she received a call from her sister to say that for the first time since his illness, he had slept through the night after the prayer session. Motivated by this news, Neh decided to 'stand for prayer line on Sunday in church' (to come forward with her petition and be prayed for) and to pray for him again. In addition, she would meet the pastor after the service and request a special prayer. A week later, her brother called to thank her and tell her that he had been blessed and his illness had disappeared. After a fortnight, he called to inform her that all was well and he had resumed his business activities. To her, this simply meant that 'the power of God is alive'.[28]

Photo 9.6 Photos posted in church by relations for prayer sessions, including that of Neh's brother

---

[28] Interview, Cape Town: 20/05/2011.

The appropriation of communication technologies (mobile phone and the Internet) is beyond the conception of its designers. This was central in restoring Neh's brother's health on the one hand and, on the other, in transmitting rumours about witchcraft to her by members of her family. Given that ICTs have removed existing borders between states and witchcraft is capable of crossing borders, so too communication technologies are being used to transgress borders to fight against evil powers at home. Migrants have found refuge and emotional comfort in the church (Sabar 2010) and the church in turn serves as a spiritual bridge between them and families back home.

Disengagement and threats of witchcraft/curses

In spite of the trumpeting of transnationalism and the untimely death of assimilation as a result of advances in communication technologies, it would be too hasty to conclude that improved communication technologies have changed migrants' relationships with the home country to one of increased involvement and engagement (see Panagakos & Horst 2006: 112). However, increases in transnational/transcultural activities are punctuated by deliberate abstention from relating to home or momentarily disconnecting from the family fold, which goes against the canons of transculturalism. Interestingly, such disconnections give emotional satisfaction to those who choose to temporally sever links with kin and kith. For some, distance may be the desired goal to keep the family at bay or to settle in without continuous calls for remittances. ICTs thus have a dual role: to connect and to disconnect. In a phone call to Nsoh (from Mankon), I enquired after his parents back home but he told me that he had not communicated with them for more than five months. He went on to explain rather incoherently and with a tone of frustration in his voice that:

> I do not like calling home because of the harassment and excessive demands for money each time I call them. They don't care about how I am living here, I earn just enough to live on (R 500) a week, I am still struggling to sort myself out as I made some mistakes in the beginning. If they want me to send them money, it means they are indirectly asking me to get involved in a dubious lifestyle; that is not what I want. When I left home, they sent me to come and hustle, but it is difficult and they don't want to hear me out. That is why I often change SIM cards and numbers just to remain out of touch.[29]

Unlike others who navigate between SIM cards, Nsoh chooses to change his number when he can no longer stand his parents' incessant demand for money or calls for him to return home. He is often out of contact for long periods of time before calling again. Remembering what Emile's mother (see Footnote 10) cautioned about excessive financial demands by parents, Nsoh's attitude is a consequence of her prediction: disengagement or engaging in illicit activities to satisfy

[29] Phone conversation with Nsoh, Cape Town: 29/12/2012.

the pressing demands from home. For Nsoh, the mobile phone is a curse (Horst 2006), sent by the devil and is an invasion of privacy.

Migrants have developed various tactics and coping strategies, one of which is the notion of 'present absence' and involves navigating between SIM cards by giving numbers that are rarely in use and thus are unreachable from home. These are the SIM cards they only use when calling home, after which they are taken out of the phone until they are needed again. In instances where migrants do not juggle SIM cards, they simply decline calls from home or call back when beeped. It thus become a game of cat and mouse and the mobile phone becomes 'both a tool for keeping in touch and for artfully avoiding that social obligation' Lamoureaux 2011: 40; see also Powell 2012: 15-17). This notwithstanding, relatives go to great lengths to get money from their migrants' kin with threats of curses or being indicted by tradition (*kontri fashion go catch you*) for having turned a blind eye to their requests (see Nyamnjoh 2005: 251-53). Mike's case illustrates this:

> His father is a retired gendarme and has four wives. The oldest wife has a son in Cape Town, while Mike, whose mother is the second wife, has two brothers – one in Johannesburg and one with him in Cape Town. When their father fails to get money from them, he directs his frustrations towards their respective mothers and accuses them of inciting the children not to look after him or send money to educate his younger children. In this respect, he sends messages to them via their mothers that '*if they no want give ma own things them wey I still dey alive make I chop, make man no come when ee hear say I die like sick wey ee kill me go kill them'*.[30] According to tradition in the Bamenda Grassfields, this is a terrible curse for any parent to say to a child and the child concerned will do anything in his/her power to avert it. A concise translation of this sentence is difficult but I will paraphrase it: 'I would want to have material comforts while I am still alive, more than when I am dead, for the cause of my death will be a lack of these comforts and anyone who attempts to display wealth at my funeral will die of the disease that I suffered from'. Sending a message via their mothers is well calculated as the father is sure the women will plead with their children to avoid such a serious curse by responding to his demands. Mike's mother recounted how, after such rants from his father, she called her son repeatedly and pleaded with him to send his father some money. At times Mike is adamant that he has been of assistance to his father and feels that his father should be understanding if he (Mike) is hard up too. However not even his mother is informed when the father receives money from one of his sons but he is calm until he needs more. From being a retired gendarme, Mike's father has changed into being a 'gendarme of tradition, who seeks to ensure or enforce poverty alleviation traditions as a kind of culture tax on the haves by the have-nots to redistribute wealth' (Nyamnjoh 2005: 251). Having failed to get his way with his sons, he resorts to threats of curses and knows the special relationship mothers have with their sons. And of course the mobile phone has become the medium of communication and transmission between mothers and sons, who are implored to avert a terrible curse.

In the absence of a curse, home-based relations use tradition as a form of extortion to enrich themselves in the event of death when a migrant comes for a funeral. When a parent passes away in Mankon, each child is obliged by tradition to provide a white cloth (*wabab*) to symbolically show that they buried the de-

[30] Conversation with Mike's mother and step-mother in Pinyin: 12/10/2011.

ceased. This has on many occasions been converted into cash. In addition, all the married daughters and nieces of the deceased are expected to give a token sum of money to the successor of the compound, known as *á-beh nifvu*, for him to use for any traditional ceremony that is required. Too often advantage is taken of visiting migrants and they are asked for large sums of money and the latter, afraid of being indicted by *kontri fashion*, pay up the amount requested. In the case of this researcher, when my father passed away, I was duly informed of this custom and I phoned our successor who quoted me one of these exaggerated sums and insisted that the amount needed to be paid as soon as I arrived home. At moments like this, most family members are keen to 'protect' kin abroad by detailing all the 'law things' that tradition demands of them, while those on the receiving end call to emphasize them as these are rare periods for them to enrich themselves in the name of tradition. In this connection, the mobile phone becomes the tool/-town crier to communicate the message. Again as Nyamnjoh (*Ibid.*: 251) aptly puts it: 'There are lots of traditions that are not really traditions but a means of some of the less privileged in our community to "beg" without losing face, for resources to help them out of poverty'.

In an era when the mobile phone and the Internet have compressed geographic distance and rendered 'mobiles' to being 'over there' and 'reachable', not all migrants are comfortable with this perception and some consciously and/or temporarily suspend communication with relations at home. Irene's case is worth mentioning. Since she moved to Cape Town from Gabon she has lost touch with her family at home and has none of their numbers. According to her:

> I don't have any of their numbers because I don't want to stress myself. I want to sort myself out first ... Business is slow and I am restarting, that is why I don't want to call. I have my aunt who is like my mum now, and I would never go for a month without calling her, but for a year now I have not called my aunt even my brother Moses, I have not spoken to him for a year now.[31]

Irene's coping strategy of self-alienation, or 'objectification' as Miller (2010: 59) puts it, is the 'way we enhance our capacity as human beings ... hence within the process of objectification, self-alienation is essential'. This does, however, go against the norm of transnational and transcultural habitus that teach people to know that the extended family are those who must not be neglected and one must keep the family unity intact.

In extreme cases, the family is divided following the death of both parents, with accusations and counter-accusations of selfishness and mismanagement of business, leading to a fractured family. Paul's story sheds light on this:

> Paul originates from Mankon and is a successful businessman in Cape Town, with a branch in Cameroon that was managed by his younger brother but is currently managed by his twin

[31] Interview, Cape Town: 07/08/2011.

sister. Since 2012, Paul has spent more time in Cameroon than in Cape Town, and his wife manages their business in Cape Town. When he was informed about his mother's illness, he called and instructed that she should be taken to a referral hospital where she was treated for lung cancer. Upon her discharge, the family took her to a traditional healer against his wishes and when he tried to dissuade them from doing so, he was met with insults 'Before you were even born, mum was sick and we took care of her. Don't you think that just because you have a little money, you can order us around now.'[32] Unable to dissuade them, Paul had to travel to Cameroon to take her back to hospital, and a few weeks later she passed away. Paul acknowledges that things have not been the same since.

> My younger brother and I were like very close, we did things together but since my mum passed away, it's been very-very difficult to deal with him and my other siblings. It's been really really difficult. It's like my mum was a shield or something, I couldn't see their real colours when my mum was alive. She's died and everybody is showing me what they are made of. I'm really struggling with them. I have been home on two occasions and I've tried to call family meetings to try and bring everybody together. I tried to raise issues. I'm not mad at them any longer about what they did because it won't help. She's gone, she's gone. I've tried to make peace. We are not like at daggers drawn, but we are just not getting along the way we used to.[33]

The relationship took a nosedive when Paul's elder brother sold a piece of land Paul bought for him at less than cost price and also when the latter set out to build on a piece of land he (Paul) had bought adjacent to their compound. The elder brother claimed the land was a gift from Paul to their father and that Paul therefore had no rights to it. During another interview session with Paul, the brother tried calling him repeatedly but Paul would not take the call saying that 'I know why he is calling, to fight with me over my land. There is no way I can let go of that land, even the person who sold the land acknowledges that it is rightfully mine, as well as my elder sister. He says my elder sister is siding with me because I give her money.'[34] This dispute divided the family into two camps, with two people in favour of Paul and five against him.

Such family feuds are not uncommon and this is one of the fallouts of migration, with some wanting to take undue advantage of their migrant kin. But it also shows the extreme consequences of migration and mobility, and the adverse effects they can have on families and migrants who end up with deep emotional wounds that take a long time to heal.

Discussion and conclusion

Trans-border and transcultural relations have grown as a result of migrants trying to maintain connections and relations with the home village and to come to terms with the emotions and trauma of separation and absence. This chapter has gone beyond debates on transnationalism to incorporate the notion of emotions and belonging. Moving away from considerations of the community, the focus has

---

32 Interview, Cape Town: 01/03/2011.
33 Interview, Cape Town: 20/05/2011.
34 Interview, Cape Town: 03/06/2011.

been on individual migrants' connections with the family in an attempt to understand the emotional challenges involved and the related aspects of witchcraft that go with them. Emotions form an integral part of transnational/transcultural habitus. If these are guided by social relationships, then emotions are regulated by and respond to social relationships.

This speaks more to the idea of upholding the values and societal norms that migrants have learnt during childhood and in their adult lives and emerges in the way they relate to their family back home, the remittances they send and of course the funeral engagements that show their adeptness in maintaining transcultural practices. By the same token, we should recognize that transnational families and relationships are not a new phenomenon and they existed before the arrival of ICTs.

Creating a home-away-from-home means that migrants have adopted the tendency to develop a 'Janus-like' view of transnational migration (Guarnizo 1997). Like the god Janus, migrants have a dual visage that enables them to transgress borders irrespective of their loci. For the majority of those in the host society, such trans-border connections provide them opportunities to consolidate ties with the home village, such as through the objects they take with them and their activities (photos and memorabilia, co-presence and visits) that allow them to remember their attachment to family and home. Through this form of double engagement, individuals hope to become a whole.

Funerals have provided us here with a lens through which we can understand migrants' emotional trauma in times of loss. Virtual and corporeal mobility becomes an option given the importance of this ritual to migrants, as was seen in the previous chapter. Loudspeakers on phones and camera apps have increasingly made it possible for migrants to be virtually present at funerals, like John who asked for the phone be put on loudspeaker so he could follow his mother's burial. Through virtual, corporeal and financial participation, migrants also gain a sense of emotional satisfaction for having done what was required of them culturally. While relationships between the home and host countries are mutually reinforcing and soothing, they tend to be normative in nature because this is the expected norm within the social network in the home village rather than a question of individual affection.

Religious practices, like funerals, create space for migrants to engage simultaneously in the home and host countries. Through telephonic prayer sessions as well as chain prayers and the warding off witchcraft, Pentecostalism has reinvented itself and has become more relevant to migrants. Boundaries have become blurred and fluid as members navigate social fields to provide a cathartic effect for both migrant and family.

Photos too become a linking device, as do the memorabilia and virtual co-presence that members seek to keep a sense of family. This duality (an absent presence) is censored by what I refer to as a 'code of honour' – what others think and say about them – which invariably drives migrants to seek to relate with home more than ever before, or minimally. It is partly this ethos that invigorates the spirit of transculturalism and transnationalism. Emotions form an integral part of transnational/transcultural habitus, and if these are guided by social relationships, then emotions are regulated by and respond to social relationships. This also alludes to the perception of success and failure, as successful migrants are seen as those who maintain family cohesion, while those who are considered to have failed stay away from the family. However this is not a logical conclusion, given that even some of those who are considered to be successful have severed links with their families.

Much as some migrants would like to minimize relations with home, ICTs does not always offer a path that allows them to completely escape these contradictions and the fear of witchcraft. Instead, they have found themselves trapped in their fears and 'provide them with explanations of misfortune, and explanations that address the "why" more than the "how" questions' (Moore & Sanders 2001: 6). Notions of witchcraft and belief systems in many ways compel migrants' transculturalism but also play an important role in the way migrants view their worlds as well as the ways in which they navigate their daily lives in both the host and home countries. For some, the mobile phone and the Internet are God-sends whereas those who develop strategies to sever links with kin and kith see it as 'Devil-sent' (Nyamnjoh 2005; Horst 2006) given the speed at which it has helped to spread information about witchcraft from home to the host country and perpetuate belief systems – indictments of tradition – used by the less wealthy to extort material accumulation and monetary gains. Witchcraft, as such, is not frozen in time but should be seen as dynamic, 'resistant to changes, and capable of reinventing and refashioning itself to suit new situations' (Moore & Sanders 2001:10-11), especially in the light of new technologies.

From this premise, we can ascertain that while transcuturalism has played a significant role in linking the home and host countries, we must guard against seeing ICTs as a panacea for a successful dual lifestyle. This would seem *prima facie*. The new communication technologies, much as they connect, equally serve to liberate migrants from the constant demands of the family as they have harnessed it to monitor calls and to know when and which calls to take. Or they are simply not reachable.

10

Conclusion: Mobilities, transcultural communities and transcultural habitus

This study set out to understand the workings of the Pinyin and Mankon mobile communities in the context of accelerated mobility and ICTs. In an attempt to show how they function, I conceptualized mobilities in Chapter 3, ICTs in Chapter 4 and networks in Chapter 5 given that these concepts underscore the social fabric of every society, including mobile communities. These chapters (3, 4 and 5) equally laid the groundwork that underpinned the remainder of the Chapters (Chapters 6, 7, 8 and 9).

Taming wild geese: ICTs and networks enhanced mobilities

Chapter 3 focused on the mobility patterns of the Pinyin and Mankon communities and how they have evolved for the current migrant. It highlighted the multifaceted forms of social, virtual, cultural and emotional mobilities and how dynamic they now are as a result of ICTs.

Communication technologies however have a nuanced role in mobility. They connect and link up communities and families but functionally, they are a form of control; a leash on migrants dragging them back into the family circle if they begin to stray too far away. And some do break loose from this form of control by (dis)connecting with the family fold. Migrants navigate between SIM cards, thus establishing connections with the family on their own terms. What this also presupposes is that they have taken control of their agentic power. It is mobile technology that is the connecting factor here and it has become a tool that enhances disconnection. In this respect, ICTs give migrants the possibility to disconnect from family. Fearful of bad mouthing, they are aware that it runs against the moral ethos and endangers the social fabric of society. Such a move could thus only be temporal and normalcy is resumed when finances permit. While disconnection may be frowned upon, it posits migrants amid a higher echelon of actor than technology because of their ability to appropriate and tame it. It gives them room to reconfigure their agency, to tame ICT and enhance mobility, which is a central conclusion of this study.

In this regard and adding to Licoppe's connected presence (2004), I include the notion of '(dis)connected complexities' to show the vagaries of connections and disconnections that exist amongst migrants and families.

Migration of women from both communities came much later compared to their male counterparts. The women came mainly for spousal reunion, with few exceptions invited by their sisters or for studies. Trade was their major preoccupation, unlike the pitcher-men; the women are sedentary, running stalls from flea markets.

Passport mobility, 'lines' and *tuyaus* perpetuated by the migration industry and that are at the migrants' disposal are continuously enhanced by mobile and wireless communications. This has also resulted in chain migration, as there is the increasing need to relocate family members to maximize assistance for the home country. As a result, migrants map a coordinated and chain migratory route that is difficult to break thanks to the pre-migratory networks and host society networks that have been forged. In the process, they are tearing down nation-state borders with little regard for state hegemony and making a farce of selective migration. Chapter 3 underscores the importance of an historical approach to understand current trends in mobility as well as the intricacies of mobility among the Pinyin and Mankon communities.

The varied forms of mobilities (cultural social, virtual), as practised by the above communities, have come to represent an historical progression from a traditional rural society to that of an urban, mobile society where space has become unbounded. Similarly, accelerated mobilities in migrants' lives have led to the recasting of identity in terms of flexibility and adaptability, and the holding onto of values that they consider define their identity (Chapters 7, 8 and 9).

Flexible mobilities and migrants' dual lives have come at a cost to nation-states given that mobiles are gradually recasting the sovereignty and autonomy of nation-states and minimizing the state's ability to provide effective border controls. This by no means suggests that borders have become redundant but rather that flexible mobility is constantly looking to make nonsense of borders: they have become contested spatiality between migrants and states. In a way, immigration policies have not optimally attained their objectives of borderland control and exclusion because they have often viewed migration and mobility as a problem. While details about the Pinyin and Mankon communities in the Netherlands are not as detailed as those in Cape Town, they nevertheless offer insight into the difficulties of migrating to Europe as a result of the stringent legislation that is in place to fence off unwanted migrants from Fortress Europe, although migrants are still able to access 'lines' that beat the system. By contrast, mobility exposes the porosity of South Africa's borders. Their fluidity and porosity are the consequence of regulations aimed at controlling the borders and a disjuncture between state policies and migration realities in sending and receiving countries. Migration and mobility, as such, should not be seen as a problem, described mostly in terms of disaster attributes or considered stasis. It is a dynamic process that is informed by different variables (information, networks, finances and serendipity). Much as these variables have given mobility a new dynamism, it is nevertheless hinged on internalized embodied practices. In this respect, policies need to take into account the development strategies that seek to manage migration flows in ways that serve the national interest, i.e. those of the sending country as well as those of the migrant, with the aim of achieving a win-win-win solution.

Chapter 4 addresses the evolution of communication technologies from its embryonic to its wide availability and affordability. I have shown how these societies, especially Pinyin, have leapfrogged from no telephonic connections straight into the era of mobile phones. Although communication technologies gingerly made their way to South Africa, they were mired by problems of affordability. As a result, we are confronted with the ingenious ways in which migrants have appropriated these technologies, especially fixed and mobile phones, to connect to their home country. The arrival of ICTs, as I have shown, is characterized by their ability to revolutionize the world and they have dramatically changed how people live their lives (dualistically) and how they perceive the world as deterritorializing spaces that are compressing time. In addition to geographical spaces, I have also shown how ICTs are changing lifestyles by creating an informal economy and job opportunities, especially in the DVD (Digital Versatile Disc) sector where migrants have championed the downloading and burning of movies for commercial purposes. The Internet has been a key factor in promoting this business in Cape Town (Chapter 6).

Throughout history and prior to these visible changes, the world was made up of layers of multiple spaces that were joined but were displaced in the event of migration. The advances in transport networks, and particularly in air travel, have given migrants access to constant and multidirectional flows of mobility. In this respect, I conclude that ICTs perpetuate mobilities and re-organize social structures. This study considers how everyday life is construed in the process of mobiles re-organizing their social structures.

Underpinning these social structures are the fact that mobility and ICTs do not function in isolation but are mutually reinforcing and inextricably linked to social relations of networks. The perception of ICTs as simply connecting people, or the catalogue of advantages that other scholars maintain, obfuscates the fact that these technologies may be manipulated and become a disconnecting device with people hiding behind them. Much as ICTs reconfigure existing relationships and build new ones, but at the same time maintain existing ones in the process, ICTs (dis)connects as well. At one level, the relationship is convivial but at another it is antagonistic. The importance of Chapter 4 is its contribution to understanding the power of ICTs in enhancing contemporary migration as well as the failure of migration policies in attempts at mitigating unwanted migration and mobility because policies are yet to consider the power of ICTs and, if they do, it is difficult to challenge.

Understanding these concepts and societies demands an historical and a multi-sited approach that allows room to capture not only migrants' mobile lifestyles but also the in-betweens that occur in the process. I equally noted the developmental stages of communication technologies in both communities. While Mankon is connected to all three mobile service providers, Pinyin is still connected to one. However, the Pinyin connection is a catapult from no communication technology to having one. In addition, following the mobile communities discussed in this study through the various stages of availability of communication technologies sheds light on the appropriation and mutual shaping that occurs. Archival research was fundamental to comprehending shifts in mobilities.

From this perspective, mobility/migration is not new but predates the migration history of every society. However what is new today are the forms and dynamism that it has taken as a result of advances in ICTs and transport networks. This has led to new mobile societies characterized by mobile flows and miniaturized mobilities.

Chapter 5 looked at how mobile communities are able to draw on the plethora of networks at their disposal, including strong and weak ties. Like ICTs, social networks are the driving force behind migration and mobility, especially in the pre-migration and actual migration process. It shows how social networks impact on migration at all stages in the process and how these networks constantly shift to accommodate new networks in order to be relevant. My aim was to discover whether mobilities have affected network ties and, if so, how? If they have not, then are these ties continuing in the norm that migrants have been brought up to respect (social cohesion)? As a result of the formation of network societies thanks to ICTs, it follows that social networks are a dynamic process due the emergence of flexible communities that reconfigure themselves and reproduce complex structures of communication. It is important to note that these networks encourage people to be mobile because they are hierarchized as people do not have equal access to the various forms of economic, cultural and social capital. Although social capital is informed by history and is accumulated over time, it has become an almost exclusive right of the elite. This is a common theme running through all the chapters. Status and social hierarchy in both communities are differentiated along generational, gender, socioeconomic and cultural parameters. Kinship ties are important corporate networks that sponsor migration and provide migrants with the possibility of double engagement as well as people to whom migrants can entrust their affairs (Chapters 6 and 7).

Although we talk of kinship ties, these are streamlined along gender roles. Despite being away from the family, female migrants continue to subscribe to their traditional role as carers as ascribed by society and care intensify with migration. ICTs, and in particular the mobile phone, have thus become the tools and links through which care is given.

Away from kinship ties (strong ties), I also focused on weak ties and, through this network, I try to understand the large number of Pinyin migrants in Cape Town compared to other ethnic communities. Weak ties include legal ties, such as lawyers and officials from the Department of Home Affairs charged with issuing refugee permit that assist migrants legalize their status. Migration syndicates that perpetuate migration are also perceived as weak ties. These weak ties appear as practical bridges that link various nodes and are ultimately stronger than strong ties, being defined by their shifting and bridging qualities. These weak ties, I maintain, connect better than strong ties.

Like mobilities, social interaction and linkages between migrants and family are not a new phenomenon but what is new is the manner in which network societies have built up their lives around ICTs and are a set of interconnected nodes that process flows. It therefore means that advances in new technologies are redefining the way migrants link up with each other and with friends and family, and how networks are forged and maintained. Earlier migrants have become the communication signposts and gatekeepers for subsequent migrants, with the latter being fed information about migration and given assistance in settling in after having migrated. While face-to-face networking is still privileged, virtual networking is incontestably the norm. As a result of the new mobile culture, mobile communities have developed an octopus-like ability to spread their tentacles in an attempt to create and link multiple networks.

Chapters 3, 4 and 5 complement earlier studies on the evolution of mobility and ICTs in marginal localities by the Mobile Africa Revisited research project,[1] of which this study is part. Its strength is its rootedness in an historical approach that is the basis for understanding contemporary mobilities; first by looking into the past. In addition, Chapter 3 is directed at those at the helm of migration policies and provides suggestions on how best to address migration and mobilities.

---

1 This is a comparative study of the relationship between new communication technologies and social spaces in Chad, Mali, Cameroon, Angola, Sudan and Senegal.
See http://mobile africarevisited.wordpress.com.

Chapters 3, 4 and 5 are the conceptual chapters that enhance our understanding of the rest of the ethnographic chapters. Together, they argue that, contemporary migration and mobility are inextricably linked with ICTs and networks and networking. While ICTs have changed the form and style of migration and mobility, earlier migrants continue to be the signposts and gatekeepers of would-be migrants. Also, advances in communication technologies have reinforced communities and individual interaction, especially interactions with greater weak ties.

Mobility and trade

Chapters 6, 7, 8 and 9 offer in-depth analysis of the research findings. I draw largely from the three preceding chapters, namely Chapters 3, 4 and 5, to anchor my analysis with the final chapters (Chapters 6, 7, 8 and 9). Chapter 6 examines migrants' economic activities, particularly pitching (hawking) and other forms of income-generating activities. I also explain how, through the rapid development of information technology, migrants have been able to do business differently from their parents' generation due to increased opportunities, improved and reduced mobility as the mobile phone provides them with mobility. This difference is possible as a result of the extensive network they can garner because of the new ICTs at their disposal. I was interested in the extent to which ICTs, mobility and networks propel the economic activities of the mobile community.

The highlight of chapter 6 is, firstly, the way Pinyin migrants do business in Cape Town. It would appear that this is informed by the historical skills inherited and internalized from a long history of mercantile trading, particularly by those who witnessed or accompanied parents to weekly village markets and from the stories they heard from their parents. Those from Mankon, however, tend to focus on the trade that they have learnt. Secondly, the chapter highlights the navigation between informal and formal business and the blurriness between legal and illegal business. This is because of the see-saw nature of activities that makes it impalpable to make a distinction between informal/formal and illegal/illegal, and this is difficult to analyze within the framework of dualism. In addition, those involved in business rarely consider it to be in any way illegal until they come into contact with the law. In general, migrants shrewdly circumvent the legal authorities. The same cannot be said of those in the Netherlands who do not have similar opportunities to operate in the informal economy because of the strict rules regarding asylum seekers. Those with educational qualifications seek formal employment. The navigation between informal/formal and illegal/legal shows the extent to which migrants' economic activities are informed by mobility, ICTs and networking.

Contrary to other studies that have announced the premature death of informal business due to lack of capital, I argue that migrants often receive support and/or loans from family and personal and ethnic enclaves to set up businesses. I proceed to show how, through wealth accumulation, migrants have again been defined and ordered. Success in this regard has played an important role in the hierarchization of society, especially to those for whom success has come as a result of 'hardwork' and not through fraudulent means (drug pushing). In summary, the informal economy and its actors is a melting pot of false designer goods, as well as a point of convergence for the global and local with marketing insights.

Mobilities, identity and belonging

Chapter 7 explored the emergence of home-village associations and their role in maintaining social cohesion within communities in the host country. It also looked at how they serve as a platform for projecting migrants' identities. It examined the ways the Pinyin and Mankon migrants are crafting lives and communities and how they maintain ties with other associations and form Internet-mediated associations with diaspora groups and home-based elites.

The chapter departs from the premise that the associations include those who belong to one ethnic group and exclude those who do not belong as well as members of the host society. I maintain that associations should be understood within the framework of agency in the production of structures because structure is both the medium and the outcome of action. Such structures define belonging, identity and exclusion, and are a means by which action can be enabled. Associations as such play a major role in the societal fabric and seek to establish social cohesion amongst members and provide a bridge between the host community and the home-village community. Although being a member of these associations is seen as a logical conclusion to proving one's belonging to the home village, belonging is also perceived by migrants as an insurance policy. Due to the internal rules and regulations of the associations, they have taken on a parental role towards members, which is a pull factor in both the host and home countries and connects migrants to the spatially dispersed. The common factor that binds them is their sense of place. Place for the migrants in this study is the home village because of a common shared identity and culture, as well as a way of engaging with the ideology of home. Place is where they hope to retire or return to posthumously.

As regards the Cape Town communities, their identification is first and foremost Pinyin and Mankon, given that this is primarily based on their sense of place and a collation of ethnic identities through which they are linked to 'home', through a shared meaning and culture. This commonality can also be explained by the fact that migrants are more driven by the need to help develop their villages to prepare a place for their homecoming. Home is constantly work in process.

As such, mediated public spheres enable participants to move from a shared imagination to collective action thanks to the 'Internet-mediated network public'. This chapter therefore acknowledged that although migrants are trying to recreate home whenever and wherever they may be, home is still deeply rooted in where they come from.

In this respect, associational formations are complex and multifaceted phenomenon that prioritize identities and belonging and are constantly being negotiated through inclusion and exclusion, similarity and difference. The identities and belonging of Pinyin and Mankon communities are best understood as a relational and social phenomenon captured in the complexities (connections, disconnections and difference), contradictions and messiness of their everyday realities.

Continuing with in-depth analysis of how associations maintain and express belonging, Chapter 8 takes a closer look at the various ritual practices that connect them to home and reinforce notions of home and belonging. The overriding concern in this chapter was to examine how mobile communities, although far from home, continue to uphold and participate in the various ritual practices and ceremonies in their host and home countries simultaneously. Also, looking at how rituals are dynamic and reconfigured as a result of mobility and new communication technologies offers insight into the workings of society. Contrary to other studies that see home not as the place where one happens to have been born but rather as the place where one becomes oneself, home to these communities is where one comes from. And they are rooted in the present as much as in the past considering the rituals they still practise. This chapter focused on rituals as connecting devices that create the glue that connects mobile communities across borders as well as with the home village. By connecting them to the home village, they are simultaneously disconnected from the host society because rituals are a verbal form of expression and a means of attaining a higher sense of identity.

Paying particular attention to rituals of birth, marriage and death, I recognize that the ritual that disconnects the most is that of death because it encapsulates and symbolizes the society. Rituals surrounding births and marriages are dynamic as migrant communities adapt and borrow aspects from the host society to make sense of their stay in the host society and also as a result of mobility. What is referred to as invariance is informed by a recognized and accepted pattern or a *bricolage* constructed of cultural elements from both the homeland and the receiving nation. While these ceremonies are also premised on the notion of belonging, they are informed by the idea of wealth and success, which invariably results in a hierarchization in society. The practising of rituals by mobile communities contradicts earlier studies that hold that rituals are likely to vanish among migrants that are not living in ethnic enclave neighbourhoods. While the

Pinyin and Mankon communities in the Netherlands cannot be said to live in such enclaves, this does not stop them from preforming various rituals.

Besides being rooted in cultural expressions, rituals are also informed by fear. Anxiety may be related to the community saying bad things about the bereaved (bad mouthing) or about a lack of a proper funeral and/or the fear of bad luck for the same reason. Migrants and mobile communities thus strive to give their members a funeral that can be seen as fitting for a 'bush faller die'.

Rituals become inescapable and are interwoven in the societal fabric, becoming a cultural and social force that dictates their wellbeing and belonging and at the same time revealing a sense of identity. The ritual of death, while it also depicts wealth, is primarily about belonging. Away from home, the mobile community wants to be seen as conforming to ritual practices. In this respect, ICTs become evidence of ritual autopsy and proof that the event was done according to cultural norms. Films and technology (DVDs and photos) have become a new form of autopsy that can be proof to either the families or the migrant that the ritual was done well. Being mobile and yet taking part in these practices calls for their reproduction at specific times.

A combination of mobility, advances in ICTs and networking has helped to discard the Weberian prediction that modernization, industrialization and the pursuit of material comfort would bring about a decline in ritual and a loss of value. Instead, what was considered to be the demise of ritual, namely mobility, advances in ICTs and networking, has increased participation in the very same rituals because they are what set communities apart from each other and underline their collectiveness. What distinguishes the mobile communities in South Africa from those in the Netherlands are the identities they have chosen to project. For those in South Africa, it is primarily Pinyin-ness and Mankon-ness, and their Grassfielder/Bamenda identity, while those in the Netherlands project their Cameroonian-ness. The argument put forward here is that the emphasis on Pinyin-ness and Mankon-ness is due to the association/ethnic enclaves that bind them together. People see themselves first as Pinyin or Mankon, secondly as a person from Bamenda in the Grassfields, and only then as Cameroonian. By contrast, migrants in the Netherlands have no established ethnic support mechanism but have found one in their Cameroonian identity and project this first. This can also be understood from the perspective of porous/regulated borders, where there is a continuous flow of Pinyin and Mankon migrants from Cameroon to Cape Town, while this is uncommon in the Netherlands where the borders are heavily fortified.

Emotions: An expression of belonging and cultural submission

In Chapters 6, 7 and 8, the focus was on communities while Chapter 9 shifted away from the collective and moved to the individual. The aim was to show how individuals deal with challenges of emotional separation from their family and the notion of witchcraft. The chapter debates how migrants, as individuals, are able to sustain multifaceted relationships back home and cope with separation. What accounts for this double engagement? Answering these questions call for a juxtaposition of the concepts of double engagement, transculturalism and transcultural habitus to unpack the ways in which migrants and families back home are engaged in each other's lives. I have tried to construe their social fabric by showing how migration is an on-going emotional journey for both the migrant and for those left behind in the home country. As individual migrants, they want to connect and relate to their families back home on a personal level and to share their emotions of missing kin, separation, of losing kin (death) and the emotions of reunion but they also have to cope with being strangers in the host country. These emotions are expressed in their relationship with their families and through their attachment to their culture and cultural practices because 'it is tradition and must be obeyed'. The increased tendency towards a dual relationship is informed by the fact that, having moved out of their original family circle, migrants lose their sense of togetherness and become individuals and strangers in the host country, which heightens their quest to link up with family back home. This quest has witnessed a rise in transcultural relationships, where the focus is more on the home village than on the nation-state. As such, they partake in family celebrations and rituals as well as holding on to objects that symbolize their connection with home to emphasize their belonging in family circles. This denotes an embodiment of their individual identity. What is predominant here is the fact that, despite the migrants' mobility pattern, they are self-conscious and are glued to their cultural practices.

This gives them the moral and emotional satisfaction of maintaining a sense of family despite being away because emotions are socially constructed. As a result of missing kin, migrants seek alternative ways to cope with separation and the feeling of being disconnected from family. They turn to photos and memorabilia as coping mechanisms to bring family close and also to the things they have taken with them to the host country. Long-distance relationships are carefully natured and choreographed, with family reunions and imminent visits much anticipated. These are events packed with emotions that give migrants and family a sense of co-family. Social interaction and relationships of this sort are important and emotions abound binding migrants to large-scale social and cultural structures. By the same token, emotions can also drive people apart and push them to tear down social structures.

In summary, individuals try to adjust and readjust transcultural practices to suit specific situations owing to the dialectical processes of simultaneous departures and arrivals, and absences and the presence of members of the community. The transnational habitus thus links South Africa, the Netherlands, Pinyin and Mankon as a web of multifocal arena of emotional and sociocultural fields. In this light, migrants' (virtual) mobility within transnational social and physical fields allows them to re-engage with their communities of origin in new ways given their insatiable craving to know what is happening and to ease the pain of separation. Keeping in touch with and being connected to the home village is measured more through relationships, attitudes and behaviour over time than through mere claims or physical presence. From this perspective, we could assert that it is by leaving their families that migrants gain a stronger presence in their home through their ability to improve the lives of those left behind by sending remittances.

Trust

What seems to keep both associations glued together is a notion of trust. The sheer fact of coming from either Pinyin or Mankon has an impact on the way migrants relate to each other. They see themselves first as brothers and sisters in relation to other ethnic groups and there is a need for trust. This is also seen in the way financial transactions are organized. It takes only one phone call for a person to be given a loan in PIFAM and there is no question of any form of collateral being requested. People are held to their word and they will repay the loan (Chapter 6) because it is all done on faith and trust that the money will be repaid as promised.

It is for such reasons that members prefer *njangi* than the bank given its relaxed nature regarding borrowing and the fact that money can be easily transferred to the home country bypassing all the laws concerning money transfers. However trust at times comes at a price as it is sometimes broken by debtors, and the matter is then referred to the Conflict Resolution Committee.

In the same way, migrants are willing to trust friends or relations back home with their credit cards to carry out financial transactions at home. Although ICT has made it possible for them to monitor authorized or unauthorized financial transactions, there is no guarantee that there will not be any mismanagement of funds as this can happen anywhere. However, these worries are held in check by the trust that migrants, friends and family are willing to put in each other.

Those engaged in business between Cape Town and Johannesburg do so solely on the basis of trust. Members send money to friends and relations to buy goods for them and these are then sent as registered baggage by train or with one of the bus agencies. Although M-pesa (mobile money) has made its mark in

Kenya and Uganda, the Pinyin and Mankon communities are wary of mobile banking in case their phones are stolen and most settle for ATM (automated teller machine) transactions.

Trust does not refer only to financial transactions but also to the way migrants and families speak well or badly of each other. Migrants seek to follow the moral ethos that is expected of them by the migrant community and their families, i.e. to support their family financially and earn a decent living in the host country. Diverting from this breaches any trust invested in the migrant and may lead to bad mouthing, which can weigh down on them emotionally. In addition, the frequency of communication between migrants and families makes for convivial relations and provides an opportunity for them to be a part of each other's lives. However, lengthy spells with no communication can lead to misgivings concerning the migrant, such as Penn who went for more than a year without communicating with his parents.

A word on methods

This research is part of the Mobile Africa Revisited project that has combined historical and anthropological methods to address how people in Africa are appropriating new ICTs and how they did so in the past. Such an approach redirects the debate towards placing more emphasis on agency in historically specific contexts. This study has maintained this tradition, with a keen focus on life histories and archival research.

Life histories became an integral component in establishing social practices in the past and these link them to the present and ascertain whether there has been continuity or disjuncture in these patterns. It thus allowed me to juxtapose mobility, trade, networks and social relations from the past with those of today's migrants and to come up with an analytical framework of transformation and continuity. Complementing life histories with archival research has resulted in interesting information on the Pinyin bush trade and trade between East and West Cameroons. Can archival information about the Pinyin man who was imprisoned be linked to the information I received from an informant about his father being imprisoned for tax evasion and the sale of contraband goods?

I also chose to use themes such as confidentiality and photos that were prominent during data collection as entrance to my methodology. Interview sessions and conversations were moments that were at times marked by significant confidential information that was embedded in an outpouring of emotions. When such moments are attained, it created a significant bond between the researcher and the informant. This in no way suggests that there was no bonding between other informants. An historical perspective was important in understanding the continuities in the past that make contemporary mobilities intelligible. Through this approach we can account for a pattern of linear mobilities and migration between the current mobile communities and their parents. Joe's case is a good example. As a young boy, he accompanied his father to various bush markets and he has since followed in the footsteps of his father in his mobility and trade. But what differentiated them is Joe's adoption of ICTs, which has enabled him to reach out to more clients and create multiple networks.

This research is built on a triangular study that spans Cameroon, South Africa and the Netherlands. A multi-sited approach was the logical path to follow as it led to the investigation of networks and the mobile community. As a result, I described the network of relationships locally and trans-locally and how mobile communities are joined or torn apart by their appropriation of ICTs. I also embarked on investigating the patterns of connections between mobile communities and the home village and amongst various mobile communities, and considered the challenges that occur and how these are dealt with. Despite the merits of this approach and its ability to follow migrants in different fields, engagement in one area prevents one from participating in important happenings in others. Equally, depending on who is telling the story, the latter may be different from one informant to the next.

A reflexive approach was used as a tool to bring me closer to the informants. The choice of a reflexive approach in part helped to understand my own personal migration and mobility trajectories and to better understand those of my informants as well as to be informed about the ethical decisions and stance that were made in the field and from the field.

Using comparative analysis, I was able to chart the various mobile trajectories of the Pinyin and Mankon communities as well as concerns concerning identities. This also shed light on the priorities of Internet-mediated forums.

Part of my approach was well embedded in the use of ICTs as a way of keeping up with the multi-sitedness of the study. Going virtual thus proved useful in making possible my absence and presence at various sites. Phone calls, texts, emails and Skype conversations complemented the quest for data collection but I was also able to follow discussions on the Pinyin News Forum and Mankon Forum that enhanced my knowledge about how Internet-mediated technology is being appropriated. In the context of accelerated mobility, ICTs have thus become a means of keeping up with mobile communities but, for researchers, they are fast becoming a research tool. Through their use, I kept abreast of happenings at different sites. Photos were important communicators throughout the research and I often carried them back and forth for families. It was always a moment of mixed feelings – nostalgia, joy, guilt and resolution when migrants and families received photos of each other. Inasmuch as these photos are virtual representations of missing kin, one cannot ignore the emotional pain that communities go through because of their separation, and photos tend to heighten the emotional challenges. In this respect, whilst families and mobile kin are connected by ICTs in their lives, they still acknowledge the geographic distance that separates them and yearn for face-to-face meetings. I agree with John Ury's notion of 'meetingness' (Ury 2003: Chapter 9) as a way of fully bridging this gap.

Frequent air travel made it possible for me to offer a courier service to the research communities. I would carry photos from Emile in Holland to his mother in Pinyin and would take foodstuffs from her to her daughter in Cape Town. I offered the perfect bridge between her and her children in Europe and Africa. This also impacted on the data collection as the recurrent visits created a relationship of trust that made it possible for regular encounters when stories emerged.

In conclusion, contrary to seeing a disintegration of societies, a breakdown in relationships and/or a dramatic transformation in societies due to advances in ICTs and mobilities, I argue that ICTs and mobilities have, in fact, complemented social relational interaction and provided mobiles with opportunities to partake in cultural practices that express their Pinyin-ness and Mankon-ness.

It should be remembered that disintegration and estrangement are part of the nuances of technologies that are akin to social relationships and are dynamic. ICTs serve as a connecting factor, especially as networking among families has increased due to improvements and transformations in communication technologies. Pinyin and Mankon migrants are still as rooted in the past as they are in the present. They are born into a community with a sense of home, moral ethos and cultural pride but live in a context of accelerated ICTs and mobility that is fast changing the way people live their lives. Acknowledging that technologies and mobilities have enhanced livelihoods in these communities, I maintain that they have not resulted in a change in people's affection for their cultural values and practices. Instead, ICTs have helped them to become more connected and increased their zeal to be transcultural migrants. In other words, mobile communities have appropriated technologies to bridge the distances that separate them from other family members and to remain true to their cultural practices by actively participating in the various activities that give them a sense of family and belonging, as well as alleviating emotional trauma. In the process, this has brought home closer to them. The empirical question at the start of this study was about ascertaining how people in these mobile worlds create and navigate their social fabric and make sense of their individual being. In an attempt to find an answer to this question, this study has contributed towards our understanding of mobile communities by taking into account their multidimensional trajectories, as shown in the different chapters. Innovatively embracing transculturalism and transcultural habitus as concepts to comprehend the societal fabric has brought out the in-betweens that would have been overlooked in the course of data collection.

In addition, the focus by migrants is increasingly less about transnationalism than transculturalism, although studies still focus on the former, thus reifying migrants discourse as national. In the context of the globalization of fluid and blurred borders, mobility should be less about (trans)nationalism than about 'villageism' and the pursuit of belonging. As a result of the move from (trans)-nationalism to villageism, it is important for the social sciences to increasingly reflect on this paradigm shift.

As far as policy orientation goes, I am optimistic that this study will influence the vision and thinking of those who develop migration policies. I trust that they will see migrants as people who want to stay firmly connected to their cultural roots and the spaces they have created for themselves, where difference should be celebrated rather than ignored. As Greenblatt (2010: 6) put it so well, cultural mobility is a thing to be celebrated as 'enhanced cultural mobility would foster new cosmopolitanism, an unregulated free trade in expression and feelings, and an epoch of global respect'. It is only when we begin to view identity as being mobile and as an often unstable relation of difference that we can begin to understand the full dimension of mobility.

From this study, it has emerged that policies put in place by states could either be seen as 'policy failures' or 'securitization'. These measures probably epitomize the failure to consider migration as a dynamic social process and the very essence of global governance of migration indicates the complexity of the subject and the difficulty of finding a comprehensive policy that will mitigate migration because nation states are yet to comprehend the full scale of transnationalism where sovereign national states are crisscrossed and undermined by transnational actors with varying prospects of orientation, identities and networks. By the same token, states and policy makers have underestimated the agency of migrants to circumvent and circumnavigate border regimes and state legislature, showing the limitations of nation-states in addressing mobility. In this regard, policies need to take into account the development strategies that seek to manage migration flows in the sending country as well as those of the migrant in ways that serve national interests. They should aim to find a 'win-win-win' solution.

With regard to recommendations for future research, a coordinated and longitudinal study of migration syndicates would give insight into the continuous flow of migrants, the role of weak and strong ties in the process, as well as the appropriation of ICTs to enhance or mitigate the migration process.

References

ADEPOJU, A. (1994), 'The African family and survival strategy'. *African Anthropology* 1(1&2): 5-8.

ADEPOJU, A. (2005*), Creating a borderless West Africa: Constraints and prospects for intra-regional migration.* Draft article for the Migration without Borders Series, UNESCO.

ADEY, P. (2010), *Mobility*. London/New York: Routledge.

ALBRECHT, S. (2006), 'Whose voice is heard in online deliberation?: A study of participation and representation in political debates on the internet'. *Information, Communication and Society* 9(1): 62-82.

ANDERSON, B. (2006), *Imagined communities*. London: Verso.

ANDERSSON, J.A. (2001a), 'Mobile worker, urban employment and "rural" identities: Rural-urban networks of Buhera migrants, Zimbabwe'. In: M.de Bruijn, R. van Dijk & D. Foeken, eds, *Mobile Africa: Changing patterns of movement in Africa and beyond.* Leiden: Brill, pp. 89-106.

ANDERSSON, J.A. (2001b), 'Reinterpreting the rural-urban connection: Migration practices and socio-cultural dispositions of Buhera workers in Harare'. *Journal of the African Institute* 71(1): 82-112.

ANNAN, K. (2006), Address by Mr Kofi Annan to the high-level dialogue of the United National General Assembly on international migration and development, New York. 14 September 2006.

ARDENER, S. (1964), 'The comparative study of rotating credit associations'. *Journal of the Royal Anthropological Institute* XCIV: 201-209.

ARDENER, S. (1995), 'Women making money go round: ROSCAs revisited'. In: S. Ardener & S. Burman, eds, *Money-go-rounds*. Oxford/Washington: Berg, pp. 1-20.

AWASOM, S.Y. (2005), 'Towards a gendered development discourse in Africa: Visible women, invisible men'. Paper presented at the 11th CODESRIA General Assembly, Maputo, 6-9 December.

BACIGALUPE, G. & M. CAMARA (2012), 'Transnational families and social technologies: Reassessing immigration psychology'. *Journal of Ethnic and Migration Studies* 38(9): 1425-1438.

BAKEWELL, O. (2007), 'The meaning and use of identity papers: Handheld and heartfelt nationality in the borderlands of North-West Zambia'. International Migration Institute, Working Paper no. 5.

BALDASSAR, L. (2007), 'Transnational families and the provision of moral and emotional support: The relationship between truth and distance'. *Identities: Global Studies in Culture and Power* 14(3): 385-409.

BALDASSAR, L. (2008), 'Missing kin and longing to be together: Emotions and the construction of co-presence in transnational relationships'. *Journal of Intercultural Studies* 29(3): 247-266.

BANERJEE, B. (1983), 'Social networks in the migration process: Empirical evidence of chain migration in India'. *The Journal of Developing Area* 17(2): 185-196.

BASCH, L., N. GLICK SCHILLER & C. SZANTON-BLANC (1994), *Nations unbound: Transnational projects, Post-colonial predicaments and deterritorialised nation-states*. Utrecht: Gordon Beach Publishers.

BECK, U. (2002), 'The cosmopolitan society and its enemies'. *Culture and Society* 19: 17-44.

BIRD, F. (1980), 'The contemporary ritual milieu'. In: R. Browne, ed., *Rituals and ceremonies in contemporary cultures*. Bowling Green, OH: Bowling Green State University Popular Press, pp. 19-35.

BLOCK, M. & J. PARRY (1982), 'Introduction: Death and the regeneration of life'. In: M. Bloch & J. Parry, eds, *Death and the regeneration of life*. Cambridge: Cambridge University Press, pp. 1-44.

BOGARD, W. (2000), 'Simmel in cyberspace: Strangeness and distance in postmodern communications'. *Space and Culture* 4(5): 23-46.

BOSWELL, C. & A. GEDDES (2011), *Migration and mobility in the European Union*. Basingstoke: Palgrave Macmillan.

BOURDIEU, P. (1977), *Outline of a theory of practice*. Translation by R. Nice. Cambridge: Cambridge University Press.

BOURDIEU, P. (1986), 'The forms of capital'. In: J.G. Richardson, ed., *Handbook of theory and research in the sociology of education*. New York: Greenwood, pp. 241-258.

BOURDIEU, P. (1990a), *Photography: A middle-brow art*. Stanford, California: Stanford University Press.

BOURDIEU, P. (1990b), *The logic of practice*. Cambridge: Polity Press.

BOURDIEU, P. (2012), *Picturing Algeria*. New York: Columbia University Press.

BOURDIEU, P. & L. WACQUANT (1992), *An invitation to reflexive sociology*. Chicago: University of Chicago Press.

BOYD, M. (1989), 'Family and personal networks in international migration: Recent developments and new agendas'. *International Migration Review* 23(3): 638-670.

BRASS, P.R. (2000), 'Foucault steals political science'. *Annual Review of Political Science* 3: 305-330.

BRATTON, M. (2013), 'Briefing: Citizens and cell phones in Africa'. *African Affairs* 112(447): 304-319.

BRETTELL, C.B. (2006), 'Introduction: Global spaces/local places: Transnationalism, diaspora, and the meaning of home'. *Identities: Global Studies in Culture and Power* 13(3): 327-334.

BRETTELL, C.B. (2008), 'Theorizing migration in anthropology: The social construction of networks, identities, communities, and globalscapes'. In: C.B. Brettell & J.F. Hollifield, eds, *Migration theory: Talking across disciplines*. Second edition. New York/London: Routledge, pp. 113-159.

BRETTELL, C.B. & J.F. HOLLIFIELD (2000), Migration theory: Talking across disciplines. New York: Routledge.

BRINKERHOFF, J. (2009), *Digital diaspora: Identity and transnational engagement*. Cambridge: Cambridge University Press.

BRYCESON, D. (2002), 'Transnational families in the twenty-first century'. In: D. Bryceson & U. Vuorela, eds, *The transnational family: New European frontiers and global networks*. Oxford: Berg, pp. 3-30.

CALHOUN, C. (1991), 'Indirect relationships and imagined communities: Large scale social integration and the transformation of everyday life'. In: P. Bourdieu & J. Coleman, eds, *Social theory for a changing society*. Boulder/San Francisco/Oxford: Westview Press, pp. 95-121.

CALHOUN, C. ([2000]2003), 'Pierre Bourdieu'. In: G. Ritzer, ed., *Blackwell companion to the major social theorists*. Cambridge, MA: Blackwell, pp.1-35.

CALLON, M. & J. LAW (2004), 'Introduction: Absence – presence, circulation and encountering in complex space'. *Environment and Planning D: Society and Space* 22: 3-11.

CASTLES, M. (1996), *The information age: Economy, society and culture. Volume 1: The rise of network society*. Oxford/Malden, MA: Blackwell Publishers.

CASTLES, M. (1997), *The information age: Economy, society and culture. Volume 2: The power of dentity*. Oxford/Malden, MA: Blackwell Publishers.

CASTLES, M. (1998), *The information age: Economy, society and culture. Volume 3: The end of millenium*. Oxford/Malden, MA: Blackwell Publishers.

CASTLES, M. (2004), 'Informationalism, networks, and the network society: A theoretical blueprint'. In: M. Castells, ed., *The network society: Cross-cultural perspective*. Cheltenham, UK/Northampton, MA, USA: Edward Elgar Publishers. pp. 3-45.

CASTLES, M. & A. PORTES (1989), 'World underneath: The origins, dynamics, and effects of the informal economy'. In: Portes, A., C. Manuells, & L. Benton, eds, *The Informal economy: Studies in advanced and less developed countries*. Baltimore/London John Hopkins University Press, pp. 11-37.

CASTLES, M., M. FERNANDEZ-ARDEVOL, L.J. QIU, & A. SEY (2007), *Mobile communication and society: A global perspective*. Cambridge: MIT Press.

CASTLES, S. (2002), 'Migration and community formation under conditions of globalization'. *International Migration Review* 36(4): 1143-1168.

CASTLES, S. (2004), 'The factors that make and unmake migration policies'. *International Migration Review* 38(3): 852-884.

CASTLES, S. & S. MILLER (2003), *The age of migration: International population movement in the modern world*. Third edition. New York: Palgrave.

CERULO, K.A. (1997), 'Identity construction: New issues, new directions'. *Annual Review of Sociology* 23: 385-409.

CHASE, I.D. (1980), 'Social process and hierarchy formation in small groups: A comparative perspective'. *American Sociological Review* 45(6): 905-924.

CHENEAU-LOQUAY, A. (2010), 'L'Afrique au soliel de la revolution des telecommunications: Les grandes tendances de la diffusion des TIC'. *Afrique Contemporaine* 234: 95-112.

CHESSON, M.S. (2001), 'Social memory, identity and death: An introduction'. In: M.S. Chesson, ed., *Social memory, identity and death: Anthropological perspectives in mortuary rituals*. Arlington, Va., Archaeological papers of the American Anthropological Association, pp. 1-10.

CHILESHE, J. (1992), *Nothing wrong with Africa, except ...* 2nd edition. New Delhi: Viokas Publishing House.

CHILVER, E.M. (1961), 'Nineteenth-century trade in the Bamenda grassfields, Southern Cameroon'. *Afrika und Ubesee* XLV: 233-258.

CHILVER, E.M. (1988), 'The political economy of Cameroon: Historical perspective'. Privately circulated.

CHILVER, E.M. & P.M. KABERRY (1970), 'Chronology of the Bamenda grassfields'. *Journal of African History* 11. (2): 249-257.

CHOLDIN, H.M. (1973), 'Kinship in the migration process'. *International Migration Review* 7(2): 163-175.

CIEKWAWY, D. & P. GESCHIERE (1988), 'Containing witchcraft: Conflicting scenario in Post-Colonial Africa'. *African Studies Review* 41(3): 1-14.

CLIFFORD, J. (1997), *Routes: Travels and translation in the late twentieth century*. Cambridge, MA: Harvard Universty Press.

COLLINS, P. & A. GALLINAT (2010), 'The ethnographic self as resource: An Introduction'. In: P. Collins& A. Gallinat-, eds, *The ethnography of the self as a resource*. New York/ Oxford: Berghahn Book, pp. 1-22.

COLLINS, R. (1993), 'Emotional energy as the common denominator of rational action'. *Rationality and Society* 5: 203-230.

COMOROFF, J. & J. COMOROFF (1993), 'Introduction'. In: J. Comoroff & J. Comoroff, eds, *Modernity and its malcontents: Ritual and power in postcolonial Africa*. Chicago: University of Chicago Press, pp. xi-xxvii.

COMOROFF, J. & J. COMOROFF (2009), *Ethnicity, INC.* South Africa: University of KwaZulu-Natal Press.

CORNIA, G., R. VAN DER HOEVEN & T. MKANDAWIRE, eds (1992), *Africa's recovery in the 1990s*. New York: St Martin's Press.

CRESSWELL, T. (2010), 'Towards a politics of mobility'. *Environment and Planning D: Society and Space* 28: 17-31.

CRISP, J. (1999), 'Policy challenges of the new diasporas: Migrant networks and their impact on asylum flows and regimes'. Geneva. UNHCR Working Paper no 7.

CRUSH, J. & D.A. MCDONALD (2000), 'Transnationalism, African immigration and new spaces in South Africa: An introduction'. *Canadian Journal of African Studies* 34(1): 1-19.

DAVIES, D.J. (1997), *Death ritual and belief: The rhetoric of funerary rites*. London & Washington: Cassell.

DE BRUIJN, M. (1998), 'The translation of Anthropological data: How to approach the chaos of fieldwork'? In: E. van Dongen & S. van Londen, eds, *Anthropology of difference, essays in honour of Professor Arie de Ruijter*. Utrecht: ISOR, pp. 68-85.

DE BRUIJN, M. (2007), 'Mobility and society in the Sahel: An exploration of mobile margins and global governance'. In: H.P.Hahn & G. Klute, eds, *Cultures of migration: African perspectives*. Berlin: LIT Verlag, pp. 109-128.

DE BRUIJN, M. (2008), '"The telephone has grown legs": Mobile communication and social change in the margins of African society'. *Inaugural address*. Faculteit der Letteren: Universiteit Leiden.

DE BRUIJN, M., R. VAN DIJK & D. FOEKEN (2001), 'Mobile Africa: An Introduction'. In: M. de Bruijn, R. van Dijk & D. Foeken, eds, *Mobile Africa: Changing patterns of movement in Africa and beyond*. Leiden: Brill, pp. 1-7.

DE BRUIJN, M., M., F. NYAMNJOHY & I. BRINKMAN (2009), 'Introduction: Mobile communications and new social spaces in Africa'. In M. de Bruijn, F. Nyamnjoh, I. Brinkman, eds, *The mobile phones: The new talking drums of everyday Africa*. Bamenda/Leiden: Langaa RPCIG/African Studies Centre, pp. 11-22.

DE BRUIJN, M. & I. BRINKMAN (2010), 'Communicating Africa. Researching mobile in communities: Communication technologies, and social transformation in Angola and Cameroon'. In: *La famille transnationale dans tous ses etats,* special issue of *Autrepart: Revue de Sciences sociales au Sud* 57/58: 41-57.

DE BRUIJN, M. & R. VAN DIJK (2011), 'Connecting and change in African societies: Examples of "ethnographies of linking" in anthropology'. *Anthropologica* 54: 45-59.

DE BRUIJN, M., I. BRINKMAN & F. NYAMNJOH (2013), 'Introduction: Mobile margins and the dynamics of communication'. In: M. de Bruijn, I. Brinkman & F. Nyamnjoh, eds, *Side@ways: Mobile margins and the dynamics of communication in Africa*. Bamenda/Leiden: Langaa RPCIG/African Studies Centre, pp. 1-16.

DE BRUIJN, M. & R. VAN DIJK (2013), 'Introduction: Connectivity and the postglobal moment: (Dis)connections and social change in Africa'. In: M. de Bruijn, & R.van Dijk, eds, *The social life of connectivity in Africa*, pp. 1-20.

DE CERTEAU, M. (1984), *The practice of everyday life*. Berkeley/Los Angeles: University of California Press

DELANCEY, M.W. (1974), 'Plantation and migration in the Mt. Cameroon region'. In: H.F.Illy, ed., *Kamerun: Sturcturen und Probleme der socio-ökonomischen Entwicklung*. Mainz: Hase & Koehler Verlag, pp. 181-236.

DELANCEY, M.W. (1988), 'The expansion of coffee production in Bamenda and bananas in the South West'. In: P. Geschiere, & P. Konings, eds, *Conference on the political economy of cameroon – historical perspectives*. Leiden: African Studies Centre, pp. 303-312.

DELANCEY, V. (1988), 'The impact of the credit union movement on production and accumulation in the agricultural sector in Cameroon'. In: P. Geschiere, & P. Konings, eds, *Conference on the political economy of cameroon – historical perspectives*. Leiden: African Studies Centre, pp. 211-248.

DE SOTO, H. (1990), *The other path: The invisible revolution in the Third World*. London, Tauris

DERCON, S. (2002), *Risk-sharing and endogenous network formation*. WIDER, Helsinki Discussion Paper no. 57.

DE WEERDT, J. & A.WEINER (1985), 'Inalienable wealth'. *American Ethnologist* 12(2): 210-227.

DONNER, J. (2008), 'Shrinking fourth world? Mobiles, development and inclusion'. In: J.E. Katz, ed., *Handbook of mobile communication studies*. Cambridge: MIT Press, pp. 29-42.

DUKE, K. (1996), 'The resettlement experiences of refugees in the UK: Main findings from an interview study'. *New Community* 22(3): 461-478.

DYCK, N. (2010), 'Remembering and the ethnography of children's sports'. In: P. Collins & A. Gallinat, eds, *The ethnography of the self as a resource*. New York/Oxford: Berghahn Books, pp. 150-164.

ELLIOTT, A. & J. URRY (2010), *Mobile lives*. London/New York: Routledge.

ESCOBAR, A. (1994), 'Welcome to cyberia: Notes on the Anthropology of cyberculture'. *Current Anthropology* 35(3): 211-231.

ETZIONI, A. (2004), 'Holidays and rituals: Neglected seedbeds of virtue, Introduction'. In: A. Etzioni & J. Bloom, eds, *We are what we celebrate: Understanding holidays and rituals*. New York/ London: New York University Press, pp. 3-40.

ETZO, S. & G. COLLENDER (2010), 'Briefing the mobile phone 'revolution' in Africa: Rhetoric or reality'? *African Affairs* 109(437): pp. 659-668.

EVANS, M. (2010), 'Primary patriotism, shifting identity: Hometown associations in Manyu Division, South-West Cameroon'. *Africa* 80(3): pp. 397-425.

EYOH, D. (1998), 'Through the prism of a local tragedy: Political liberalisation, regionalism and elite struggles for power in Cameroon'. *Africa: Journal of International African Institute* 68(3): pp. 338-359.

EYOH, D. (1999), 'Community, citizenship, and the politics of ethnicity in post-colonial Africa.' In: P.T. Zeleza & E. Kalipeni, eds, *Sacred spaces and public quarrels: African cultural and economic landscapes*. Trenton, N.J.: Africa World Press, pp. 271-300.

FAIST, T. (2000), *The volume and dynamics of international migration and transnational social spaces*. Oxford: Oxford University Press.

FAUBION, J.D. (1984), 'Governmentality.' In: J.D. Faubion, ed., *Michel Foucault: Power. Essential works of Michel Foucault, 1954-1984, vol. 3*. London: Penguin Books, pp. 201-222.

FAWCETT, J.T. (1989), 'Networks, linkages, and migration systems. *International Migration Review* 1(3): pp. 671-680.

FAWCETT, J.T. & F. ARNOLD (1987), 'Explaining diversity: Asian and Pacific immigration systems'. In: J.T. Fawcett & B.V. Corino, eds, *Pacific bridges: The new immigration from Asia and the Pacific Islands*. New York: Centre for Migration Studies, pp.453-473.

FEATHERSTONE, D., R. PHILLIPS & J. WATERS (2007), 'Introduction: Spatialities of transnational networks'. *Global Networks* 7(4): 383-391.

FELDMAN-SAVELSBERG, P. & T. NDONKO (2010), 'Urbanities and urban villages: Comparing 'home' among elite and non-elite Bamileke women's hometown associations'. *Africa* 80(3): 371-396.

FERGUSON, J. (1999), *Expectations of modernity: Myths and meanings of urban life in Zambian Copperbelt*. Berkeley: University of California press.

FEYDER, S. (2009), *Think positive, make negatives: Black popular photography and urban identies in Johannesburg townships, 1920-1960*. Unpublished Master's Thesis. Leiden University.

FIELD, J. (2003), *Social capital*. London/New York: Routledge.

FISHER, C. (1992), *America calling: A social history of the telephone to 1940*. Berkeley: University of California Press.

FISIY, C. & P. GESCHIERE (1991), 'Sorcery, witchcraft and accumulation: Regional variations in the South and West Cameroon'. *Critique of Anthropology* 11(3): 251-278.

FISIY, C. & P. GESCHIERE (2001), 'Witchcraft development and paranoia in Cameroon: Interaction between popular, academic and state discourses'. In: H. Moore & T. Sanders, eds, *Magical interpretations, material realities: Modernity, witchcraft and the occult in postcolonial Africa*. London: Routledge, pp. 226-246.

FITZGERALD, D. (2000), 'Negotiating extra-territorial citizenship: Mexican migration and the transnational politics of community'. La Jolla, CA: Center for Comparative Immigration Studies, Monograph Series no. 2.

FLEISCHER, A. (2006), 'Family obligations and migration: The role of kinship in Cameroon.' Rostock: MPIDR, Working Paper no. 047.

FON TANNI LL, G.A. (2010), Brief monograph of the Pinyin People: A presentation to the Senior Divisional Officer Mezam by His Royal Majesty the Fon of Pinyin Clan.

FÖRSTER, T. (2001), 'Seeing and observing: Ethnography after post-modernism'. *Sozialer Sinn* 3: 569-574.

FREEMAN, R. (2006), 'People flows in globalisation'. *The Journal of economic Perspectives* 20(2): 145-170.

FRIEDMAN, J. (1990), 'Being in the world: Globalisation and localization'. In M. Featherstone, ed., *Global culture: Nationalism, globalization and modernity*. London: Sage, pp. 311-328.

GAO, J. (2006), 'Organized international asylum-seeker networks: Formation and utilization by Chinese students'. *International Migration Review* 40(2): 294-317.

GARDENER, K. (2002), 'Death of a migrant: Transnational death rituals and gender among British Sylhettis'. *Global Networks* 2(3): pp. 191-204.

GARDENER, K. & R. GRILLO (2002), 'Transnational households and rituals: an overview'. *Global Networks* 2(3): 179-190.

GEERTZ, C. (1973), *The interpretation of cultures*. New York: Basic Books.

GEERTZ, C. (2005), 'Deep play: Notes on the Balinese cockfight'. *Daedalus* 134(4): 56-86.

GESCHIERE, P. (2005), 'Funerals and belonging: Different patterns in South Cameroon'. *African Studies Review* 48(2): 45-64.

GESCHIERE, P. & J. GUGLER (1998), 'The urban-rural connection: Changing issues of belonging and identification'. *Africa: Journal of International African Institute* 68(3): 309-319.

GESCHIERE, P. & F. NYAMNJOH (2000), 'Capitalism and autochthony: The seesaw of mobility and belonging' *Public Culture*12(30): 423-452.

GESCHIERE, P. & F. NYAMNJOH (2001), 'Capitalism and autochthony: The seesaw of mobility and belonging'. In J. Comaroff& J. Comaroff, eds, *Millennial capitalism and culture of neoliberalism*. Durham/London: Duke University Press, pp. 159-190.

GIDDENS, A. (1990), *Consequences of modernity*. Oxford: Polity Press.

GIDENGIL, E. & B. O'NEIL (2006), 'Removing rose colored glasses: Examining theories of social capital through a gendered lens'. In: B. O'Neil & E. Gidengil, eds, *Gender and social capital*. New York/London: Routledge, pp. 1-14.

GILLIS, J.R. (2004), 'Gathering together: Remembering memory through ritual'. In: A. Etzioni, & J. Bloom, eds, *We are what we celebrate: Understanding holidays and rituals*. New York/London: New York University Press, pp. 89-103.

GLICK SCHILLER, N., L. BASCH & C.S. BLANC (1992), 'Transnationalism: A new analytical framework for understanding migration'. In: N. Glick Schiller, L. Basch, C.S. Blanc, eds, *Towards a transnational perspective on migration: Race, class, ethnicity and nationalism reconsidered*. New York: Annals of the New York Academy of Sciences, pp. 1-24.

GLICK SCHILLER, N., L. BASCH & C.S. BLANC (1995), 'From immigrant to transmigrant: Theorizing transnational migration'. *Anthropological Quarterly* 68(1): 48-63.

GOODY, J. (1961), 'Religion and ritual: The definitional problem'. *The British Journal of Sociology* 12(2): 142-164.

GORDON, S. (1988), Review of N. Frijda (1986), 'Emotion'. *Contemporary Sociology* 17(3): 403-404.

GORDON, S. (1990), 'Social structure effects on emotions'. In: T.D. Kemper, ed., *Research agendas in the sociology of emotions*. Albany: State University of New York Press, pp. 145-179.

GOUGH, K. (1959), 'The Nayars and the definition of marriage'. *The Journal of the Royal Anthropological Institute* 89(1): 23-34.

GRAHAM, F. (2010), *M-Pesa: Kenya's mobile wallet revolution.* http://www.bbc.co.uk/news/business-11793290 Accessed: 13/12/11.

GRANOVETTER, M. (1973), 'The strength of weak ties'. *American Journal of Sociology* 78(6): 1360-1380.

GRANOVETTER, M. (1985), 'Economic action and social structure: The problem of embeddedness'. *American Journal of Sociology* 91(3): 481-510.

GREENBLATT, S. (2010), 'Cultural mobility: An introduction'. In: S. Greenblatt, I. Zupanov, R. Meyer-Kalkus, H. Paul, P. Nyiri & F. Pannewick, eds, *Cultural mobility*. Cambridge: Cambridge University Press, pp. 1-23.

GRILLO, R. & V. MAZZUCATO (2008), 'Africa< >Europe: A double engagement'. *Journal of Migration Studies* 34(2): 175-198.

GROSH, B. & G. SOMOLKAE (1996), 'Mighty Oaks from little acorns: Can microenterprises serve as the seedbed for industrialisation'? *World Development* 24: 1879-1890.

GUARNIZO, L.E. (1997), 'The emergence of a transnational social formation and the mirage of return migration among Dominican transmigrants'. *Identities* 4(2): 281-322.

GUARNIZO, L.E. (2003), 'The economics of transnational living'. *International Migration Review* 37(3): 666-699.

GUARNIZO, L.E., I. SANCHEZ, & E.M. ROACH (1999), 'Mistrust, fragmented solidarity, and transnational migration: Colombians in New York and Los Angeles'. *Journal of Ethnic and Migration Studies* 22(2): 367-396.

GUGLER, J. (1971), Life in a dual system: Eastern Nigerians in Town. *Cahiers d'Etudes Africaines* 11(3): 400-421.

GUGLER, J. (1997), 'Life in a dual system: revisited: Urban-rural ties in Enugu, Nigeria, 1961-1987'. In: J. Gugler, ed., *Cities in the developing world: Issues, theories and policies*. Oxford: Oxford University Press, pp. 62-73.

GUGLER, J. (2002), 'The son of a hawk does not remain abroad: The urban-rural connection in Africa'. *African Studies Review* 45(1): 21-41.

GUILLEMIN, M. & L. GILLAM (2004), 'Ethics, reflexivity, and "ethics important moments" in research'. *Qualitative Inquiry* 10(2): 261-280.

GUPTA, A. (1992), 'The song of nonaligned world: Transnational identities and the reinscription of space in late capitalism'. *Cultural Anthropology* 7(1): 6-23.

GUPTA, A. & J. FERGUSON (1997), 'Discipline and practice: "The field" as site, method, and location in Anthropology'. In: A. Gupta & J. Ferguson, eds, *Anthropological locations: Boundaries and grounds of a field science*. Berkeley/Los Angeles: University of California Press, pp. 1-46.

GUYER, J.I. (2004), '*Marginal gains: Monetary transactions in Atlantic Africa*'. Chicago/ London: University of Chicago Press.

HAGAN, J.M. (1998), 'Social networks, gender, and immigrant ncorporation: Resources and constraints'. *American Sociological Review* 63(1): 55-67.

HAGAN, J. & R.H. EBAUGH (2003), 'Calling upon the Sacred: Migrants' use of religion in the migration process'. *International Migration Review* 37(4): 1145-1162.

HAHN, H.P. (2012), 'The appropriation of bicycles in West Africa: Pragmatic approaches to sustainability'. *Transfers* 2(2): 31-48.

HAHN, H.P. & G. KLUTE (2007), 'Cultures of migration: Introduction. In: H.P. Hahn & G. Klute, eds, *Cultures of migration: African perspectives*. Berlin: LIT Verlag, pp. 9-27.

HAHN, H.P. & L. KIBORA (2008), 'The domestication of mobile telephone: Oral society and new ICT in Burkina Faso'. *Journal of Modern African Studies* 46(1): 87-109.

HAMMERSLEY, M. & P. ATKINSON (2007), *Ethnography: Principles in practice* (3rd edition). London/New York: Routledge.

HANEMAAYER, A. (2009), 'A grounded theory approach to engaging technology on the paintball field'. In: P. Vannini, ed., *Material culture and technology in everyday life: Ethnographic approaches*. New York: Peter Lang, pp. 157-168.

HANNERZ, U. (2003), 'Being there ... and there ... and there! Reflections on multi-site ethnography'. *Ethnography* 4(2): 210-216.
HANSEN, K.T. (2010), 'Changing youth dynamics in Lusaka's informal economy in the context of economic liberalization'. *African Studies Quarterly* 11(2&3): 13-27.
HART, K. (1973), 'Informal income opportunities and urban employment in Ghana'. *Journal of Modern African Studies* 11: 61-89.
HEIDEGGER, M. (1993), *Basic writings*. London: Routledge.
HICKEY, S. (2004), ''*Hometown associations' as social movements for citizenship: A case study from Northwest Cameroon'*. A revised version of a paper presented at the 47th Annual Meeting of the African Studies-US Association, New Orleans LA, 1114 November.
HIRSCHMAN, C. (2007), 'The role of religion in the origins and adaptation of immigration groups in the United States'. In: A. Portes, & J. DeWind, eds, *Rethinking migration: New theoretical and empirical perspectives*. New York/Oxford: Berghahn Books, pp. 391-418.
HOCHSCHILD, A.R. (1979), 'Emotion work, feeling rule and social structure'. *American Journal of Sociology* 85: 551-575.
HOLLAND, R. (1999), 'Reflexivity'. *Human Relations* 52(4): 463-484.
HOLLIFIELD, J. (2004), 'The emerging migration states'. *International Migration Review* 38(3): 885-913.
HOLMES, M. (2010), 'The emotionalization of reflexivity'. *Sociology* 44(1): 139-154.
HOLNESS, S., E. NEL & T. BINNS (1999), 'The changing nature of informal street trading in post-apartheid South Africa' *Urban Forum* 10(2): 285-302.
HOLY, L. & M. STUCHLIK (1983), *Actions norms and representations: Foundations of anthropological inquiry*. Cambridge: Cambridge University Press.
HOPKINS, N., S. REICHER, & K. HARRISON (2006), 'Young people's deliberations on geographic mobility: Identity and cross-border relocation'. *Political Psychology* 27(2): 227-245.
HORST, H. (2006), 'The blessings and burdens of communication: Cell phones in Jamaican transnational social fields'. *Global Networks* 6(2): 143-159.
HORST, H. & D. MILLER (2005) 'From kinship to link-up: Cell phones and social networking in Jamaica'. *Current Anthropology* 46(5): 755-778.
HORST, H. & D. MILLER (2006), *The cell phone: An anthropology of communication*. Oxford: New York. Berg.
HOUSEMAN, M. (1995), 'Kinship networks in Africa: Past and future'. *African Anthropology* 2(1): 77-90.
IKENGA-METUH, E. (1985), 'Religious symbolisms in Igbo life crisis rituals'. *Africana Marburgensia* 18(2): 59-79.
ILAHIANE, H. & J. SHERRY (2008), 'Joutia: Street vendor entrepreneurship and the informal economy of information and communication technologies in Morocco'. *Journal of North Africa Studies* 13(2): 243-255.
IMBER-BLACK, E. & J. ROBERTS (1992), *Rituals for our times: Celebrating, healing, and changing our lives and our relationship*. New York: HarperPerenial.
INTERNET WORLD STATS (2007), *Internet usage statistics for Africa: African Internet usage and population stats*. http://www.internetworldstats.com/stats1.htm (Accessed, 20/03/2012).
ITU (2004), *African telecommunication indicators 2004*. www.itu.int (Accessed, 09/03/2012).
ITU (2008), *African Telecommunication/ICT indicators 2008: At a crossroads*. http://www.itu.int/ITU-D/ict/papers/2008/AFTI2008_VG.pdf (Accessed, 21/03/2012).
ITU (2009), *Information society statistical profiles 2009: Africa*. http://www.itu.int/ITU-D/ict/papers/2009/RPM_AF09.pdf (Accessed, 21/03/2012; 17/04/2013).
JACKSON, K.M. (2005), 'Rituals and patterns in children's lives: Introduction'. In: K.M. Jackson, ed., *Rituals and patterns in children's lives*. Wisconsin: University of Wisconsin Press, pp. 3-13.

JENKINS, H. (2006), *Convergence culture: Where old and new media collide*. New York: New York University Press.

JINDRA, M. (2005), 'Christianity and the proliferation of ancestors: Change in hierarchy and mortuary ritual in the Cameroon Grassfields'. *Journal of International African Institute* 75(3): 356-377.

JUMA, C., C. TORORI, & C. KIRIMA (1993), *The adaptive economy: Economic crisis and technological innovation*. Nairobi: African Centre for Technology Studies Press.

KABERRY, P.M & E.M. CHILVER (1961), 'An outline of the traditional political system of Bali-Nyonga, Southern Cameroons'. *Africa: Journal of International African Institute* 31(4): 355-371.

KARP, I. (1986), 'Agency and social theory: A review of Anthony Giddens new rules of sociological methods: A positive critique of interpretative sociologies; central problems in social theory: Action, structure and contradiction in social analysis; A contemporary critique of Historical materialism. Vol. 1. Power, property and the state'. *American Ethnologist* 13(1): 131-137

KATZ, J.E. & M. AAKHUS (2002), 'Introduction: Framing the issues'. In: J.E. Katz & M. Aakhus, eds, *Perpetual contact: Mobile communication, private talk, public performance*. Cambridge: Cambridge University Press, pp. 1-13.

KEARNEY, M. (1991), 'Borders and boundaries of state and self at the end of empire'. *Journal of Historical sociology* 5(1): 52-74.

KERLIN, M.D. (2000), 'New agents of socio-economic development: Guinea-Bissauan hometown associations in Portugal'. *South European Society and Politics* 5(3): 33-55.

KIEN, G. (2009), 'Actor-network theory: Translation and material culture'. In: P. Vannini, ed., *Material culture and technology in everyday life: Ethnographic approaches*. New York: Peter Lang, pp. 26-44.

KIVISTO, P. (2001), 'Theorizing transnational immigration: A critical review of current effort'. *Ethnic and Racial Studies* 24(4): 549-577.

KLASS, D. (2001), 'Continuing bonds in the resolution of grief in Japan and North America'. *American Behavioral Scientist* 44: 742-763.

KNOKE, D. & S. YANG (2008), '*Social network analysis* (2nd edition). Los Angeles/London: Sage Publications, Inc.

KONINGS, P. (2001), 'Mobility and exclusion: Conflicts between autochthons and allochthons during political liberalisation in Cameroon'. In: M. de Bruijn, R. van Dijk & D. Foeken, eds, *Mobile Africa: Changing patterns of movement in Africa and beyond.* Leiden: Brill, pp. 169-194.

KONINGS, P. (2008), Autochthony and ethnic cleansing in the post-colony: The 1966 Tombel disturbances in Cameroon. *The International Journal of African Historical Studies* 41(2): 203-222.

KONINGS, P. & F.B. NYAMNJOH (2000), Construction and deconstruction: Anglophones or autochtones? *The African Anthropologist* 7(1): 5-32.

KOPYTOFF, I. (1987), *The African frontier: The reproduction of traditional African societies*. Bloomington/Indianapolis: Indiana University Press.

KOSER, K. (1997), 'Social networks and the asylum cycles: The case of Iranians in The Netherlands'. *International Migration Review* 31(3): 591-611.

KOSINSKI, A.L. & M.R. PROTHERO (1975), 'Introduction: The study of migration'. In: A.L. Kosinski & M.R. Prothero, eds, *People on the move: Studies on internal migration*. London: Methuen & Co Ltd, pp. 23-38.

LAMOUREAUX, S. (2011), *Messages in a mobile: Mixed-messages, tales of missing and mobile communities at the University of Khartoum*. Bamenda/Leiden: Langaa RPCIG/African Studies Centre.

LATOUR, B. (1988), 'Part two: Irreductions'. In: A. Sheridan & J. Law, eds, *The pasteurization of France*. Cambridge, M.A: Harvard University Press, pp. 158-236.

LATOUR, B. (1996), 'On actor-network theory: A few clarifications'. *Soziale Welt* 47(4): 369-381.

LEACH, C.W. & L.Z. TIEDENS (2004), 'A world of emotion, Introduction'. In: L.Z. Tiedens & C.W. Leach, eds, *The social life of emotions*. Cambridge: Cambridge University Press, pp. 1-16.

LEE, R. (2012), 'Introduction: Themes of the study of death and loss in Africa'. *African Studies* 71(2): 163-173.

LEVITT, P. (2006), 'God has no passport: Defining the boundaries of belonging'. *Harvard Divinity Bulletin* 45-57.

LEVITT, P. & N. GLICK SCHILLER (2003), 'International perspective on transnational migration: An introduction'. *International Migration Review* 37(4): 565-575.

LEVITT, P. & N. GLICK SCHILLER (2004), 'Conceptualizing simultaneity: A transnational social field perspective on society'. *International Migration Review* 38(3): 1002-1039.

LEVITT, P. & B.N. JAWORSKY (2007), 'Transnational migration studies: Past developments and future trends'. *Annual Review of Sociology* 33: 129-156.

LEWIS, C. (2007), 'South Africa: The Internet war'. In: E.J. Wilson III & K. R. Wong, eds, *Negotiating the net in Africa: The politics diffusion*. Boulder London: Lynne Publishers, pp. 105-135.

LICOPPE, C. (2004), ''Connected' presence: The emergence of a new repertoire for managing social relationships in a changing communications technoscape'. *Environment and Planning D: Society and Space* 22: 135-156.

LINDELL, I. (2010), 'Between exit and voice: Informality and the spaces of popular agency'. *African Studies Quarterly* 11(2&3): 1-11.

LING, R. (2004), *The mobile connection: The cell phone's impact on society*. San Francisco: Morgan Kaufmann.

LING, D. & J. DONNER (2009), *Mobile communication: Digital media and society*. Cambridge: Polity Press.

LOWNDES, V. (2006), 'It's not what you've got but what you do with it: Women, social capital and political participation'. In: B. O'Neil & E. Gidengil, eds, *Gender and social capital*. New York/London: Routledge, pp. 213-240.

LURY, C. (1997), 'The objects of travel'. In: C. Rojek & J. Urry, eds, *Touring cultures: Transformations of travel and theory*. London: Routledge, pp. 75-95.

MACGAFFEY, J. (1987), *Entrepreneurs and parasites: The struggle for indigenous capitalism in Zaire*. Cambridge: Cambridge University Press.

MACGAFFEY, J. (1998), 'Creatively coping with crisis: Entrepreneurs in the second economy in Zaire (the Democratic Republic of Congo)'. In: A. Spring & B. McDade, eds, *African entrepreneurship: Theory and reality*. Florida: University of Florida Press, pp. 37-50.

MACGAFFEY, J. (2005), Review of J.I. Guyer (2004), 'Marginal gains: Monetary transaction in Atlantic Africa'. *Journal of Modern African Studies* 43(2): 328-329.

MACGAFFEY, J. & R. BAZENGUISSA-GANGA (2000), *Congo – Paris: Transnational traders on the margins of the Law*. Oxford: James Currey.

MACCLANCY, J. & A. FUENTES (2011) 'Centralizing fieldwork'. In: J. MacClancy & A. Fuentes, eds, *Centralizing fieldwork: Critical perspectives from primatology, Biology and Social Anthropology*. New York/Oxford: Berghahn Books, pp. 1-26.

MACDONALD, S. (2011), 'Making ethics'. In: M. Melhuus, J.P. Mitchell & H. Wulff, eds, *Ethnographic practice in the present*. New York/Oxford: Berghahn Books, pp. 80-94.

MAHARAJ, B. & V. MOODLEY (2000), 'New African immigration to Durban region'. *Canadian Journal of African Studies* 34(1): 149-160.

MAITLAND, D. (1984), *The missing link: Report on the independent commission for Worldwide Telecommunications Development*. Geneva: International Telecommunication Union.

MALINOWSKI, B. (1922), 'Introduction: Subject, method and scope of this enquiry'. In: *Argonauts of the Western Pacific: An account of native enterprise and adventure in the Archipaelagoes of Melanesia New Guinea*: Routledge, pp.1-25.

MALLARD, A. (2004), 'From the telephone to the economic exchange: How small businesses use the telephone in their market relations'. *Environment and Planning D: Society and Space* 22: 117-134.

MAND, K. (2002), 'Place, gender and power in transnational Sikh marriages'. *Global Networks* 2(3): 233-248.

MANGA-EDIMO, R.M. (2010), 'Les TIC, nouvelles formes d'action politique: Le cas des diasporas Camerounaise'. *Afrique Contemporaine* 234: 127-140.

MARCUS, G. (1995), 'Ethnography in/of the World System: The emergence of multi-sited ethnography'. *Annual Review of Anthropology* 24: 95-117.

MASSEY, D.S., R. ALARCON, J, DURAND & H. GONZÁLEZ (1987), *Return to Aztlan: The social process of international migration from Western Mexico*. Berkeley: University of California Press.

MASSEY, D.S. (1992), *Space, place and gender*. Cambridge: Polity Press.

MAUTHNER, N. & A. DOUCET (2003), Reflexive accounts and accounts of reflexivity in qualitative data analysis. *Sociology* 37(3): 413-431.

MAVHUNGU, K. (2012), *Witchcraft in post-colonial Africa: Beliefs, techniques and containment strategies*. Bamenda: Langaa RPCIG.

MAZZUCATO, V. (2008), 'The double engagement: Transnationalism and integration. Ghanaian migrants' lives between Ghana and The Netherlands'. *Journal of Ethnic and Migration Studies* 34(2): 199-216.

MAZZUCATO, V. (2003), Asente transnational relations: historical negotiation or changing structure of social relationships? Paper presented at the CERES Summer School, Utrecht, 23- 26 June.

MAZZUCATO, V. M. KABKI & L. SMITH (2006), 'Transnational migration and economy of funerals: Changing practices in Ghana'. *Development and Changes* 37(5): 1047-1072.

MBEMBE, A. (2001), 'At the edge of the world: Boundaries, territoriality, and sovereignty in Africa'. *Public Culture* 12(1): 259-284.

MBUY, T.H. (2012), *The faith of our ancestors: New perspectives in the study and understanding of African traditional religion*. Bamenda: Archdiocesan Print Media.

MELUCCI, A. (1989), *Nomads of the present: Social movements and individual needs in contemporary society*. Philadelphia, PA: Temple University Press.

MERCER, C., B. PAGE, & M. EVANS (2008), *Development in the African Diaspora: Place and the politics of home*. London: Zed Books.

MERCER, C., B. PAGE, & M. EVANS (2009), 'Unsettling connections: Transnational networks, development and African home associations'. *Global Networks* 9(2): 141-161.

MERCER, C. & B. PAGE (2010), 'African home associations in Britain: Between political moral and conviviality'. *African Diaspora* 3: 111-131.

MERLINGEN, M. (2006), 'Foucault and world politics: Promises and challenges of extending Governmentality theory to the European and beyond'. *Millennium: Journal of International Studies* 35(1): 181-196.

METCALF, P. & R. HUNTINGTON (1991), *Celebrations of death: The Anthropology of mortuary ritual*. Second edition. Cambridge: Cambridge University Press.

MEYER, B. (1998), 'Commodities and the power of prayer: Pentecostal attitudes towards consumption in contemporary Ghana'. *Development and Change* 29: 751-776.

MILLER, D. (2008), *The comfort of things. Portrait 7: Home and homeland.* Cambridge: Polity Press, pp. 73-82.

MILLER, D. (2011), *Stuff.* Cambridge: Polity Press.

MILLER, D. & D. SLAER (2000), *The Internet: An ethnographic approach.* Oxford/ New York: Berg.

MILLS, D. (2005), Made in Manchester?: Methods and myths in disciplinary history: Social analysis. *The International Journal of Cultural and Social Practice* 49(3): 129-143.

MITCHELL, J.P. (2011), 'Introduction'. In: M. Melhuus, J.P. Mitchell & H. Wulff, eds, *Ethnographic practice in the present*. New York/Oxford: Berghahn Books, pp. 1-15.

MITRA, A. (2005), 'Creating immigrant identities in cybernetic space: Examples from a non-resident Indian website'. *Media Culture Society* 27(3): 371-390.

MOORE, H.L. & T. SANDERS (2001), 'Magical interpretations and material realities: An introduction'. In: H.L. Moore & T. Sanders, eds, *Magical interpretations, material realities: Modernity, witchcraft and the occult in Postcolonial Africa*. London: Routledge, pp. 1-27.

MUTCH, A. (2002), 'Actors and networks or agents and structures: Towards a realist view of information systems'. *Organization* 9(3): 477-496.

NAIDO, M. (2012), 'Editorial: Mobilities and transnational lives'. *Alternation* 19(1): 1-6.

NAVILLE, P. (1963), 'Technical elites and social elites'. *Sociology of Education* 37(1): 27-29.

NCHEAWAH, Z.K. & S.A. TANIBANG (2004), *A brief history of the Mankon people*. Paper presented at Akongnu-Mankon Cultural festival, Douala.

NEDELCU, M. (2012), 'Migrants' new transnational habitus: Rethinking migration through a cosmopolitan lens in the digital age'. *Journal of Ethnic and Migration Studies* 38(9): 1339-1356.

NDIPINWI, R.M. (2010), *Pinyin - Bali Nyonga Conflict 1904*. Unpublished thesis in partial fulfilment of the requirements for the award of Higher Teacher's Diploma (DIPES I) in History, University of Yaoundé I, Cameroon.

NGWA, L & W. NGWA, eds, (2006), *From dust to snow: Bush-faller*. Princeton: Horeb Communications.

NIESWAND, B. (2011), *Theorising transnational migration: The status paradox of migration*. New York & London: Routledge.

NIESWAND, B. (2005), 'Charismatic Christianity in the context of migration: Social status, the experience of migration and the construction of self among Ghanaian migrants in Berlin'. In: A. Adogame & C. Weisskoppel, eds, *Religion in the context of African migration*. Bayreuth: Pia Thielmann & Eckhard Breitinger, pp. 243-265.

NIGER-THOMAS, M. (1995), 'Women's access to and the control of credit in Cameroon: The Mamfe case'. In: S. Bauman & S. Ardener, eds, *Money-go-rounds: The importance of rotating credit and savings association for women*. Oxford: Berg, pp. 95-110.

NJUBI, F.N. (2002), 'African intellectual in the belly of the beast: Migration, identity and the politics of exile'. *African Issues* 30(1): 70-75.

NKWI, W. (2011), *Kfaang and its technologies: Towards a social history of mobility in Kom, Cameroon, 1928-1998*. Leiden: African Studies Centre.

NORBECK, E. (1963), 'African rituals of conflict'. *American Anthropologist* 65: 1254-1279.

NUMFO, B. (1996), *The content and form of Mankon rites of passage songs*. Unpublished dissertation in partial fulfilment of the requirement of the Post-Graduate Teacher's Diploma in English (DIPES II), University of Yaoundé I, Cameroon.

NYAMNJOH, F.B. (1999), 'Cameroon: A country united by ethnic ambition and difference'. *African Affairs* 98(390): 101-118.

NYAMNJOH, F.B. (2001), 'Delusions of development and the enrichment of witchcraft discourses in Cameroon.' In: H.L. Moore & T. Sanders, eds, *Magical interpretations, material realities: Modernity, witchcraft and the occult in Postcolonial Africa*. London: Routledge, pp. 28-49.

NYAMNJOH, F.B. (2005), 'Images of Nyongo amongst Bamenda Grassfielders in Whiteman Kontri'. *Citizenship Studies* 9(3): 241-269.

NYAMNJOH, F.B. (2006), *Insiders and outsiders: Citizenship and xenophobia in Contemporary Southern Africa*. Dakar: CODESRIA.

NYAMNJOH, F.B. (2007), 'Theorizing agency in and on Africa: The questions are key'. In: R. van Dijk, M. de Bruijn & J. Gewald, eds, *Strength beyond structure: Social and historical trajectories of agency in Africa*. Leiden: Brill, pp. 334-338.

NYAMNJOH, F.B. (2009), *Married but available*. Bamenda: Langaa RPCIG.

NYAMNJOH, F.B. (2011), 'Cameroonian bushfalling: Negotiation of identity and belonging in fiction and ethnography'. *American Ethnologist* 38(4): 701-713.

NYAMNJOH, F.B. (2012), Fiction and reality of mobility in Africa. *Keynote address* presented at Stanford Forum for African Studies, Stanford, 26- 27 October.

NYAMNJOH, F. & M. ROWLANDS (1998), 'Elite association and the politics of belonging in Cameroon'. *Journal of International African Institute* 68(3): 320-337.

NYAMNJOH, F. & B. PAGE (2002), 'Whiteman kontri and the enduring allure of modernity among Cameroonian youth'. *African Affairs* 101: 607-634.

NYAMNJOH, H.M. (2010), *'We get nothing from fishing': Fishing for boat opportunities amongst Senegalese fisher migrants.* Bamenda & Lieden: Langaa RPCIG & African Studies Centre.

NYAMNJOH, H.M (forthcoming), ''Penetrating the unseen' The role of religion and spiritual practices in the Senegalese boat migration process.' In: A. Adogame, ed., *Imagining the religious 'other': The public face of African new religious movements in diaspora.* London: Ashgate.

NYAMNJOH, H.M. & M. ROWLANDS (2013), 'Do you eat achu here? Nurturing as a way of life in a Cameroon diaspora'. *Critical African Studies* 5(3): 140-152.

NYMAN, J. (2009), *Home, identity, and mobility in contemporary diasporic fiction.* Amsterdam/New York: Rodopi.

O'BYRNE, J. D. (2001), 'On passports and border controls'. *Annals of Tourism Research* 28(2): 399-416.

OIARZABAL, P.J. & U. REIPS (2012), 'Migration and diaspora in the age of information and communication technologies'. *Journal of Ethnic and Migration studies* 38(9): 1333-1338.

OLSEN, B. (2003), 'Material culture after text: Re-membering things'. *Norwegian Archaeological Review* 36(2): 87-104.

OONK, G. (2005), 'Global Indian diasporas: Exploring trajectories of migration and theory'. In: G. Oonk, ed., *Global Indian diasporas: Exploring trajectories of migration and theory.* Amsterdam: Amsterdam University Press, pp. 9-30.

OLWIG, K.F. (2009), 'A proper funeral: Contextualizing community among the Caribbean migrants'. *Journal of Royal Anthropological Institute* 15: 520-537.

O'REILLY, K. (2009), *Key concepts in ethnography.* London: Sage Publications.

OSENI, A.L. (2013), 'Informal economy and entrepreneurial development'. *West Africa Insight* 4(5): 3-5.

PAGE, B. (2007), 'Slow going: The mortuary, modernity and the home-town association in Bali-Nyonga, Cameroon'. *Africa* 77(3): 419-441.

PAGE, B., M. EVANS & C. MERCER (2009), 'Introduction: African transnationalisms and diaspora networks'. *Global Networks* 9(2): 137-140.

PANAGAKOS, A. & H. HORST (2006), 'Return to cyberia: Technology and the social worlds of transnational migrants'. *Global Networks* 6(2): 109-124.

PARHAM, A.A. (2005), 'Internet, place, and public sphere in diaspora communities'. *Diaspora* 14(2/3): 349-380.

PARISH, J. (2000), 'From the body to the wallet: Conceptualizing Akan witchcraft at home and abroad'. *Journal of the Royal Anthropological Institute* 6: 487-500.

PEBERDY, S. (2009), *Selecting Immigrants: National identity and South Africa's immigration policies 1920-2008.* Johannesburg: Wits University Press.

PEBERDY, S. & C. ROGERSON (2000), 'Transnational and non-South African entrepreneurs in South Africa's small, medium and micro-enterprise (SMME) economy'. *Canadian Journal of African Studies* 34(1): 20-40.

PELICAN, M. (2006), *Getting along in the grassfields: Interethnic relations and identity politics in northwest Cameroon.* Unpublished PhD thesis in der Philosophischen Fakultät, Martin-Luther-King-Universität, Halle-Wittenberg.

PELICAN, M., P. TATHAH & B. NDJIO (2008), 'Local perspectives on transnational relations of Cameroon migrants'. *African Sociological Review* 12(2): 117-127.

PERRONS, D. (2004), *Globalization and social change: People and places in a divided world.* London: Routledge.

PICKERING, W.S. (1974), 'The persistence of rites of passage: Towards an explanation'. *The British Journal of Sociology* 25(1): 63-78.
PLECK, E.H. (2004), 'Who are we and where do we come from? Rituals families and identities.' In: A. Etzioni & J. Bloom, eds, *We are what we celebrate: Understanding holidays and rituals*. New York and London: New York University Press, pp. 43-60.
PLECK, E.H. (2000), *Celebrating the family: Ethnicity, consumer culture, and family rituals*. Cambridge, MA/London: Harvard University Press.
POITIER, S. (2000), *The measure of a man: A spiritual autobiography*. New York: HarperCollins Publishers.
POPKIN, E. (1999), 'Guatemalan Mayan migration to Los Angeles: Constructing transnational linkages in the context of the settlement process'. *Ethnic and Racial Studies* 22(2): 267-289.
POROS, M.V. (2001), 'The role of migrants networks in linking labour markets: The case of Asian Indian migrants to New York and London'. *Global Networks* 1(3): 243-259.
PORTES, A. (1995), 'Transnational communities: Their emergence and significance in the contemporary world system'. Series of the Program in Comparative and International Development. Baltimore, MD: The John Hopkins University, Working Paper. no. 16.
PORTES, A. (1998), 'Social capital: Its origins and applications in modern sociology'. *Annual Reviews Sociology* 24: 1-24.
PORTES, A .(2001), 'The debates and significance of ImmigrantTransnationalism'. *Global Networks* 1(3): 181-193.
PORTES, A. (2003), 'Conclusion: Theoretical convergencies and empirical evidence in the study of immigrant transnationalism'. *International Migration Review* 37(3): 874-892.
PORTES, A., M. CASTELLS & L. BENTON (1991), *Introduction: The informal economy: Studies in advanced and less developed countries*. Baltimore/London: John Hopkins University Press, pp. 1-7.
PORTES, A. & P. LANDOLT (2000), 'Social capital: Promise and pitfalls of its role in development'. *Journal of Latin American Studies* 32(2): 529-547.
PORTES, A., L.E. GUARNIZO & W.J. HALLER (2002), 'Transnational entrepreneurs: An alternative form of immigrant economic adaptation'. *American Sociological Review* 67(2): 278-298.
PORTES, A. & J. DEWIND (2007), 'Across-Atlantic dialogue: The progress of research and theory in the study of international migration'. In: A. Portes, & J. DeWind, eds, *Rethinking migration: New theoretical and empircal perspectives*. New York/Oxford: Berghahn Books, pp. 3-26.
POWELL, C. (2012), *Me and my cell phone and other essays on technology in everyday life*. Bamenda: Langaa RPCIG
POWLES, J. (2004), ‚*Life history and personal narrative: Theoretical and methodical issues relevant to research and evaluation in refugee contexts*. UNHCR, The UN Refugee Agency: Evaluation and Policy Analysis Unit, pp. 1-24.
PRELIPCEANU, R. (2008), 'Building transnational lives: Using ICT to connect mobility and home'. Series of the Research Network, Working Paper no. 19/2008.
PROCTER, D. (2004), 'Victorian days: Performing community through local festivals'. In: A. Etzioni, & J. Bloom, eds, *We are what we celebrate: Understanding holidays and rituals*. New York/London: New York University Press, pp. 131-148.
PRONK, J.P. (1993), 'Migration: The Nomand in each of us'. *Population and Development Review* 19(2): 323-327.
PUTNAM, R.D., L.M. FELDSTEIN & D. COHEN (2003), 'Introduction'. In: R.D. Putnam, L.M. Feldstein & D. Cohen, eds, *Better together: Restoring the American community*. New York/London: Simon & Schuster, pp. 1-10.
RAGIN, C.C. (1987), *The comparative method: Moving beyond qualitative and quantitative strategies*. Berkeley: University of California Press.
RAPPAPORT, R.A. (1997), *Ritual and religion in the making of humanity*. Cambridge: Cambridge University Press.

RAPPORT, N. (2010), 'The ethics of particpant observation: Personal reflection on fieldwork in England'. In: P. Collins & A. Gallinat, eds, *The ethnography of the self as a resource*. New York/Oxford: Berghahn Books, pp. 78-94.

RICCIO, B. (2011), 'Rehearsing transnational citizenship: Senegalese associations, co-development and simultaneous inclusion'. *Africa Diaspora* 4: 97-113.

BER-SCHLOSSER, D., G. DE MEUR, B. RIHOUX & C.C. RAGIN (2009), 'Qualitative comparative analysis (QCA) as an approach'. In B. Rihoux & C.C. Ragin, eds, *Configurational comparative methods: Qualitative comparative analysis and related techniques*. Los Angeles/London: Sage Publications, pp. 1-18.

ROBERTS, B.R., R. FRANK, & F. LOZANO-ASCENCIO (1999), 'Transnational migrant communities and Mexican migration to the US'. *Journal of Ethnic and Migration studies* 22(2): 238-266.

ROSE, J. & M. JONES (2005), 'The double dance of agency: A socio-theoretic account of how machines and humans interact'. *International Journal on Communication, Information Technology and Work* 1(1): 19-37.

ROWLANDS, M. (1979), 'Local and long distance trade and incipient state formation in the Bamenda Plateau in the late 19th century'. *Paideuma* 25: 1-19.

ROWLANDS, M. (1988a), 'Notes on labour supply in the Grassfields'. In: P. Geschiere & P. Konings, eds, *Conference on the political economy of Cameroon – historical perspectives*. Leiden: African Studies Centre, pp. 175-180.

ROWLANDS, M. (1988b), 'The material culture of success: Households and consumption in Bamenda'. In: P. Geschiere, & P. Konings, eds, *Conference on the historical economy of Cameroon – historical perspectives*. Leiden: African Studies Centre, pp. 503-523.

ROWLANDS, M. (1992), 'Accounting for personal success in Bamenda'. In: J. Maw & J. Picton, eds, *Concepts of the body/self in Africa*. Wien: Beiträge zur Afrikanistik Band 43, pp. 117-134.

ROWLANDS, M. (1993), 'Accumulation and the cultural politics of identity in the Grassfields'. In: P. Geschiere, & P. Konings, eds, *Itinéraires d'accumulation au Cameroun*. Paris: Karthala; Leiden: African Studies Centre, pp. 71-97.

ROWLANDS, M. (1995), 'Looking at financial landscapes: A contextual analysis of ROSCA in Cameroon'. In: S. Bauman & S. Ardener, eds, *Money-go-rounds: The importance of rotating credit and savings association for women*. Oxford: Berg, pp. 111-124.

ROWLANDS, M. (2005), 'Value and the cultural transmission of things.' In: W. van Binsbergen & P. Geschiere, eds, *Commodification: Things, agency, and identities*. Münster: LIT Verlag, pp. 267-281.

RUSSELL, H. (1999), 'Friends in low places: Gender, unemployment and sociability'. *Work, Employment and Society* 13: 205-224.

RYAN, L. (2008), 'Navigating the emotional terrain of families "here" and "there": Women, migration and the management of emotions'. *Journal of Intercultural Studies* 29(3): 299-313.

SABAR, G. (2010), 'Witchcraft and concepts of evil amongst African migrant workers in Israel'. *Canadian Journal of African Studies* 44(1): 110-141.

SACK, R. (1986), *Human territoriality: Its theory and history*. Cambridge: Cambridge University Press.

SAWYERR, H. (1974), 'African concepts of life and death'. In: J.S. Pobee, ed., *Religion, morality, and population dynamics*. Legon: University of Ghana, pp. 78-86.

SCHANK, R.C. (1990), *Tell me a story: A new look at real and artificial memory*. New York: Scribner's.

SCHILBRACK, K. (2004), 'Introduction: On the use of philosophy in the study of rituals.' In: K. Schilbrack, ed., *Thinking through rituals: Philosophical perspectives*. New York/London: Routledge, pp. 1-30.

SCHROVER, M., J. VAN DER LEUN, L. LUCASSEN & C. QUISPEL (2008), *Illegal migration and gender in a global and historical perspective*. Amsterdam: Amsterdam University Press.

SCOT, J. (1985), *Weapons of the weak: Everyday forms of peasant resistance*. New Haven, CT: Yale University Press.

SEGALL, A. (2001), 'Critical ethnography and the invocation of voice: From the field/in the field – single exposure, double standard'? *International Journal of Qualitative Studies in Education* 14(4): 579-592.

SEGATTI, A. (2011), 'Migration to South Africa: Regional challenges versus national instruments and interests'. In A. Segatti & L.B. Landau, eds, *Contemporary migration to South Africa: A regional development issue*. Washington: World Bank, pp. 9-29.

SEGATTI, A. & L.B. LANDAU (2011), 'Immigration legislation and policy in South Arica and the Southern African Development Community'. In A. Segatti & L.B. Landau, eds, *Contemporary migration to South Africa: A regional development issue*. Washington: World Bank, pp. 137-144.

SHINDONDOLA, H. (2007), 'Chasing an illusion and being trapped in the social and economic obligations: The case of Cameroonians in Johannesburg'. Paper presented at the Anthropology and Development Studies seminar, University of Johannesburg, 9 March.

SHELLER, M. (2004), 'Mobile publics: beyond the network perspective'. *Environment and Planning D: Society and Space* 22: 39-52.

SICHONE, O. (2008), 'Xenophobia and xenophilia in South Africa: African migrants in Cape Town'. In: P. Werbner, ed., *Anthropology and the new cosmopolitanism: Rooted, feminist and vernacular perspectives*. Oxford: Berg, pp. 309-332.

SKRBIŠ, Z. (2008), 'Transnational families: Theorising migration, emotions and belonging'. *Journal of Intercultural Studies* 29(3): 231-246.

SLATER, D & J. TACHHI (2004), *Research: ICT innovations for poverty reduction*. New Dheli: UNESCO.

SMITH, D.J. (2006), 'Cell phones: Social inequality, and contemporary culture in Nigeria'. *Canadian Journal African Studies* 40(3): 496-523.

SMITH, E.M. (1976), 'Networks and migration resettlement: Cherchez la femme'. *Anthropological Quarterly* 49(1): 20-27.

SMITH, L. (2007), *Tied to migrants: Transnational influences on the economy of Accra, Ghana*. Leiden: African Studies Centre.

SMITH, L. & V. MAZZUCATO (2009), 'Constructing homes, building relationships: Migrant investments in houses'. *Royal Dutch Geographical Society KNAG* 100 (5): 662-673.

SMITH, M.P. (2001), *Transnational urbanism locating globalisation*. Oxford: Blackwell Publishers.

SMITH, R. (1998), 'Transnational localities: Community, technology and the politics of membership within the context of Mexico and U.S. migration.' In: M.P. Smith & L. E. Guarnizo, eds, *Transnationalism from below*. New Brunswick, NJ: Transaction Publishers, pp. 196-238.

SPRING, A. & B. MCDADE (1998), 'Entrepreneurship in Africa: Traditional and contemporary paradigms'. In: A. Spring & B. McDade, eds, *African entrepreneurship: Theory and reality*. Florida: University of Florida Press, pp. 1-32.

SVAŠEK, M. (2010), 'On the move: Emotions and human mobility'. *Journal of Ethnic and Migration Studies* 36(6): 865-880.

TASI, L.N. (2009), 'An introduction to Pinyin political history: 1690-1995'. Unpublished manuscript.

TAZANU, P.M. (2012), *Being available and reachable new media and Cameroonian transnational sociality*, Bamenda: Langaa RPCIG.

TED CONFERENCES (2009), Chimamanda, A. N. *The danger of a single story* http://www.ted.com/talks/chimamanda_adichie_the_danger_of_a_single_story.html Accessed: 12/03/2012.

THOMAS, J.J. (1992), *Informal economic activity*. Ann Arbor: University of Michigan Press.

THOMAS, W. & F. ZNANIECKI ([1918]1974), *The polish peasant in Europe and America*. New York: Dover Publications.

THOMPSON, E.C. (2009), 'Mobile phones, communities and social networks among foreign workers in Singapore'. *Global Networks* 9(3): 359-380.

TRAGER, L. (1998), 'Home-town linkages and local developments in south-western Nigeria: Whose agenda? What impact'? *Africa* 68(3): 360-382.

TSAGAROUSIANOU, R. (2004), 'Rethinking the concept of diaspora: Mobility, connectivity and communication in a globalised world'. *Westminister Papers in Communication and Culture* 1: 52-65.

TURNER, J.A. (2009), *The Philippines: Mobilities, identities, globalization.* New York/London: Routledge.

TURNER, J.H. & J. STETS (2005), *Sociology of emotions.* Cambridge: Cambridge University Press, Chapter 1, pp. 1-25.

TURNER, S. (1995), 'Torture, refugee and trust.' In: E.V. Daniel & J.C. Knudsen, eds, *Mistrusting refugees.* Berkeley: University of California Press, pp. 56-72.

TURNER, V. (1967), 'Betwixt and between: The liminal periods in rites de passage'. In: V. Turner, ed., *The forest of symbols: Aspects of Ndembe ritual.* Ithaca, New York: Cornell University Press, pp. 93-111.

TURNER, V. (1969), *The ritual process: Structure and anti-structure.* Chicago: Aldine Publishing Company.

ULF, H. (1996), *Transnational connections: Culture people and places. Introduction.* London/New York: Routledge, pp. 1-13.

UNITED NATIONS (2005), *Global Commission on International Migration.*

URRY, J. (2000), *Sociology beyond societies: Mobilities for the twenty-first century.* London: Routledge.

URRY, J. (2003), Social networks, travel and talk. *British Journal of Sociology* 54: 155-175.

URRY, J. (2004), 'Connections'. *Environment and Planning D: Society and Space* 22: 27-37.

URRY, J. (2007), *Mobilities.* Cambridge/Malden: Polity Press

VALE, P. (2002), 'Migration, xenophobia and security-making in post-apartheid South Africa'. *Politikon: South African Journal of Political Studies* 29(1): 7-29.

VAN BEEK, W. & T.D. BLAKELY (1994), *Introduction, religion in Africa: Experience and expression.* London: James Curry, pp. 1-20.

VAN BINSBERGEN, W. (2007), 'Manchester as the birth place of modern agency research: The Manchester School explained from the perspective of Evans-Prichard's book The Nuer'. In: M. de Bruijn, R. van Dijk & J.B. Gewald, eds, *Strength beyond structure: Social and historical trajectories of agency in Africa.* Leiden/Boston: Brill, pp. 16-61.

VAN DER GEEST, S. (2000), 'Funerals for the living: Conversations with elderly people in Kwahu, Ghana'. *African Studies Review* 43(3): 103-129.

VAN DIJK, R. (2013), 'A ritual connection: Urban youth marrying in the village in Botswana'. In: M. de Bruijn & R. van Dijk, eds, *The social life of connectivity in Africa.* New York: Palgrave Macmillan, pp. 141-159.

VAN DIJK, R. (2002), 'Religion, reciprocity and restructuring family responsibility in the Ghanaian Pentecostal diaspora'. In: D. Bryceson & U. Vuorela, eds, *The transnational family: New European frontiers and global networks.* Oxford: Berg, pp. 173-193.

VAN DIJK, R., M. DE BRUIJN & J.B. GEWALD (2007), 'Social and historical trajectories of agency in Africa: An introduction'. In: M. de Bruijn, R. van Dijk & J.B. Gewald, eds, *Strength beyond structure: Social and historical trajectories of agency in Africa.* Leiden/Boston: Brill, pp. 1-15.

VAN GENNEP, A. (1960), *The rites of passage.* London: Routledge & Kegan Paul.

VAN HEAR, N. (1998), *New Diasporas: The mass exodus, dispersal and regrouping of migrant communities.* Seattle: University of Washington Press.

VERTOVEC, S. (1999), 'Conceiving and researching transnationalism'. *Journal of Ethnic and Migration Studies* 22(2): 447-462.

VERTOVEC, S. (2001), 'Transnationalism and identity'. *Journal of Ethnic and Migration Studies* 27 (4): 573-582.

VERTOVEC, S. (2002), Migrant transnationalism and modes of transformation'. *International Migration Review* 38(3): 970-1001.
VERTOVEC, S. (2004a), 'Cheap calls: The social glue of migrant transnationalism'. *Global Networks* 4(2): 219-224.
VERTOVEC, S. (2004b), 'Trends and impacts of migrants transnationalism'. COMPAS, University of Oxford, Working Paper no. 3.
VERTOVEC, S. (2007) 'Migrant transnationalism and modes of transformation'. In: A. Portes & J. de Wind, eds, *Rethinking migration: New theoretical and empirical perspectives*. New York/Oxford: Berghahn Books, pp. 149-180.
VERTOVEC, S. (2009), *Transnationalism*. London. Routledge.
VIDACS, B. (2003), 'Football and identity in Cameroon.' In: J.G. Gros, ed., *Politics and society in critical perspectives*. New York/Toronto/Oxford: University of America, pp. 167-184.
VUEROLA, U. (2002), 'Transnational families: Imagined and real communities'. In: D. Bryceson & U. Vuorela, eds, *The transnational family: New European frontiers and global networks*. Oxford: Berg, pp. 63-82.
WARNIER, J.P. (1985), 'The Mankon palace'. *Paideuma* 31: 15-29.
WARNIER, J.P. (2012), *Cameroon Grassfields civilization*. Bamenda: Langaa RPCIG.
WARNIER, J.P. (2012), 'The revival of vernacular public autopsy in rural Cameroon'. Paper presented at the Centre for African Studies 50th anniversary conference, Edinburgh, 7 June.
WENTWORTH, W. & D. YARDLY (1994), 'Deep sociality: A bioevolutionary perspectives on the sociology of human emotions'. In D.D. Franks, W.M. Wentworth & J. Ryan, eds, *Social perspectives on emotions*. Greenwich, CT: JAI Press, pp. 1-17.
WILDING, R. (2006), ''Virtual' intimacies? Families communicating across transnational contexts'. *Global Networks* 6(2): 125-142.
WILLIAMS, L. (2006), 'Social networks of refugees in the United Kingdom: Tradition, tactics and new community spaces'. *Journal of Ethnic and Migration Studies* 32(5): 865-879.
WILLIAMS-JONES, B. & G. GRAHAM (2003), 'Actor-network theory: A tool to support ethical analysis of commercial genetic testing'. *New Genetics and Society* 22(3): 271-296.
WILSON III, E.J. & K.R. WOMG (2007), 'Negotiating the net in Africa: Introduction'. In: E.J. Wilson III & K.R. Wong, eds, *Negotiating the net in Africa: The politics diffusion*. Boulder/London: Lynne Publishers, pp. 1-16.
WILSON, G. (1939), 'Nyakyusa conventions of burial'. *Bantu Studies* 13: 1-31.
WILSON, M. (1954), 'Nyakyusa Ritual and symbolism'. *American Anthropologist* 54: 228-241.
WILSON, M. (1957), *Rituals of kinship among the Nyakyusa*. London: Oxford University Press.
WOLF, D. (2002), 'There is no place like 'home': Emotional transnationalism and the struggles of second-generation Fillipinos'. In: P. Levitt & M.C. Waters, eds, *The changing face of home: The transnational lives of the second generation*. New York: Russell Sage Foundation, pp. 255-294.
WOODS, D. (1994), 'Elites, ethnicity, and 'home town' associations in the Cote d'Ivoire: An historical analysis of state-society links'. *Journal of the International African Institute* 64(4): 465-483.
WORLD BANK (2001), *World Development Report 2000/2001: Attacking poverty*. New York: Oxford University Press.
WRIGHT, G. (1991), 'Prescribing the model home'. *Social Research* 58(1): 213-225.
YUMBA, D. (2004), 'Nurturing information and communications technology in Africa'. *Humanities Review Journal* 4: 11-17.
ZOLBERG, A.R. (2006), 'Patterns of international migration policy: A diachronic comparison.' In: M.A. Messina & G. Lahav, eds, *The migration reader: Exploring politics and policies*. London: Lynne Rienner Publishers, pp. 110-125.

www.ingramcontent.com/pod-product-compliance
Lightning Source LLC
LaVergne TN
LVHW060027210926
848481LV00013B/3803